Ciba Foundation Symposium 198

P2 PURINOCEPTORS: LOCALIZATION, FUNCTION AND TRANSDUCTION MECHANISMS

1996

JOHN WILEY & SONS

Chichester · New York · Brisbane · Toronto · Singapore

Published in 1996 by John Wiley & Sons Ltd
Baffins Lane, Chichester
West Sussex PO19 1UD, England

Telephone National (01243) 779777
International (+44) (1243) 779777

Other Wiley Editorial Offices

John Wiley & Sons, Inc., 605 Third Avenue,
New York, NY 10158-0012, USA

Jacaranda Wiley Ltd, G.P.O. Box 859, Brisbane,
Queensland 4001, Australia

John Wiley & Sons (Canada) Ltd, 22 Worcester Road,
Rexdale, Ontario M9W 1L1, Canada

John Wiley & Sons (SEA) Pte Ltd, 37 Jalan Pemimpin #05-04,
Block B, Union Industrial Building, Singapore 2057

Suggested series entry for library catalogues:
Ciba Foundation Symposia

Ciba Foundation Symposium 198
ix + 336 pages, 72 figures, 14 tables

Library of Congress Cataloging-in-Publication Data

P2 purinoceptors: localization, function, and transduction mechanisms.
p. cm.—(Ciba Foundation symposium ; 198)
Editors: Derek J. Chadwick (organizer) and Jamie A. Goode.
Symposium held at Ciba Foundation, London, 11–13 July 1995.
Includes bibliographical references and index.
ISBN 0 471 96125 6 (alk. paper)
1. Purines—Receptors—Congresses. 2. Adenosine triphosphate—Receptors—Congresses. 3. Neurotransmitter receptors—Congresses.
I. Chadwick, Derek. II. Goode, James. III. Series.
[DNLM: 1. Receptors, Purinergic P2—congresses. W3 C161F v. 198 1996/QU 58 P99 1996]
QP801.P8P2 1996
599′.0188—dc20
DNLM/DLC
for Library of Congress 96-1089
CIP

British Library Cataloguing in Publication Data

A catalogue record for this book is available from the British Library

ISBN 0 471 96125 6

Typeset in 10/12pt Times by Dobbie Typesetting Limited, Tavistock, Devon.
Printed and bound in Great Britain by Biddles Ltd, Guildford.
This book is printed on acid-free paper responsibly manufactured from sustainable forestation, for which at least two trees are planted for each one used for paper production.

P2 PURINOCEPTORS: LOCALIZATION, FUNCTION AND TRANSDUCTION MECHANISMS

The Ciba Foundation is an international scientific and educational charity (Registered Charity No. 313574). It was established in 1947 by the Swiss chemical and pharmaceutical company of CIBA Limited—now Ciba-Geigy Limited. The Foundation operates independently in London under English trust law.

The Ciba Foundation exists to promote international cooperation in biological, medical and chemical research. It organizes about eight international multidisciplinary symposia each year on topics that seem ready for discussion by a small group of research workers. The papers and discussions are published in the Ciba Foundation symposium series. The Foundation also holds many shorter meetings (not published), organized by the Foundation itself or by outside scientific organizations. The staff always welcome suggestions for future meetings.

The Foundation's house at 41 Portland Place, London W1N 4BN, provides facilities for meetings of all kinds. Its Media Resource Service supplies information to journalists on all scientific and technological topics. The library, open five days a week to any graduate in science or medicine, also provides information on scientific meetings throughout the world and answers general enquiries on biomedical and chemical subjects. Scientists from any part of the world may stay in the house during working visits to London.

Contents

Symposium on P2 purinoceptors: localization, function and transduction mechanisms, held at the Ciba Foundation, London, 11–13 July 1995

This symposium is based on a proposal made by Peter Illes

Editors: Derek J. Chadwick (Organizer) and Jamie A. Goode

Participants

M. P. Abbracchio Facoltà di Farmacia, Istituto di Scienze Farmacologiche, Università di Milano, Via Balzaretti 9, I-20133 Milano, Italy

E. A. Barnard Molecular Neurobiology Unit, Division of Basic Medical Sciences, Royal Free Hospital School of Medicine, Rowland Hill Street, London NW3 2PF, UK

J. M. Boeynaems Institute of Interdisciplinary Research, School of Medicine and Department of Medical Chemistry, Université Libre de Bruxelles, Campus Hôpital Erasme, Bâtiment C, Route de Lennik 808, B-1070 Bruxelles, Belgium

R. C. Boucher Division of Pulmonary Diseases, Department of Medicine, University of North Carolina, School of Medicine, 724 Burnett Womack Building, Chapel Hill, NC 27599-7020, USA

U. Brändle *(Ciba Foundation Bursar)* Sektion Sensorische Biophysik, Hals-Nasen-Ohren Universitätsklinik, Röntgenweg 11, D-72076 Tübingen, Germany

G. Burnstock *(Chairman)* Department of Anatomy and Developmental Biology, University College London, Gower Street, London WC1E 6BT, UK

F. Di Virgilio Institute of General Pathology, University of Ferrara, Via L Borsari 46, I-44100 Ferrara, Italy

F. A. Edwards Department of Pharmacology, University of Sydney D06, NSW 2006, Australia

T. K. Harden Department of Pharmacology, School of Medicine, University of North Carolina, CB 7365 FLOB, Chapel Hill, NC 27599-7365, USA

S. E. Hickman The Rover Laboratory of Physiology, Department of Physiology and Cellular Biophysics, Columbia University, College of Physicians and Surgeons, 630 West 168th Street, New York, NY 10032, USA

S. M. O. Hourani Receptors and Cellular Regulation Research Group, School of Biological Sciences, University of Surrey, Guildford, Surrey GU2 5XH, UK

A. P. IJzerman Division of Medicinal Chemistry, Leiden/Amsterdam Center for Drug Research, University of Leiden, PO Box 9502, NL-2300 RA Leiden, The Netherlands

P. Illes Institut für Pharmakologie und Toxikologie der Universität Leipzig, Hättelstrasse 16-18, D-04107 Leipzig, Germany

K. A. Jacobson Molecular Recognition Section, Chief Laboratory of Bioorganic Chemistry, Bdg 8A, Rm B1A-17, NIDDK, NIH, Bethesda, MD 20892, USA

C. Kennedy Department of Physiology and Pharmacology, University of Strathclyde, Royal College, 204 George Street, Glasgow G1 1XW, UK

P. Leff Department of Pharmacology, Astra Charnwood, R & D Laboratories, Bakewell Road, Loughborough LE11 0RH, UK

K. D. Lustig Department of Cell Biology, Harvard Medical School, 25 Shattuck Street, Boston, MA 02138, USA

M. T. Miras-Portugal Departamento de Bioquimica, Facultad de Veterinaria, Universidad Complutense, E-28040 Madrid, Spain

J. T. Neary Laboratory of Neuropathology, Research Service 151, VA Medical Center, 1201 NW 16 St, Miami, FL 33125, USA

R. A. North Glaxo Institute for Molecular Biology, 14 Chemin des Aulx, 1228 Plan-les-Oautes, CH-1211 Geneva, Switzerland

P. Petit Laboratoire de Pharmacologie, Institut de Biologie, Faculté de Médecine, Boulevard Henri IV, F-34060 Montpellier Cedex, France

K. Starke Pharmakologisches Institut, Albert Ludgwigs-Universität Freiburg, Hermann-Herder-Strasse 5, D-79104 Freiburg im Breisgau, Germany

A. Surprenant Glaxo Institute for Molecular Biology, 14 Chemin des Aulx, 1228 Plan-les-Oautes, CH-1211 Geneva, Switzerland

D. P. Westfall Department of Pharmacology, School of Medicine, University of Nevada, Howard Medical Building 318, Reno, NV 89557-0046, USA

J. S. Wiley Haematology Department, Austin Hospital, Studley Road, Heidelberg, Victoria 3084, Australia

M. Williams Neuroscience Discovery, Pharmaceutical Products Division, Abbott Laboratories, 100 Abbott Park Road, Abbott Park, IL 60064, USA

P2 purinoceptors: historical perspective and classification

Geoffrey Burnstock

Department of Anatomy and Developmental Biology, University College London, Gower Street, London WC1E 6BT, UK

Abstract. This article presents an overview that gives some historical perspective to the detailed papers at the cutting edge of P2 purinoceptor research that follow. I consider the proposal, first put forward by Abbracchio & Burnstock (Pharmacol Ther 64:445–475, 1994), that P2 purinoceptors should be regarded as members of two main families: a P2X purinoceptor family consisting of ligand-gated ion channels, and a P2Y purinoceptor family consisting of G protein-coupled receptors. The latest subclasses of these two families ($P2X_{1-4}$ and $P2Y_{1-5}$), identified largely on the basis of molecular cloning and expression, are tabled. Finally, I suggest some future directions for P2 purinoceptor research, including studies of the long-term (trophic) actions of purines, the evolution and development of purinoceptors and therapeutic applications.

1996 P2 purinoceptors: localization, function and transduction mechanisms. Wiley, Chichester (Ciba Foundation Symposium 198) p 1–34

Early history

The first report about the potent actions of adenine compounds was published by Drury & Szent-Györgyi (1929), and the first hint that ATP might be a neurotransmitter appeared three decades later when ATP was shown to be released during antidromic stimulation of sensory nerves supplying the rabbit ear artery (Holton 1959). In the early 1960s, my colleagues and I in Melbourne and a group in Sweden proposed the existence of autonomic nerves supplying the gastrointestinal tract that were neither adrenergic nor cholinergic (Burnstock et al 1963, Martinson & Muren 1963). In the years that followed, strenuous efforts were made to identify the transmitter in non-adrenergic, non-cholinergic (NANC) nerves supplying the gut and the urinary bladder. Perhaps surprisingly, the substance that most satisfied the criteria at that time (Eccles 1964) was adenosine 5′-triphosphate (ATP) (Burnstock et al 1970), and the word ‘purinergic’ was coined and purinergic transmission proposed (Burnstock 1972).

Implicit in the concept of purinergic transmission was the existence of postjunctional receptors for ATP, although there was considerable confusion in the literature about the variable effects of adenosine nucleotides and nucleosides on a wide variety of tissues (Burnstock 1976a). However, a step forward was taken in 1978 when, from a detailed analysis of the literature and some preliminary experiments, it was proposed (Burnstock 1978) that 'purinoceptors' could be subdivided into P1 (adenosine) purinoceptors which were coupled to adenylate cyclase and were competitively antagonized by low concentrations of methylxanthines, and P2 purinoceptors which were activated preferentially by ATP and ADP. Two of the most important implications of this purinoceptor subdivision were: (1) the importance of establishing whether in a particular situation ATP acts directly on P2 purinoceptors or via P1 purinoceptors after ectoenzymic breakdown to adenosine (Moody et al 1984); and (2) that during purinergic transmission, whereas ATP released from the nerve terminals acts on postjunctional P2 purinoceptors, adenosine generated from the extracellular breakdown of ATP acts largely via P1 purinoceptors on the nerve terminals to inhibit release of transmitter (De Mey et al 1979). Prejunctional modulation via P1 purinoceptors operates both as a negative feedback system in autoregulation in purinergic transmission and also to modulate the release of noradrenaline, acetylcholine (ACh) and other neurotransmitters (see Burnstock 1995).

In 1985, Burnstock & Kennedy proposed the first subdivision of P2 purinoceptors into P2X purinoceptors (which mediate vasoconstriction and contraction of visceral smooth muscle, with α,β-methylene ATP as a potent agonist) and P2Y purinoceptors (which mediate vasodilatation as well as relaxation of the smooth muscle of the gut, with 2-methylthioATP as a particularly potent agonist). Soon after, two further P2 purinoceptors were tentatively proposed (Gordon 1986): a P2T purinoceptor, which is ADP-selective involved in platelet aggregation; and a P2Z purinoceptor, which appears to be activated by ATP^{4-} and is prominent in macrophages, lymphocytes and mast cells. Another important landmark, following the seminal studies of Furchgott & Zawadzki (1980), was the recognition that P2Y purinoceptors on endothelial cells mediate vasodilatation via release of EDRF (endothelium-derived relaxing factor; De Mey & Vanhoutte 1981). This important discovery challenged the early hypothesis of Berne (1963) that adenosine is the local regulator of blood flow following hypoxia in heart and other vascular beds—it now seems likely that reactive hyperaemia is largely due to ATP, released from endothelial cells during hypoxia, acting on P2Y purinoceptors to release EDRF (now known to be nitric oxide, NO), resulting in vasodilatation; adenosine, produced following the breakdown of ATP, is likely to contribute to the later component of vasodilatation by direct action on P1 purinoceptors on vascular smooth muscle (Fig. 1, Hopwood et al 1989, Burnstock 1987, 1993a, Burnstock & Ralevic 1994).

The concept of co-transmission was put forward in 1976 in a review article entitled 'Do some nerve cells release more than one transmitter?' (Burnstock

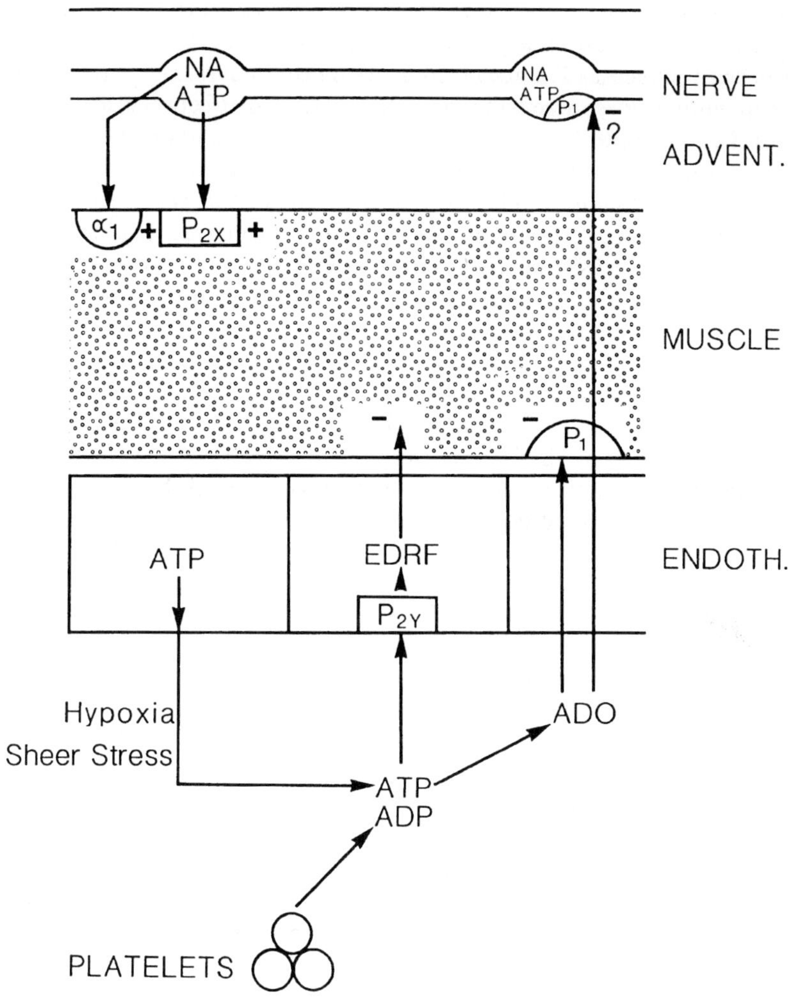

FIG. 1. A schematic representation of the interactions of ATP released from perivascular nerves and from endothelial cells (ENDOTH.). ATP is released from endothelial cells during hypoxia to act on endothelial P2Y purinoceptors leading to production of EDRF (NO) and subsequent vasodilatation (−). In contrast, ATP released as a co-transmitter with noradrenaline (NA) from perivascular sympathetic nerves at the adventitial (ADVENT.)–muscle border produces vasoconstriction (+) via P2X purinoceptors on the muscle cells. Different subclasses of P2Y purinoceptor are involved—both $P2Y_1$ and $P2Y_2$ (formerly termed $P2_U$) purinoceptors are present on most endothelial cells, while the subclass of P2Y purinoceptor found on some vascular smooth muscles has not been identified yet. Adenosine (ADO), resulting from rapid breakdown of ATP by ectoenzymes, usually produces vasodilatation by direct action on the muscle via P1 purinoceptors; it may also act via P1 purinoceptors on the perivascular nerve terminal varicosities to inhibit release of transmitter. (Figure modified from Burnstock 1987.)

1976b). This concept is now widely supported (Hökfelt et al 1986, Kupfermann 1991, Burnstock 1990b). It appears that ATP is a primitive transmitter and that it has been retained as a co-transmitter with other neurotransmitters in many different nerve types, albeit in proportions that vary between locations and species. For example, in sympathetic nerves ATP co-exists and is co-released with noradrenaline and neuropeptide Y; in some parasympathetic nerves with ACh; in some sensory-motor nerves with calcitonin gene-related peptide (CGRP) and substance P; and in NANC inhibitory nerves, together with NO and vasoactive intestinal peptide (VIP) (see Fig. 2).

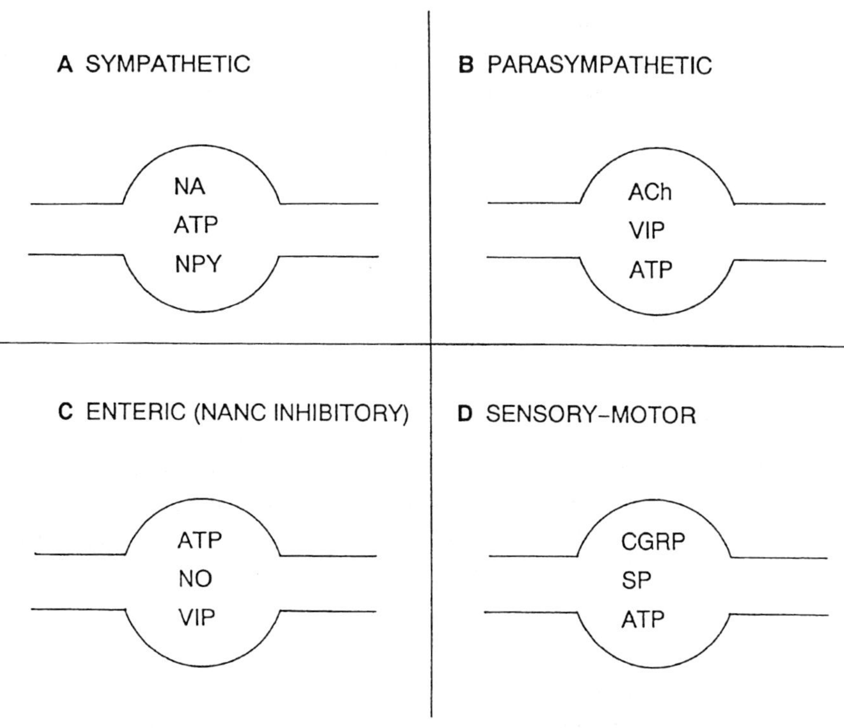

FIG. 2. Schematic representation of the chemical coding of autonomic nerves. The main transmitters found in terminal varicosities of sympathetic, parasympathetic, enteric (NANC inhibitory) and sensory–motor nerves are included, although the proportions of these transmitters varies considerably in different locations, species and at different stages of development and ageing. NA, noradrenaline; NPY, neuropeptide Y; ACh, acetylcholine; VIP, vasoactive intestinal polypeptide; CGRP, calcitonin gene-related peptide; SP, substance P.

In some recent studies, ATP has also been shown to be co-localized with γ-aminobutyric acid (GABA) in retinal nerves and perhaps released with glutamate in some hippocampal neurons (Perez & Bruun 1987, Wieraszko et al 1989, Kupfermann 1991). It is interesting that when excitatory junction potentials (EJPs) were first recorded in smooth muscle cells of the vas deferens in response to single pulses evoked by sympathetic nerves (Burnstock & Holman 1961), the authors were puzzled that depletion of noradrenaline with reserpine failed to block these responses (Burnstock & Holman 1962). It was about 15 years later before it was clear that we had been recording purinergic sympathetic co-transmission when, in response to sympathetic stimulation, EJPs were shown to be abolished by selective desensitization of the ATP receptor with α,β-methylene ATP in the vas deferens and rat tail artery (Kasakov & Burnstock 1983, Sneddon & Burnstock 1984a,b). Furthermore, local application of ATP, but not noradrenaline, mimicked EJPs in the vas deferens (Sneddon & Westfall 1984).

Expression of co-transmitters in autonomic nerves shows remarkable plasticity in development and ageing, in nerves that remain following trauma and surgery, under the influence of hormones, and in various disease situations (Burnstock 1990a). There are several examples which involve purinergic co-transmission. (1) In spontaneously hypertensive rats a significantly greater role for ATP compared with noradrenaline has been demonstrated in tail arteries (Vidal et al 1986), mesenteric arteries (Woolridge & van Helden 1990) and in various blood vessels *in vivo* (Bulloch & McGrath 1992). (2) Whereas the purinergic component of parasympathetic contraction of the rodent bladder is prominent compared to the cholinergic component (Burnstock et al 1978, Dean & Downie 1978, see Hoyle & Burnstock 1991), there has been debate about whether there is a purinergic component in the parasympathetic supply to the human bladder (Husted et al 1983, Sibley 1984, Hoyle et al 1989), even though the presence of P2 purinoceptors is well established (Hoyle et al 1989, Inoue & Brading 1991, Bo & Burnstock 1995). Interestingly, however, evidence for a substantial purinergic component in the responses to the urinary bladder of women with interstitial cystitis has been reported (Palea et al 1993) and increased purinergic responses demonstrated in the neurogenic bladder (Ruggieri et al 1990). (3) Whereas in the developing myotube, ATP acts as a co-transmitter with ACh with both transmitters opening ion channels (Kolb & Wakelam 1983, Hume & Honig 1986, Häggblad & Heilbronn 1988, Henning et al 1993), in the adult skeletal neuromuscular junction, the ATP released from adult nerves acts either as a postjunctional modulator potentiating the action of ACh or as a prejunctional modulator of ACh release via P1 purinoceptors after ectoenzymic breakdown of ATP to adenosine (Nagano et al 1992, Smith & Lu 1991, Lu & Smith 1991).

The concept of purinergic transmission was boosted when it was shown clearly that ATP was used as a fast transmitter between neurons in both autonomic ganglia (Evans et al 1992, Silinsky et al 1992) and the medial habenula (Edwards et al 1992). Since then there have been several reports of the potent actions of ATP in the CNS (see Inoue et al 1992, Harms et al 1992, Ueno et al 1992, Wieraszko & Ehrlich 1994, Ergene et al 1994, Illes et al 1995) and convincing demonstrations of the autoradiographic localization of P2X purinoceptors in different regions of the brain (Bo & Burnstock 1994).

Current status of P2 purinoceptors

Knowledge of the structure and properties of P2 purinoceptors has lagged behind information for most other neurotransmitters. The general progress of information has involved structure–activity, pharmacology, quantitation of receptor expression, ligand binding and autoradiographic localization, transduction mechanisms involving second messenger systems and ion channels, and finally the molecular biology of the receptors with cloning and sequencing. Since the subdivisions of the P2 purinoceptor into P2X, P2Y, P2T and P2Z mentioned earlier, several subclasses have been proposed, including P2U purinoceptors where ATP and UTP are equipotent (O'Connor et al 1991) and a P2D purinoceptor, selective for diadenosine polyphosphates (Pintor & Miras-Portugal 1993). It was clearly shown that P2X purinoceptors involved ligand-gated cation channels, while P2Y purinoceptors involve G protein activation (Fig. 3, Dubyak 1991). More recently, the possibility that some P2Y purinoceptors act via G_i proteins to inhibit adenylate cyclase has been raised (Harden et al 1995) and also that there may be uridine nucleotide-selective G protein-linked receptors (Lazarowski & Harden 1994).

The first P2 purinoceptors to be cloned were G protein-coupled P2Y purinoceptors: $P2Y_1$ purinoceptors were isolated from chick brain (Webb et al 1993); and a P2U purinoceptor (later designated $P2Y_2$) from neuroblastoma cells (Lustig et al 1993). A year later two ligand-gated ion channel ATP receptors were reported—one from vas deferens (Valera et al 1994) and another from rat phaeochromocytoma PC12 cells (Brake et al 1994). The P2Y and P2U purinoceptors had the typical seven-transmembrane-domain structures (Fig. 4b), while the P2X purinoceptors consisted of two transmembrane domains with a large extracellular loop rich in cysteines (Fig. 4a).

In the recent paper from the subcommittee concerned with the nomenclature of P2 purinoceptors (Fredholm et al 1994), it was emphasized that the current purinoceptor subclassification, with so many letters of the alphabet being somewhat randomly added as new receptor subtypes were discovered, was unsatisfactory. They supported, in principle, a new system of classification proposed by Abbracchio & Burnstock (1994). In this proposal, it was suggested

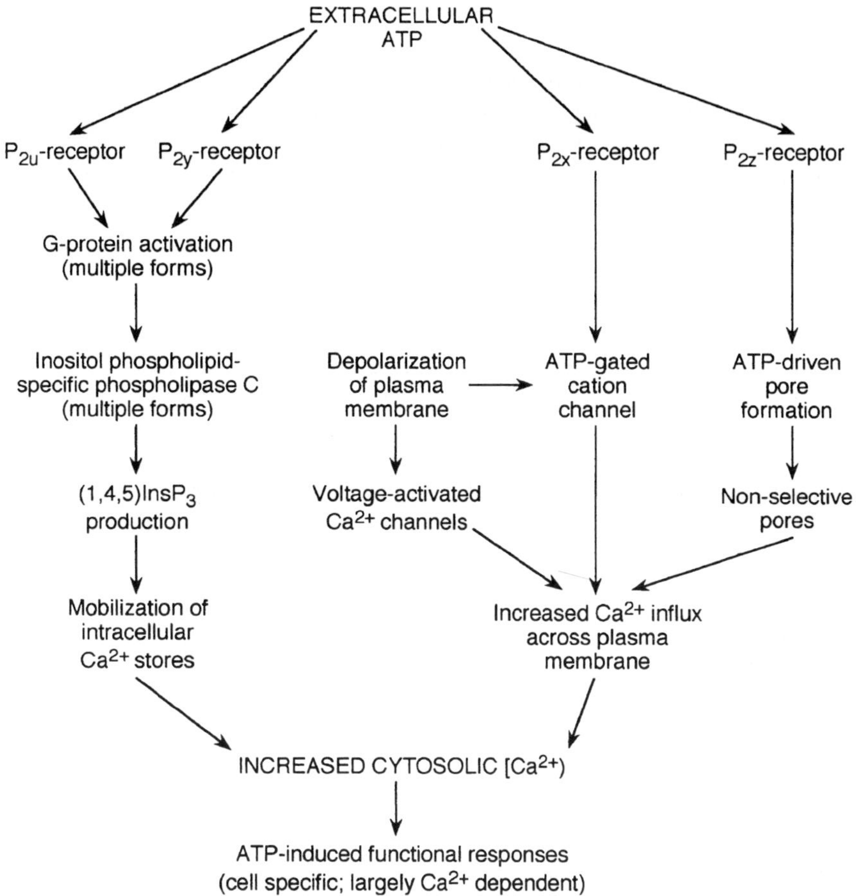

FIG. 3. Mechanisms underlying the increase in cystolic [Ca^{2+}] activated by different P2 purinoceptors for extracellular ATP. Extracellular ATP can interact with several P2 purinoceptor subtypes expressed in different cell types. Occupation of each of these receptor subtypes has been shown to induce rapid increases in cystolic [Ca^{2+}], with consequent activation of Ca^{2}-regulated cellular functions (e.g. contraction in muscle cells or exocytotic secretion in endocrine/neuroendocrine cells). Occupation of both P2Y and P2U purinoceptors primarily induces mobilization of Ca^{2+} sequestered in inositol 1,4,5-trisphosphate ($InsP_3$)-releasable, intracellular stores. Conversely, P2X and P2Z purinoceptors primarily increase Ca^{2+} influx, via a variety of channels and 'pores', across the plasma membrane. (From Dubyak 1991.)

that P2 purinoceptors should be placed in two major families, a P2X family consisting of ligand-gated cation channels and a P2Y family consisting of G protein-mediated receptors. They recognized that the P2Z purinoceptor, which opens non-selective pores, might represent a third family (Table 1). It

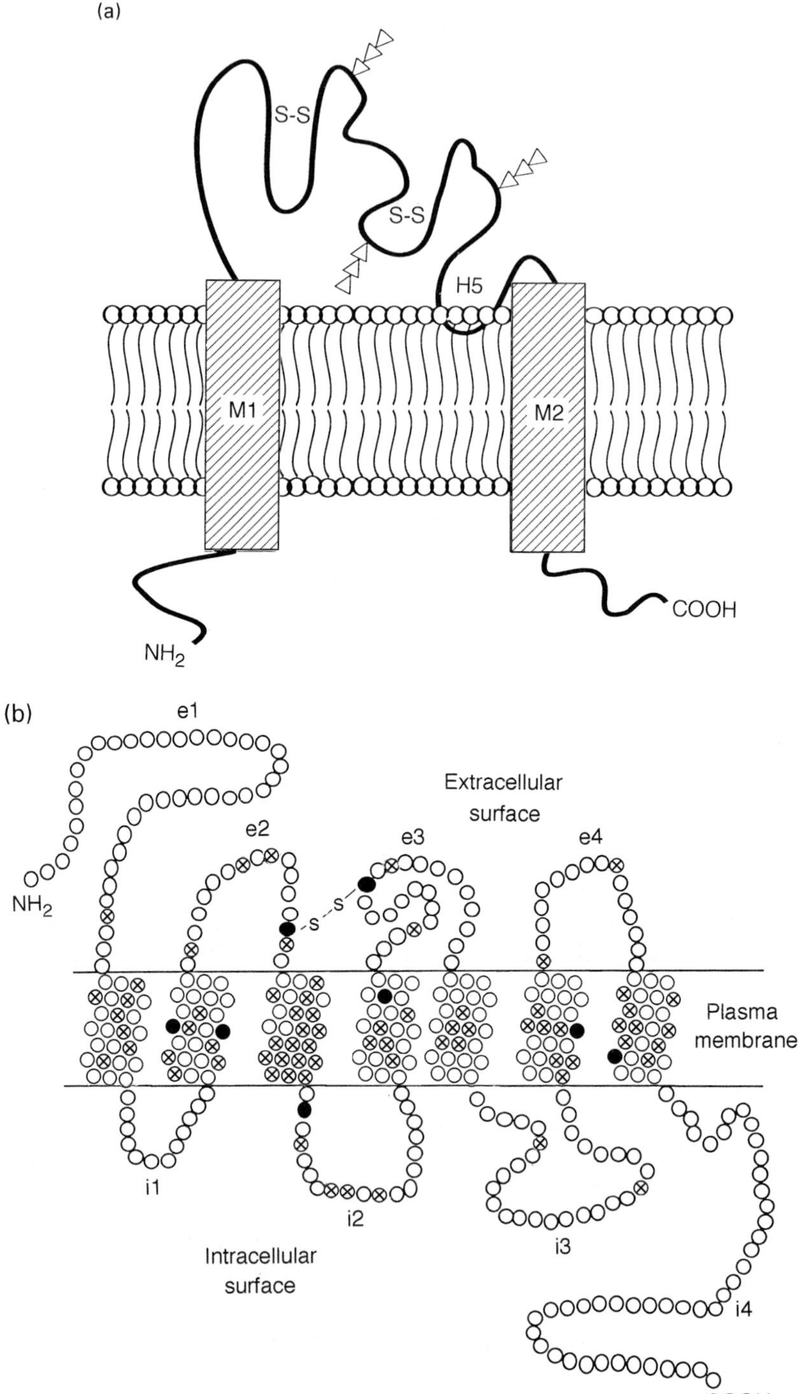
(a)
S-S
S-S
H5
M1
M2
COOH
NH_2
(b)
e1
Extracellular
surface
e2
e3
e4
NH_2
s
s
Plasma
membrane
i1
i2
i3
i4
Intracellular
surface
COOH

TABLE 1 Proposed subclassification of P2 purinoceptors

Name	*P2X purinoceptor family*	*P2Y purinoceptor family*	*P2Z*
Type	Ligand-gated channel	G protein-coupled	Non-selective pore
General agonist profile	α,β-meATP > β,γ-meATP > ATP ≈ 2-MeSATP ≈ ADP	2-MeSATP > ATP = ADP > α,β-meATP ⩾ β,γ-meATP	ATP^{4-}
Antagonists	α,β-meATP desensitization Suramin Selectively blocked by PPADS ANAPP3	Suramin Reactive blue 2	Oxidized ATP

α,β-meATP, α,β-methylene ATP; β,γ-meATP, β,γ-methylene ATP; 2-MeSATP, 2-methylthioATP; PPADS, pyridoxalphosphate-6-azophenyl-2′,4′-disulfonic acid; ANAPP3, 3-*O*-3[*N*-(4-azido-2-nitrophenyl)amino] proprionyl ATP. (Reproduced with permission from Abbracchio & Burnstock 1994.)

was pointed out that this classification brought ATP into line with most other neurotransmitters such as ACh, GABA, glutamate and 5-hydroxytryptamine (5-HT), where ligand-gated and G protein-mediated receptor subclassifications have already been established (Table 2, Burnstock 1995).

On the basis of the relative actions of a series of newly established purine analogues (Fischer et al 1993, Burnstock et al 1994, Bo et al 1994), an analysis of transduction mechanisms (Friel 1988, Dubyak 1991, Pirotton et al 1991, El-Moatassim et al 1992, Harden et al 1995) and on the structure of cloned receptors, Abbracchio & Burnstock tentatively named four subclasses of the

FIG. 4. Molecular structure of P2X and P2Y purinoceptors. (a) Model depicting a proposed transmembrane topology for $P2X_2$ protein is shown with both N- and C-termini in the cytoplasm. Two putative membrane-spanning segments (M1 and M2) traverse the lipid bilayer of the plasma membrane and are connected by a hydrophilic segment of 270 amino acids. This putative extracellular domain is shown containing two disulfide-bonded loops (S-S) and three *N*-linked glycosyl chains (triangles). The $P2X_2$ cDNA was sequenced on both strands using Sequenase (USB). (From Brake et al 1994.) (b) Schematic diagram of the sequence of $P2Y_1$ purinoceptor showing its differences from $P2Y_2$ and $P2Y_3$ purinoceptors. Crossed circles represent amino acid residues that are conserved between the three receptors; open circles represent residues that are not conserved; and filled circles represent residues that are known to be functionally important in other G protein-coupled receptors. (From Barnard et al 1994.)

TABLE 2 Comparison of fast ionotropic and slow metabotropic receptors for acetylcholine (ACh), γ-aminobutyric acid (GABA), glutamate and 5-hydroxytryptamine (5-HT) with those proposed for ATP

	Receptors	
Messenger	*Fast ionotropic*	*Slow metabotropic*
ACh	Nicotinic Muscle type Neuronal type	Muscarinic M1–M5
GABA	$GABA_A$	$GABA_B$
Glutamate	AMPA Kainate NMDA	$mGlu_1$–$mGlu_7$
5-HT	5-HT_3	$5\text{-HT}_{1A–F}$ $5\text{-HT}_{2A–C}$ 5-HT_4 $5\text{-HT}_{5A–B}$ 5-HT_6 5-HT_7
ATP	$P2X_{1–6}$	$P2Y_{1–7}$

AMPA, 2-(aminomethyl)phenylacetic acid; NMDA, *N*-methyl-D-aspartate.

P2X purinoceptor family and seven subclasses of the P2Y purinoceptor family. A number of further P2X and P2Y purinoceptor molecular sequences have subsequently been reported and the current situation is summarized in Tables 3 and 4. A new member of the P2X purinoceptor family has just been identified in our laboratory which is expressed by a subset of sensory neurons (Chen et al 1995). *In situ* hybridization and Northern blotting of the $P2X_3$ purinoceptor show that the channel transcript is present in a subset of rat dorsal root ganglion sensory neurons, some of which express nociceptor-associated markers; it is absent in other tissues we have tested, including sympathetic, enteric and CNS neurons. When expressed in *Xenopus* oocytes, the channel shows an ATP-dependent cation flux ($EC_{50} \approx 1.2\,\mu M$), and a rank order of potency of 2-methylthioATP > > ATP > α,β-methylene ATP > ATPγS > 2′-deoxy ADP > CTP ≈ ADP > > UTP ≈ β,γ-methylene ATP > GTP. $P2X_3$ is the only ligand-gated channel known to be expressed exclusively by a subset of sensory neurons. The remarkable selectivity of expression of the channel coupled with its sensory neuron-like pharmacology suggest that this channel may transduce ATP-evoked nociceptor activation. In Fig. 5, the primary sequence of the $P2X_3$ purinoceptor is aligned with the two other known members of the P2X purinoceptor family isolated from PC12 cells ($P2X_2$) and rat vas deferens ($P2X_1$).

TABLE 3 **Classification of subtypes of P2X purinoceptor family on the basis of molecular and functional characteristics**

P2 receptor subtype	*Tissue*	*Activity*	*Properties*	*References*	*Genbank/EMBL accession no.*
$P2X_1$	Vas deferens (rat)	2-MeSATP > ATP > α,β-meATP	$I_{Na/K/Ca}$	Valera et al 1994	X80477
$P2X_2$	PC12 cells (rat)	2-MeSATP > ATP (α,β-meATP inactive)	$I_{Na/K}$	Brake et al 1994	U14414
$P2X_2-1$ (short form)	Cochlea (rat)	ND	ND	Housley et al 1995	L43511
$P2X_3$	DRG cells (rat)	2-MeSATP > ATP > α,β-meATP	$I_{Na/K}$	Chen et al 1995a	—
	DRG cells (rat)	ATP > 2-MeSATP > α,β-meATP	$I_{Na/K/Ca}$	Lewis et al 1995	X91167
$P2X_4$	Hippocampus (rat)	ATP > 2-MeSATP > α,β-meATP	$I_{Na/K}$	Bo et al 1995	—
	DRG cells (rat)	ATP active, α,β-meATP inactive	$I_{Na/K/Ca}$	Buell et al 1996	—
	Neurons (rat)	ATP > > 2-MeSATP > CTP > α,β-meATP > dATP	$I_{Na/K/Ca}$	Soto et al 1996	X93565
	Neurons (rat)	—	—	Seguela et al 1996	U32497
$P2X_5$	Neurons (rat)	ATPγ5 $\geqslant$ ATP $\geqslant$ 2-MeSATP > > ADP (α,β-meATP inactive)	$I_{Na/K/Ca}$	Collo et al 1996	—
$P2X_6$	Neurons (rat)	2-MeSATP = 2-cl-ATP = ATP > ATPγ5 > > ADP (α,β-meATP inactive)	$I_{Na/K/Ca}$	Collo et al 1996	—
P2Z (or P2X-like)	Macrophage (mouse)	(i) BzATP > ATP > UTP (ii) ATP > UTP > BzATP	(i) $I_{Na/K}$ (ii) I_{Ca}	Nuttle et al 1993	—

ND, not determined; DRG, dorsal root ganglion; 2-MeSATP, 2-methylthioATP; a,β-meATP, α,β-methylene ATP; BzATP, 3′-O-(4-benzoyl)benzoyl ATP; dATP, 3′-deoxy ATP.

TABLE 4 Classification of subtypes of P2Y purinoceptor family on basis of molecular and functional characteristics

P2 receptor subtype	*Tissue*	*Activity*	*Properties*	*References*	*Genbank/EMBL accession no.*
$P2Y_1$	Brain (chick)	2-MeSATP > ATP > ADP (UTP inactive)	PLCβ/InsP_3/Ca^{2+}	Webb et al 1993	X73268
	Brain (turkey)	2-MeSATP > ADP > ATP (UTP inactive)	PLCβ/InsP_3/Ca^{2+}	Filtz et al 1994	U09842
	Brain (human)	—	—	Ayyanathan et al, unpublished	*U42029/30
	Insulinoma cells (mouse)	ND	ND	Tokuyama et al 1995	U22829
	Insulinoma cells (rat)	2-MeSATP > 2-Cl-ATP > ATP (α,β-meATP inactive)	PLCβ/InsP_3/Ca^{2+}	Tokuyama et al 1995	U22830
	Placenta (human)	ND	ND	Leon et al 1996	Z49205
	Endothelium (bovine)	2-MeSATP > ATP > > UTP	PLCβ/InsP_3/Ca^{2+}	Henderson et al 1995	X87628
$P2Y_2$	NG108-15 cells (mouse)	ATP = UTP > > 2-MeSATP	PLCβ/InsP_3/Ca^{2+}	Lustig et al 1993	L14751
	CT/43 cells (human)	ATP = UTP > > 2-MeSATP	PLCβ/InsP_3/Ca^{2+}	Parr et al 1994	U07225
	Lung (rat)	ATP = UTP	PLCβ/InsP_3/Ca^{2+}	Rice et al 1995	U09402
	Bone (human)	ND	ND	Bowler et al 1995	—
	Pituitary (rat)	ND	ND	Chen et al 1996	L46865
$P2Y_3$	Brain (chick)	ADP > UTP > ATP = UDP	PLCβ/InsP_3/Ca^{2+}	Webb et al 1996a	—
$P2Y_4$	Placenta (human)	UTP = UDP > ATP = ADP	PLCβ/InsP_3/Ca^{2+}	Communi et al 1995	—
	Brain (rat)	ND	ND	(to be confirmed)	—
$P2Y_5$	HEL cells (human)	ADP > ATP > UTP	ND	Kunapuli et al 1996	U41070
	Lymphocytes (chicken)	ATP > ADP > 2-MeSATP > > α,β-meATP = UTP†	ND	Webb et al 1996b	—
$P2Y_6$	Aortic smooth muscle (rat)	UTP > ADP = 2-MeSATP > ATP	PLCβ/Ins$P_3$$Ca^{2+}$	Chang et al 1995	D63665

ND, not determined; 2-MeSATP, 2-methylthioATP; 2-Cl-ATP, 2-chloroATP; α,β-meATP, α,β-methylene ATP; PLCβ, phospholipase Cβ, InsP_3, inositol 1,4,5-trisphosphate. *Direct submission to Genbank of long and short form of gene encoding the same receptor protein.
†Radioligand binding.

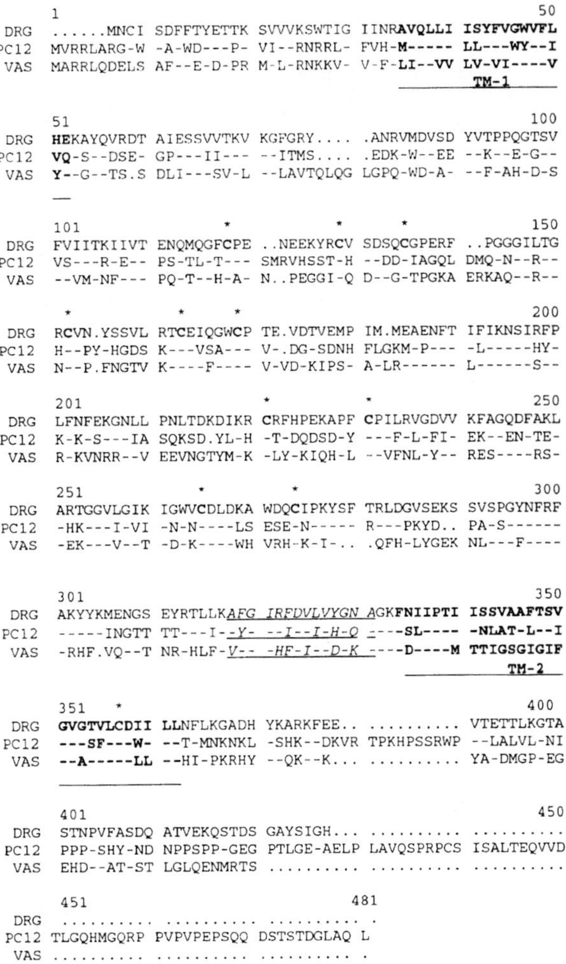

FIG. 5. Primary sequence of the $P2X_3$ purinoceptor (DRG) aligned with those of two other known members of the P2X purinoceptor family isolated from PC12 cells ($P2X_2$) and rat vas deferens ($P2X_1$). Dashes represent sequence identity, dots represent gaps. The putative transmembrane regions are bold and underlined, and the H5-related regions also present in K^+ channels are italicized and underlined. Note the conserved cysteine residues (bold) marked with asterisks. (From Chen et al 1995.)

Future directions

Long-term (trophic) actions of purines

While there are substantial studies of the short-term actions of purines in neural and neuromuscular transmission and secretion in a range of different cell types, there are relatively few studies on the long-term (trophic) actions of

TABLE 5 P2 purinoceptor antagonists

Compounds	*Receptor subtype claimed*	*Tissue*	*pA_2 value (if known)*	*References*	*Comments*
Quinidine	P2X	Taenia coli	—	Burnstock et al 1970	Low concentrations antagonize adrenergic responses; higher concentrations are needed to antagonize purinergic responses (≈300 μM)
	P2Y	Urinary bladder	—	Burnstock et al 1972	
Imidazolines (phentolamine)	P2Y	Taenia coli	—	Sathell et al 1973	Effective only at high concentrations (≈200 μM)
PIT	P2Y	Taenia coli	—	Hooper et al 1975	Causes direct relaxation of the muscle. At concentrations greater than 100 μM causes non-specific effects
			—	Spedding et al 1975	
ANAPP3	P2X	Vas deferens, urinary bladder	—	Hogaboom et al 1980	Irreversible antagonism occurs after photolysis due to covalent binding to ATP-binding sites. Binding occurs also to any ATP-metabolizing protein
			—	Westfall et al 1983	
Apamin	P2Y	Taenia coli	—	Brown & Burnstock 1981	Now best known as an inhibitor of K^+ channels
α,β-meATP	P2X	Urinary bladder, vas deferens	—	Kasakov & Burnstock 1983 Meldrum & Burnstock 1983	Repeat administrations cause selective antagonism due to desensitization of purinoceptors. Very selective for P2X purinoceptors, but not competitive
Reactive blue 2	P2Y	Colon, duodenum, caecum, portal vein, taenia coli	$pK_B = 5.0$	Kerr & Krantis 1979 Manzini et al 1985, 1986 Reilly et al 1987	Produces reasonably selective antagonism at P2Y purinoceptors at concentrations less than 50 μM
	P2X	Urinary bladder, vas deferens	—	Choo 1981 Bültmann & Starke 1994a	
	P2Z	Lymphocytes	—	Wiley et al 1993	
Suramin	P2X	Vas deferens, urinary bladder	$pA_2 = 4.6–5.0$	Dunn & Blakeley 1988 Hoyle et al 1990	Fairly selective for P2 purinoceptors, but cannot discriminate between subtypes Has diverse other biological activities (Voogd et al 1993)

(Continued)

	P2Y	Taenia coli	—	Hoyle et al 1990	
	P2T	Platelet	$pA_2 = 4.6$	Hourani et al 1992	
	P2Z	Lymphocytes	—	Wiley et al 1993	
XAMR0721	P2Y	Turkey erythrocytes	$K_i = 19 \mu M$	van Rhee et al 1994	Benzamide congener of suramin. Claimed to be selective, but has not been tested on other receptors
PPADS	P2X	Vas deferens, urinary bladder, blood vessels	$pK_B = 6.0–6.5$	Lambrecht et al 1992 Ziganshin et al 1993, 1994 McLaren et al 1994	At present the drug of choice for selective antagonism at P2X purinoceptors
	P2Y	Turkey erythrocytes	—	Boyer et al 1994	
isoPPADS	P2X	Vas deferens, vagus nerve, sympathetic ganglia	$pK_B = 6.6$	Bültmann & Starke 1994a Trezise et al 1994a Khahk et al 1994 Connoly 1995	Nearly as effective as PPADS
P5P	P2X	Vas deferens, vagus nerve, sympathetic ganglia	$pK_i = 4.9$	Trezise et al 1994b Connoly 1995	Synthesis precursor of PPADS
DIDS	P2X	Vas deferens	$IC_{50} = 3.9 \mu M$	Bültmann & Starke 1994b	Also inhibits anion transport (Cabantchik & Greger 1992)
	P2Z	Parotid acinar cells	—	Soltoff et al 1993	
Evans blue	P2X	Vas deferens	$pK_B = 5.9$	Bültmann & Starke 1993	
Trypan blue	P2X	Vas deferens	$pA_2 = 5.3$	Bültmann et al 1994	
Congo red	P2X	Vas deferens	$pK_B = 4.6$	Khakh et al 1994	
Brilliant blue	P2Z	Parotid acinar cells	—	Soltoff et al 1989	
	P2Y?	Vas deferens, urinary bladder	$pK_B = 4.5$	von Kügelgen et al 1994 Palea et al 1995	
TNP-ATP	P2Y?	Cochlear hair cells	—	Mockett et al 1994	
o-ATP	P2Z	Macrophages	—	Murgia et al 1993	
ARL 66096	P2T	Platelets	$pK_B = 8.7$	Humphries et al 1994	

PIT, 2,2′-pyridylisatogen tosylate; ANAPP3, 3-*O*-3[*N*-(4-azido-2-nitrophenyl) amino] proprionyl ATP; α,β-MeATP, α,β-methylene ATP; XAMRO721, 8-β,5-dinitrophenylene carbonylimino)-1,3,5-napthalene trisulfonate; PPADS, pyridoxalphosphate-6-azophenyl-2′4′-disulfonic acid; isoPPADS, pyridoxalphosphate-6-azophenyl-2′,5′-disulfonic acid; P5P, pyridoxal-5-phosphate; DIDS, 4,4′-diisothiocyanotostilbene-2,2′-disulfonate; ARL66096 (formerly FPL66096), 2-propylthio-D-β,γ-difluoromethylene ATP; TNP-ATP, 2′-(or 3′)-*O*-trinitrophenyl ATP; O-ATP, oxidized ATP.

purines (Burnstock 1993b, Neary et al 1996). However, there are an increasing number of reports about the roles of purines in embryonic development (Knudsen & Elmer 1987, Laasberg 1990), growth and cell proliferation (Gonzales et al 1990, Wang et al 1990, Malam-Souley et al 1993, Erlinge et al 1995). There is also recognition of the involvement of purines in programmed cell death or apoptosis (Zheng et al 1991, Murgia et al 1992, Avery et al 1992). Trophic roles for purines on the activation of glial and neuronal cells have recently been described (Rathbone et al 1992, Neary & Norenberg 1992, Abbracchio et al 1994), as have their interactions with growth factors (Neary et al 1994, Schäfer et al 1995, Abbracchio et al 1995).

The participation of purines in neuro-immune mechanisms (Steinberg & Silverstein 1987, El-Moatassim et al 1987, Murgia et al 1992) and in bone and cartilage resorption (Caswell et al 1992, Leong et al 1994, Yu & Ferrier 1993) is beginning to be explored.

Evolution and development of purinoceptors

ATP is one of the first molecules to appear in the evolution of biological systems as an energy source and as an essential component of nucleic acids (Sigel 1992). There are growing indications that ATP might also have been used early in evolution as an extracellular messenger (Burnstock 1975, 1977, 1979, Berlind 1977, Venter et al 1988, Hoyle & Greenberg 1988). For example, ATP has been reported to have potent effects on various activities of amoebae (Zimmerman 1962, Pothier et al 1987). Purine compounds elicit contractions of the pedal disc of the sea anemone (Hoyle et al 1989) and depolarize specific neurons in the nervous system of the leech (Backus et al 1994). Separate receptors for adenosine and ATP have been identified in molluscan preparations, including neurons of the snail (Yatani et al 1982), the bivalve *Mytilus* (Barraco & Stefano 1995), the heart of various marine molluscs (Hoyle & Greenberg 1988, Knight et al 1992) and proboscis muscle of *Buccinium* (Nelson & Huddart 1994). Purine compounds also relax the gastric ligament of the starfish (Hoyle & Greenberg 1988, Knight et al 1990).

The sensillae of the olfactory organ of the spiny lobster are excited by low concentrations of AMP, ADP or ATP, which act as chemoreceptors in feeding responses (Carr et al 1986, Zimmer-Faust et al 1988), and ATP stimulates the gorging responses in a variety of blood-sucking insects (Galun & Kabayo 1988, Liscia et al 1993). The effects of purine nucleosides and nucleotides have been studied in a number of preparations from lower vertebrates, including the elasmobranch and teleost fish intestine (Young 1988, Jensen & Holmgren 1985, Kitazawa et al 1990), the cardiovascular system of fish, amphibians, reptiles and birds (Flitney et al 1977, Colin et al 1979, Meghji & Burnstock 1984,

Knight & Burnstock 1995, Lennard & Huddart 1989), and fish chromatophores (Miyashita et al 1984, Namoto 1992). While there are reports about P2 purinoceptor activity in early development in a variety of tissues (e.g. Hilfer et al 1977, Smurs 1981, Levin et al 1981, Furukawa & Nomoto 1989, Laasberg 1990, Rathbone et al 1992, Zheng et al 1994, Abe et al 1995), few attempts have been made yet to characterize the purinoceptor subtypes involved (Erlinge et al 1995).

Therapeutic applications

There has been considerable interest in the therapeutic potential of adenosine acting via different subclasses of P1 purinoceptors for some time (see reviews by Daly 1982, Stone 1992, Williams 1993). Interest in the therapeutic potential of ATP acting through P2 purinoceptors is more recent (see Burnstock 1993b, Abbracchio & Burnstock 1994). It is remarkable that during the past year, serious explorations of a therapeutic role for P2 purinoceptors have been reported for cystic fibrosis (Boucher et al 1995), diabetes (Loubatiéres-Mariani et al 1995), immune and inflammatory disease situations (Di Virgilio et al 1995) and cancer (Rapaport 1993). Most recently, the discovery of a selective P2T antagonist (Humphries et al 1994) is leading to the development of an antithrombotic agent (Humphries et al 1995). Other conditions where there is interest in pursuing therapeutic goals involving P2 purinoceptors include bladder incontinence (Hoyle & Burnstock 1996), pulmonary hypertension (Rubino & Burnstock 1994), surfactant and mucosal secretion (Griese et al 1993), constipation and diarrhoea (Milner & Burnstock 1994), behavioural disorders such as epilepsy, depression and ageing-associated neurodegenerative diseases (Williams 1993), contraception and sterility (Foresta et al 1992), ischaemia (Phillis et al 1993), and wound healing (Wang et al 1990, Namiot et al 1993). Increasing interest in the therapeutic potential of purinoceptors reinforces the pressing need for the development of specific agonists and antagonists for all new purinoceptor subtypes as they become established and especially agents that will be effective *in vivo*. A summary of the various compounds claimed to be P2 purinoceptor antagonists through the years, with comments about their actions, is given in Table 5. The development and use of nucleoside transport inhibitors (Van Belle 1995) and ATPase inhibitors (Crack et al 1995) is also of considerable interest in therapeutic terms.

Final comment

There is clearly an exploding interest in the molecular identification, distribution and roles of different P2 purinoceptor subtypes and in their application to clinical medicine. The future of this field is intensely exciting and I look forward to much fruitful debate in the coming days.

Acknowledgements

My thanks to Brian King and Airat Ziganshin for their help in preparing Tables 3, 4 and 5, to Charles Hoyle and Vera Ralevic for their critical feedback, and to Daphne Christie and Annie Evans for their help with the manuscript.

References

Abbracchio M, Burnstock G 1994 Purinoceptors: are there families of P2X and P2Y purinoceptors? Pharmacol Ther 64:445–475

Abbracchio MP, Saffrey MJ, Höpker V, Burnstock G 1994 Modulation of astroglial cell proliferation by analogues of adenosine and ATP in primary cultures of rat striatum. Neuroscience 59:67–76

Abbracchio MP, Ceruti S, Burnstock G, Cattabeni F 1995 Purinoceptors on glial cells of the central nervous system: functional and pathological implications. In: Belardinelli L, Pelleg A (eds) Adenosine and adenine nucleotides: from molecular biology to integrative physiology. Kluwer Acad, Norwell, MA, p 271–280

Abe Y, Itoyama Y, Furukawa K, Akaike N 1995 ATP responses in the embryo chick ciliary ganglion cells. Neuroscience 64:547–551

Avery RK, Bleier KJ, Pasternack MS 1992 Differences between ATP-mediated cytotoxicity and cell-mediated cytotoxicity. J Immunol 149:1265–1270

Backus KH, Braum S, Lohner F, Deitmer JW 1994 Neuronal responses to purinoceptor agonists in the leech central nervous system. J Neurobiol 25:1283–1292

Barnard EA, Burnstock G, Webb TE 1994 G protein-coupled receptors for ATP and other nucleotides: a new receptor family. Trends Pharmacol Sci 15:67–70

Barraco R, Stefano GB 1995 Pharmacological evidence for the modulation of monoamine release by adenosine in the invertebrate nervous system. J Neurochem 54:2002–2006

Berlind A 1977 Cellular dynamics in invertebrate neurosecretory systems. Int Rev Cytol 49:171–251

Berne RM 1963 Cardiac nucleotides in hypoxia: possible role in regulation of coronary blood flow. Am J Physiol 204:317–322

Bo X, Burnstock G 1994 Distribution of [^{3}H]α,β-methylene ATP binding sites in rat brain and spinal cord. NeuroReport 5:1601–1604

Bo X, Burnstock G 1995 Characterization and autoradiographic localization of [^{3}H]α,β-methylene adenosine 5′-triphosphate binding sites in human urinary bladder. Br J Urol 76:297–302

Bo X, Fischer B, Maillard M, Jacobson KA, Burnstock G 1994 Comparative studies on the affinities of ATP derivatives for P2X-purinoceptors in rat urinary bladder. Br J Pharmacol 112:1151–1159

Bo X, Zhang Y, Nassar M, Burnstock G, Schöepfer R 1995 A P2X purinoceptor cDNA conferring a novel pharmacological profile. FEBS Lett 375:129–133

Boucher RC, Knowles MR, Olivier KN, Bennett W, Mason SJ, Stutts MJ 1995 Mechanisms and therapeutic actions of uridine triphosphate in the lung. In: Belardinelli L, Pelleg A (eds) Adenosine and adenine nucleotides: from molecular biology to integrative physiology. Kluwer Acad, Norwell, MA, p 525–532

Bowler WB, Birch MA, Gallagher JA, Bilbe G 1995 Identification and cloning of human P_{2U} purinoceptor present in osteoblastoma, bone, and osteoblasts. J Bone Min Res 10:1137–1145

Boyer JL, Zohn IE, Jacobson KA, Harden K 1994 Differential effects of P2-purinoceptor antagonists on phospholipase C- and adenylyl cyclase-coupled P2Y-purinoceptors. Br J Pharmacol 113:614–620
Brake AJ, Wagenbach MJ, Julius D 1994 New structural motif for ligand-gated ion channels defined by an ionotropic ATP receptor. Nature 371:519–523
Brown CM, Burnstock G 1981 Evidence in support of the P1/P2 purinoceptor hypothesis in the guinea-pig taenia coli. Br J Pharmacol 73:617–624
Buell G, Lewis C, Collo G, North RA, Surprenant A 1996 An antagonist-insensitive P_{2X} receptor expressed in epithelia and brain. EMBO J 15:55–62
Bulloch JM, McGrath JC 1992 Evidence for increased purinergic contribution in hypertensive blood vessels exhibiting cotransmission. Br J Pharmacol 107:145P
Bültmann R, Starke K 1993 Evans blue blocks P_{2X}-purinoceptors in rat vas deferens. Naunyn-Schmiedeberg's Arch Pharmacol 348:684–687
Bültmann R, Starke K 1994a P_2-purinoceptor antagonists discriminate three contraction-mediating receptors for ATP in rat vas deferens. Naunyn-Schmeidebergs Arch Pharmacol 349:74–80
Bültmann R, Starke K 1994b Blockade by 4,4′-diisothiocyanatostilbene-2,2′-disulfonate (DIDS) of P_{2X}-purinoceptors in rat vas deferens. Br J Pharmacol 112:690–694
Bültmann R, Trendelenburg M, Starke K 1994 Blockade of P_{2X}-purinoceptors by trypan blue in rat vas deferens. Br J Pharmacol 113:349–354
Burnstock G 1972 Purinergic nerves. Pharmacol Rev 24:509–581
Burnstock G 1975 Ultrastructure of autonomic nerves and neuroeffector junctions: analysis and drug action. In: Daniel EE, Paton DM (eds) Methods in pharmacology, vol 3: Smooth muscle. Plenum, New York, p 113–137
Burnstock G 1976a Purinergic receptors. J Theor Biol 62:491–503
Burnstock G 1976b Do some nerve cells release more than one transmitter? Neuroscience 1:239–248
Burnstock G 1977 Purine nucleotides and nucleosides as neurotransmitters or neuromodulators in the central nervous system. In: Usdin E, Humburg DA, Barshas JD (eds) Neuroregulators and psychiatric disorders. Oxford University Press, New York, p 470–477
Burnstock G 1978 A basis for distinguishing two types of purinergic receptor. In: Straub RW, Bolis L (eds) Cell membrane receptors for drugs and hormones: a multidisciplinary approach. Raven, New York, p 107–118
Burnstock G 1979 Speculations on purinergic nerves in evolution. In: Olive G (ed) Advances in pharmacology and therapeutics, vol 8: Drug–action modification—comparative pharmacology. Pergamon, Oxford, p 267–273
Burnstock G 1987 Local control of blood pressure by purines. Blood Vessels 24:156–160
Burnstock G 1990a Changes in expression of autonomic nerves in aging and disease. J Auton Nerv Syst 30:25S-34S
Burnstock G 1990b Co-transmission. Arch Int Pharmacodyn Ther 304:7–33
Burnstock G 1993a Hypoxia, endothelium and purines. Drug Dev Res 28:301–305
Burnstock G 1993b Physiological and pathological roles of purines: an update. Drug Dev Res 28:195–206
Bunstock G 1995 Receptors for ATP at peripheral neuroeffector junctions. In: Belardinelli L, Pelleg A (eds) Adenosine and adenine nucleotides: from molecular biology to integrative physiology. Kluwer Acad, Norwell, MA, p 289–295
Burnstock G, Holman ME 1961 The transmission of excitation from autonomic nerve to smooth muscle. J Physiol 155:115–133
Burnstock G, Holman ME 1962 Effect of denervation and reserpine treatment on transmission at sympathetic nerve endings. J Physiol 160:461–469

Burnstock G, Kennedy C 1985 Is there a basis for distinguishing two types of P_2 purinoceptor? Gen Pharmacol 16:433–440

Burnstock G, Ralevic V 1994 New insights into the local regulation of blood flow by perivascular nerves and endothelium. Br J Plast Surg 47:527–543

Burnstock G, Campbell G, Bennett M, Holman ME 1963 Inhibition of the smooth muscle of the taenia coli. Nature 200:581–582

Burnstock G, Campbell G, Satchell D, Smythe A 1970 Evidence that adenosine triphosphate or a related nucleotide is the transmitter substance released by non-adrenergic inhibitory nerves in the gut. Br J Pharmacol 40:668–688

Burnstock G, Dumsday B, Smythe A 1972 Atropine resistant excitation of the urinary bladder: the possibility of transmission via nerves releasing a purine nucleotide. Br J Pharmacol 44:451–461

Burnstock G, Cocks T, Crowe R, Kasakov L 1978 Purinergic innervation of the guinea-pig urinary bladder. Br J Pharmacol 63:125–138

Burnstock G, Fischer B, Hoyle CHV et al 1994 Structure–activity relationships for derivatives of adenosine 5′-triphosphate as agonists at P_2 purinoceptors: heterogeneity within P_{2X} and P_{2Y} subtypes. Drug Dev Res 31:206–219

Cabantchik ZI, Greger R 1992 Chemical probes for anion transporters of mammalian cell membranes. Am J Physiol 262:803C–827C

Carr WES, Gleeson RA, Ache BW, Milstead ML 1986 Olfactory receptors of the spiny lobster: ATP-sensitive cells with similarities to P_2-type purinoceptors of vertebrates. J Comp Physiol A 158:331–338

Caswell AM, Leong WS, Russell RGG 1992 Interleukin-1β enhances the response of human articular chondrocytes to extracellular ATP. Biochim Biophys Acta 1137:52–58

Chang K, Hanaoka K, Kumada M, Takuwa Y 1995 Molecular cloning and functional analysis of a novel P_2 nucleotide receptor. J Biol Chem 270:26152–26158

Chen CC, Akopian AN, Sivilotti L, Colquhoun D, Burnstock G, Wood JN 1995 A P_{2X} purinoceptor expressed by a subset of sensory neurons. Nature 377:428–431

Chen ZP, Krull N, Xu S, Levy A, Lightman SL 1996 Molecular cloning and functional characterization of a rat pituitary G protein-coupled ATP receptor. Endocrinology, in press

Choo LK 1981 The effect of reactive blue, an antagonist of ATP, on the isolated urinary bladder of guinea-pig and rat. J Pharm Pharmacol 33:248–250

Colin D, Kirscg R, Leray C 1979 Haemodynamic effects of adenosine on gills of the trout (*Salmo gairdneri*). J Comp Physiol 130:325–330

Collo G, North RA, Kawashima E et al 1996 Cloning of $P2X_5$ and $P2X_6$ receptors, and the distribution and properties of an extended family of ATP-gated ion channels. J Neurosci, in press

Communi D, Pirotton S, Parmentier M, Boeynaems J-M 1995 Cloning and functional expression of a human uridine nucleotide receptor. J Biol Chem 270:30849–30852

Connolly GP 1995 Differentiation by pyridoxal 5-phosphate, PPADS and isoPPADS between responses mediated by UTP and those evoked by α,β-methylene-ATP on rat sympathetic ganglia. Br J Pharmacol 114:727–731

Crack BE, Pollard CE, Beukers MW et al 1995 Pharmacological and biochemical analysis of FPL 67156, a novel, selective inhibitor of ecto-ATPase. Br J Pharmacol 114:475–481

Daly JW 1982 Adenosine receptors: targets for future drugs. J Med Chem 25:197–207

De Mey J, Vanhoutte PM 1981 Role of the intima in cholinergic and purinergic relaxation of isolated canine femoral arteries. J Physiol 316:347–355

De Mey J, Burnstock G, Vanhoutte PM 1979 Modulation of evoked release of noradrenaline in canine saphenous vein via presynaptic receptors for adenosine but not ATP. Eur J Pharmacol 55:401–405

Dean DM, Downie JW 1978 Interaction of prostaglandins and adenosine 5′-triphosphate in non-cholinergic neurotransmission in rabbit detrusor. Prostaglandins 16:245–251

Di Virgilio F, Ferrari D, Munerati M et al 1995 The P2Z receptor and its regulation of macrophage function. In: Belardinelli L, Pelleg A (eds) Adenosine and adenine nucleotides: from molecular biology to integrative physiology. Kluwer Acad, Norwell, MA, p 329–335

Drury AN, Szent-Györgyi A 1929 The physiological activity of adenine compounds with special reference to their action upon the mammalian heart. J Physiol 68:213–237

Dubyak GR 1991 Signal transduction by P2-purinergic receptors for extracellular ATP. Am J Respir Cell Mol Biol 4:295–300

Dunn PM, Blakeley AGH 1988 Suramin: a reversible P_2-purinoceptor antagonist in the mouse vas deferens. Br J Pharmacol 93:243–245

Eccles JC 1964 The physiology of synapses. Springer-Verlag, Berlin

Edwards FA, Gibb AJ, Colquhoun D 1992 ATP receptor-mediated synaptic currents in the central nervous system. Nature 359:144–147

El-Moatassim C, Dornand J, Mani JC 1987 Extracellular ATP increases cytosolic free calcium in thymocytes and initiates the blastogenesis of the phorbol 12-myristate 13-acetate-treated medullary population. Biochim Biophys Acta 927:437–444

El-Moatassim C, Dornand J, Mani JC 1992 Extracellular ATP and cell signaling. Biochim Biophys Acta 1134:31–45

Ergene E, Dunbar JC, O'Leary DS, Barraco RA 1994 Activation of P_2-purinoceptors in the nucleus tractus solitarius mediates depressor responses. Neurosci Lett 174:188–192

Erlinge D, You J, Wahlestedt C, Edvinsson L 1995 Characterisation of an ATP receptor mediating mitogenesis in vascular smooth muscle cells. Eur J Pharmacol 289:135–149

Evans RJ, Derkach V, Surprenant A 1992 ATP mediates fast synaptic transmission in mammalian neurons. Nature 357:503–505

Filtz TM, Li Q, Boyer JL, Nicholas RA, Harden TK 1994 Expression of a cloned P2Y purinergic receptor that couples to phospholipase C. Mol Pharmacol 46:8–15

Fischer B, Boyer JL, Hoyle CHV et al 1993 Identification of potent selective P_{2Y} purinoceptor agonists: structure–activity relationships for 2-thioether derivatives of adenosine 5′-triphosphate. J Med Chem 36:3937–3946

Flitney FW, Lamb JF, Singh J 1977 Effects of ATP on the hypodynamic frog ventricle. J Physiol 273:50P–52P

Foresta C, Rossato M, Di Virgilio F 1992 Extracellular ATP is a trigger for the acrosome reaction in human spermatozoa. J Biol Chem 257:19443–19447

Fredholm BB, Abbracchio MP, Burnstock G et al 1994 Nomenclature and classification of purinoceptors. Pharmacol Rev 46:143–156

Friel DD 1988 An ATP sensitive conductance in single smooth muscle cells from the rat vas deferens. J Biol Chem 268:8787–8792

Furchgott RF, Zawadzki JV 1980 The obligatory role of endothelial cells in the relaxation of arterial smooth muscle by acetylcholine. Nature 288:373–376

Furukawa K, Nomoto T 1989 Postnatal changes in response to adenosine and adenine nucleotides in rat duodenum. Br J Pharmacol 97:1111–1118

Galun R, Kabayo JP 1988 Gorging response of *Glossima palpalis* to ATP analogues. Physiol Entomol 13:419–423

Gonzales FA, Wang DJ, Huang N-N, Heppel LA 1990 Activation of early events of the mitogenic response by a P2Y-purinoceptor with covalently bound 3′-O-(4-benzoyl)-benzoyladenosine 5′-triphosphate. Proc Natl Acad Sci USA 87:9717–9771

Gordon JL 1986 Extracellular ATP: effects, sources and fate. Biochem J 233:309–319

Griese M, Gobran LJ, Rooney SA 1993 Signal-transduction mechanisms of ATP-stimulated phosphatidylcholine secretion in rat type II pneumocytes: interactions between ATP and other surfactant secretagogues. Biochim Biophys Acta 1167:85–93

Häggblad J, Heilbronn E 1988 P2-purinoceptor-stimulated phosphoinositide turnover in chick myotubes: calcium mobilization and the role of guanyl nucleotide-binding proteins. FEBS Lett 235:133–136

Harden TK, Boyer JL, Nicholas RA 1995 P_2-purinergic receptors: subtype-associated signaling responses and structure. Ann Rev Pharmacol Toxicol 35:541–579

Harms L, Finta EP, Tschöpl M, Illes P 1992 Depolarization of rat locus coeruleus neurons by adenosine 5′-triphosphate. Neuroscience 48:941–952

Henderson DJ, Elliot DG, Smith GM, Webb TE, Dainty IA 1995 Cloning and characterisation of a bovine P_{2Y} receptor. Biochem Biophys Res Commun 212:648–656

Henning RH, Duin M, Hertog A, Nelemans A 1993 Activation of the phospholipase C pathway by ATP is mediated exclusively through nucleotide type P2-purinoceptors in C2C12 myotubes. Br J Pharmacol 110:747–752

Hilfer SR, Palmatier BY, Fithian EM 1977 Precocious evagination of the embryonic chick thyroid in ATP-containing medium. J Embryol Exp Morphol 42:163–175

Hogaboom GK, O'Donnel JP, Fedan JS 1980 Purinergic receptors: a photoaffinity analog of adenosine triphosphate is a specific adenosine triphosphate antagonist. Science 208:1273–1276

Hökfelt TB, Fuxe K, Pernow B 1986 Coexistence of neuronal messengers: a new principle in chemical transmission. Elsevier, New York

Holton P 1959 The liberation of adenosine triphosphate on antidromic stimulation of sensory nerves. J Physiol 145:494–504

Hooper M, Spedding M, Sweetman AJ, Weetman DF 1974 2,2′-pyridylisatogen tosylate: an antagonist of the inhibitory effects of ATP on smooth muscle. Br J Pharmacol 50:458P–459P

Hopwood AM, Lincoln J, Kirkpatrick KA, Burnstock G 1989 Adenosine 5′-triphosphate, adenosine and endothelium-derived relaxing factor in hypoxic vasodilatation of the heart. Eur J Pharmacol 165:323–326

Hourani SMO, Hall DA, Nieman CJ 1992 Effects of the P2-purinoceptor antagonist, suramin, on human platelet aggregation induced by adenosine 5′-diphosphate. Br J Pharmacol 105:453–457

Housely GD, Greenwood D, Bennett T, Ryan AF 1995 Identification of a short form of the P2xR1-purinoceptor subunit produced by alternative splicing in the pituitary and cochlea. Biochem Biophys Res Commun 212:501–508

Hoyle CHV, Burnstock G 1991 ATP receptors and their physiological roles. In: Stone TW (ed) Adenosine in the nervous system. Academic Press, London, p 43–76

Hoyle CHV, Burnstock G 1995 Purines. In: Bittar EE, Bittar N (eds) Principles of medical biology. JAI Press, Connecticut

Hoyle CHV, Greenberg MJ 1988 Actions of adenylyl compounds in invertebrates from several phyla: evidence for internal purinoceptors. Comp Biochem Physiol 90:113C–122C

Hoyle CHV, Chapple C, Burnstock G 1989 Isolated human bladder: evidence for an adenine dinucleotide acting on P_{2X}-purinoceptors and for purinergic transmission. Eur J Pharmacol 174:115–118

Hoyle CHV, Knight GE, Burnstock G 1990 Suramin antagonizes responses to P_2-purinoceptor agonists and purinergic nerve stimulation in the guinea-pig urinary bladder and taenia coli. Br J Pharmacol 99:617–621

Hume RI, Honig MG 1986 Excitatory action of ATP on embryonic chick muscle. J Neurosci 6:681–690

Humphries RG, Tomlinson W, Ingall AH, Cage PA, Leff P 1994 FPL 66096: a novel, highly potent and selective antagonist at human platelet P_{2T}-purinoceptors. Br J Pharmacol 113:1057–1063

Humphries RG, Robertson MJ, Leff P 1995 A novel series of P2T purinoceptor antagonists: definition of the role of ADP in arterial thrombosis. Trends Pharmacol Sci 16:179–181

Husted S, Sjögren C, Andersson K-E 1983 Direct effects of adenosine and adenine nucleotides on isolated human urinary bladder and their influence on electrically induced contractions. J Urol 130:392–398

Illes P, Sevcik J, Finta EP, Fröhlich R, Nieber K, Nörenberg W 1995 Modulation of locus coeruleus neurons by extra- and intracellular adenosine 5′-triphosphate. Brain Res Bull 35:513–519

Inoue R, Brading AF 1991 Human, pig and guinea-pig bladder smooth muscle cells generate similar inward currents in response to purinoceptor activation. Br J Pharmacol 103:1840–1841

Inoue K, Nakazawa K, Fujimori K, Watano T, Takanaka A 1992 Extracellular adenosine 5′-triphosphate-evoked glutamate release in cultured hippocampal neurons. Neurosci Lett 134:215–218

Jensen J, Holmgren S 1985 Neurotransmitters in the intestine of the atlantic cod, *Gadus morhua*. Comp Biochem Physiol 82:81C–89C

Kasakov L, Burnstock G 1983 The use of the slowly degradable analog, α,β-methylene-ATP, to produce desensitization of the P_2-purinoceptor: effect on non-adrenergic, non-cholinergic responses of the guinea-pig urinary bladder. Eur J Pharmacol 86:291–294

Kerr DIB, Krantis A 1979 A new class of ATP antagonist. Proc Aust Physiol Pharmacol Soc 10:156P

Khakh BS, Michel A, Humphrey PPA 1994 Estimates of antagonist affinities at P2X purinoceptors in rat vas deferens. Eur J Pharmacol 263:301–309

Kitazawa T, Hoshi T, Chugun A 1990 Effects of some autonomic drugs and neuropeptides on the mechanical activity of longitudinal and circular muscle strips isolated from the carp intestinal bulb (*Cyprimus carpio*). Comp Biochem Physiol 97:12C–24C

Knight GE, Burnstock G 1995 Responses of the aorta of the garter snake (*Thamnophis sirtalis parietalis*) to purines. Br J Pharmacol 114:41–48

Knight GE, Hoyle CHV, Burnstock G 1990 Glibenclamide antagonises the responses to ATP, but not adenosine or adrenaline, in the gastric ligament of the starfish *Asterias rubens*. Comp Biochem Physiol 97:363C–367C

Knight GE, Hoyle CHV, Burnstock G 1992 Effects of adenine nucleosides and nucleotides on the isolated heart of the snail *Helix aspersa* and the slug *Arion ater*. Comp Biochem Physiol 101:175C–181C

Knudsen TB, Elmer WA 1987 Evidence for negative control of growth by adenosine in the mammalian embryo: induction of $Hm^{\times}/+$ mutant limb outgrowth by adenosine deaminase. Differentiation 33:270–279

Kolb H-A, Wakelam MJO 1983 Transmitter-like action of ATP on patched membranes of cultured myoblasts and myotubes. Nature 303:621–623

Kunapuli SP, Akbar GKM, Webb T, Matsumoto M, Mills DCB, Barnard EA 1996 Cloning and characterisation of a novel P2 purinoceptor from human erythro leukemia cells. In: Structure and function of P2-purinoceptors. Satellite meeting of experimental biology, Atlanta, GA, 7–9 April 1995 (abstr)

Kupfermann I 1991 Functional studies of cotransmission. Physiol Rev 71:683–732

Laasberg T 1990 Ca^{2+}-mobilizing receptors of gastrulating chick embryo. Comp Biochem Physiol 97:9C–12C

Lambrecht G, Friebe T, Grimm U et al 1992 PPADS, a novel functionally selective antagonist of P2 purinoceptor-mediated responses. Eur J Pharmacol 217:217–219

Lazarowski ER, Harden TK 1994 Identification of a uridine nucleotide-selective G-protein-linked receptor that activates phospholipase C. J Biol Chem 269:11830–11836

Lennard R, Huddart H 1989 Purinergic modulation in the flounder gut. Gen Pharmacol 20:849–853

Léon C, Vial C, Cazenave J, Gachet C 1996 Cloning and sequencing of a human endothelial $P2Y_1$ purinoceptor. Gene, in press

Leong WS, Russell RGG, Caswell AM 1994 Stimulation of cartilage resorption by extracellular ATP acting at P2-purinoceptors. Biochim Biophys Acta 1201:298–304

Levin RM, Malkowicz B, Jacobowitz D, Wein AJ 1981 The ontogeny of the autonomic innervation and contractile response of the rabbit urinary bladder. J Pharmacol Exp Ther 219:250–257

Lewis C, Neidhardt S, Holy C, North RA, Buell G, Surprenant A 1995 Coexpression of $P2X_2$ and $P2X_3$ receptor subunits can account for ATP-gated currents in sensory neurons. Nature 377:432–435

Liscia A, Cingar R, Barbarossa JT, Esu S, Muroni P, Galun R 1993 Electrophysiological responses of labral apical chemoreceptors to adenine nucleotides in *Culex pipiens*. J Insect Physiol 39:261–265

Loubatiéres-Mariani M-M, Petit P, Chapal J, Hillaire-Buys D, Bertrand G, Ribes G 1995 Effects of purinoceptor agonists on insulin secretion. In: Belardinelli L, Pelleg A (eds) Adenosine and adenine nucleotides: from molecular biology to integrative physiology. Kluwer Acad, Norwell, MA, p 337–345

Lu Z, Smith DO 1991 Adenosine 5′-triphosphate increases acetylcholine channel opening frequency in rat skeletal muscle. J Physiol 436:45–56

Lustig KD, Shiau AK, Brake AJ, Julius D 1993 Expression cloning of an ATP receptor from mouse neuroblastoma cells. Proc Natl Acad Sci USA 90:5113–5117

Malam-Souley R, Campan M, Gadeau A-P, Desgranges C 1993 Exogenous ATP induces a limited cell cycle progression of arterial smooth muscle cells. Am J Physiol 264:783C–788C

Manzini S, Maggi CA, Meli A 1985 Further evidence for involvement of adenosine 5′-triphosphate in non-adrenergic, non-cholinergic relaxation of the isolated rat duodenum. Eur J Pharmacol 113:399–408

Manzini S, Hoyle CHV, Burnstock G 1986 An electrophysiological analysis of the effect of reactive blue 2, a putative P_2-purinergic receptor antagonist, on inhibitory junction potentials of rat caecum. Eur J Pharmacol 127:197–204

Martinson J, Muren A 1963 Excitatory and inhibitory effects of vagus stimulation on gastric motility in the cat. Acta Physiol Scand 57:309–316

McLaren GJ, Lambrecht G, Mutschler E, Bäumert HG, Sneddon P, Kennedy C 1994 Investigation of the actions of PPADS, a novel P_{2X}-purinoceptor antagonist, in the guinea-pig isolated vas deferens. Br J Pharmacol 111:913–917

Meghji P, Burnstock G 1984 The effect of adenyl compounds on the heart of the dogfish, *Scyliorhinus canicula*. Comp Biochem Physiol 77:295C–300C

Meldrum M, Burnstock G 1983 Evidence that ATP acts as a co-transmitter with noradrenaline in sympathetic nerves supplying the guinea-pig vas deferens. Eur J Pharmacol 92:161–163

Milner, P, Burnstock G 1994 Neurotransmitters in the gut. In: Kamm M, Lennard-Jones JE (eds) Constipation and related disorders: pathophysiology and management in adults and children. Wrightson Biomedical, Bristol, p 41–49

Miyashita Y, Kumazawa T, Fujii R 1984 Receptor mechanisms in fish chromatophores. VI: Adenosine receptors mediate pigment dispersion in guppy and catfish melanophores. Comp Biochem Physiol 77:205C–210C

Mockett BG, Housley GD, Thorne PR 1994 Fluorescence imaging of extracellular purinergic receptor sites and putative ecto-ATPase sites on isolated cochlear hair cells. J Neurosci 14:6992–7007

Moody CJ, Meghji P, Burnstock G 1984 Stimulation of P_1-purinoceptors by ATP depends partly on its conversion to AMP and adenosine and partly on direct action. Eur J Pharmacol 97:47–54

Murgia M, Pizzo P, Zanovello P, Zambon A, Di Virgilio F 1992 *In vitro* cytotoxic effects of extracellular ATP. ATLA-Alt L 20:66–70

Murgia M, Hanau S, Pizzo P, Rippa M, Di Virgilio F 1993 Oxidised ATP. An irreversible inhibitor of the macrophage purinergic P_{2Z} receptor. J Biol Chem 271:8199–8203

Nagano O, Földes FF, Nakatsuka H et al 1992 Presynaptic A1-purinoceptor-mediated inhibitory effects of adenosine and its stable analogs on the mouse hemidiaphragm preparation. Naunyn-Schmiedebergs Arch Pharmacol 346:197–202

Namiot Z, Marcinkiewicz M, Jaroszewicz W, Stasiewicz J, Gorski J 1993 Mucosal adenosine deaminase activity and gastric ulcer healing. Eur J Pharmacol 243:301–303

Namoto S 1992 Effects of purine compounds and forskolin on melanophores of the medaka, Oryzias latipes: evidence for an adenosine-A_2 receptor. Comp Biochem Physiol 103:391C–398C

Neary JT, Norenberg MD 1992 Signaling by extracellular ATP: physiological and pathological considerations in neuronal–astrocytic interactions. Prog Brain Res 94:145–151

Neary JT, Whittemore SR, Zhu Q, Norenberg MD 1994 Synergistic activation of DNA synthesis in astrocytes by fibroblast growth factors and extracellular ATP. J Neurochem 63:490–494

Neary JT, Rathbone MP, Cattabeni F, Abbracchio MP, Burnstock G 1996 Trophic actions of extracellular nucleotides and nucleosides on glial and neuronal cells. Trends Neurosci 19:13–18

Nelson ID, Huddart H 1994 Neuromodulation in molluscan smooth muscle: the action of 5-HT, FMRF amide and purine compounds. Gen Pharmacol 25:539–552

Nuttle LC, El-Moatassim C, Dubyak GR 1993 Expression of the pore-forming P2Z purinoceptor in *Xenopus* oocytes injected with poly$(A)^+$ RNA from murine macrophages. Mol Pharmacol 44:93–101

O'Connor SE, Dainty IA, Leff P 1991 Further subclassification of ATP receptors based on agonist studies. Trends Pharmacol Sci 12:137–141

Palea S, Artibani W, Ostardo E, Trist DG, Pietra C 1993 Evidence for purinergic neurotransmission in human bladder affected by interstitial cystitis. J Urol 150:2007–2012

Palea S, Pietra C, Trist DG, Artibani W, Calpista A, Corsi M 1995 Evidence for the presence of both pre- and postjunctional P2-purinoceptor subtypes in human isolated urinary bladder. Br J Pharmacol 114:35–40

Parr CE, Sullivan DM, Paradiso AM et al 1994 Cloning and expression of a human P_{2U} nucleotide receptor, a target for cystic fibrosis pharmacotherapy. Proc Natl Acad Sci USA 91:3275–3279

Perez MTR, Bruun A 1987 Colocalization of (^{3}H)-adenosine accumulation and GABA immunoreactivity in the chicken and rabbit retinas. Histochemistry 87:413–417

Phillis JW, Smith-Barbour M, Perkins LM, O'Regan MH 1993 Acetylcholine output from the ischemic rat cerebral cortex: effects of adenosine agonists. Brain Res 613:337–340

Pintor J, Miras-Portugal MT 1993 Diadenosine polyphosphates (Ap_xA) as new neurotransmitters. Drug Dev Res 28:259–262

Pirotton S 1995 The action of ATP on vascular endothelium: transduction signals and molecular characterisation of receptors. ASPET colloquium, Atlanta, GA

Pirotton S, Verjans B, Boeynaems J-M, Erneux C 1991 Metabolism of inositol phosphates in ATP-stimulated vascular endothelial cells. Biochem J 277:103–110

Pothier F, Forget H, Sullivan R, Couillard P 1987 ATP and the contractile vacuole in *Amoeba proteus*: mechanism of action of exogenous ATP and related nucleotides. J Exp Zool 243:379–387

Rapaport E 1993 Anticancer activities of adenine nucleotides in tumour bearing hosts. Drug Dev Res 28:428–431

Rathbone MP, Christjanson L, De Forge S et al 1992 Extracellular purine nucleosides stimulate cell division and morphogenesis: pathological and physiological complications. Med Hypoth 37:232–240

Reilly WM, Saville VL, Burnstock G 1987 An assessment of the antagonistic activity of reactive blue 2 at P_1- and P_2-purinoceptors: supporting evidence for purinergic innervation of the rabbit portal vein. Eur J Pharmacol 140:47–53

Rice WR, Burton FM, Fiedeldey DT 1995 Cloning and expression of the alveolar type II cell P2U-purinergic receptor. Am J Respir Cell Mol Biol 12:27–32

Rubino A, Burnstock G 1994 Recovery after dietary vitamin E supplementation of impaired endothelial function in vitamin E-deficient rats. Br J Pharmacol 112:515–518

Ruggieri MR, Whitmore KE, Levin RM 1990 Bladder purinergic receptors. J Urol 144:176–181

Satchell D, Burnstock G, Dann P 1973 Antagonism of the effects of purinergic nerve stimulation and exogenously applied ATP on the guinea-pig taenia coli by 2-substituted imidazolines and related compounds. Eur J Pharmacol 23:264–269

Schäfer K, Saffrey MJ, Burnstock G 1995 Trophic actions of 2-chloroadenosine and bFGF on cultured myenteric neurones. NeuroReport 6:937–941

Seguela P, Haghihi A, Soghomonian J-J, Cooper E 1996 A novel neuronal P2XATP receptor ion channel with widespread distribution in the brain. J Neurosci 16:448–455

Sibley GNA 1984 A comparison of spontaneous and nerve-mediated activity in bladder muscle from man, pig and rabbit. J Physiol 354:431–443

Sigel H 1992 Have adenosine 5′-triphosphate (ATP^{4-}) and related purine-nucleotides played a role in early evolution? ATP, its own 'enzyme' in metal ion facilitated hydrolysis. Inorg Chim Acta 200:1–11

Silinsky EM, Gerzanich V, Vanner SM 1992 ATP mediates excitatory synaptic transmission in mammalian neurons. Br J Pharmacol 106:762–763

Smith DO, Lu Z 1991 Adenosine derived from hydrolysis of presynaptically released ATP inhibits neuromuscular transmission in the rat. Neurosci Lett 122:171–173

Smuts MS 1981 Rapid nasal pit formation in mouse embryos stimulated by ATP-containing medium. J Exp Zool 216:409–414

Sneddon P, Burnstock G 1984a Inhibition of excitatory junction potentials in guinea-pig vas deferens by α,β-methylene-ATP: further evidence for ATP and noradrenaline as cotransmitters. Eur J Pharmacol 100:85–90

Sneddon P, Burnstock G 1984b ATP as a co-transmitter in rat tail artery. Eur J Pharmacol 106:149–152

Sneddon P, Westfall DP 1984 Pharmacological evidence that adenosine triphosphate and noradrenaline are co-transmitters in the guinea-pig vas deferens. J Physiol 347:561–580

Soltoff SP, McMillian MK, Talamo BR 1989 Coomassie brilliant blue G is a more potent antagonist of P2 purinergic responses than reactive blue 2 (cibacron blue 3GA) in rat parotid acinar cells. Biochem Biophys Res Commun 165:1279–1285

Soltoff SP, McMillian MK, Talamo BR, Cantley LC 1993 Blockade of ATP binding site of P_2 purinoceptors in rat parotid acinar cells by isothiocyanate compounds. Biochem Pharmacol 45:1936–1940

Soto F, Garcia-Guzman M, Gomez-Hernandez JM, Hollman M, Karschin C, Stühmer W 1996 P2Xoc4: an ATP-activated ionotropic receptor cloned from rat and human brain. Proc Natl Acad Sci USA, in press

Spedding M, Sweetman AJ, Weetman DF 1975 Antagonism of adenosine 5′-triphosphate-induced relaxation by 2,2′-pyridylisatogen in the taenia of guinea-pig caecum. Br J Pharmacol 53:575–583

Steinberg TH, Silverstein SC 1987 Extracellular ATP^{4-} promotes cation fluxes in the J774 mouse macrophage cell line. J Biol Chem 267:6451–6454

Stone T 1992 Therapeutic potential of adenosine. Scrip Magazine, July/August, p 41–43

Tokuyama Y, Hara M, Jones EMC, Fan Z, Bell GI 1995 Cloning of rat and mouse P_{2Y} purinoceptors. Biochem Biophys Res Commun 211:211–218

Trezise DJ, Bell NJ, Khakh BS, Michel AD, Humphrey PPA 1994a P2 purinoceptor antagonist properties of pyridoxal-5-phosphate. Eur J Pharmacol 259:295–300

Trezise DJ, Kennedy I, Humphrey PPA 1994b The use of antagonists to characterize the receptors mediating depolarization of the rat isolated vagus nerve by α,β-methylene adenosine 5′-triphosphate. Br J Pharmacol 112:282–288

Ueno S, Harata N, Inoue K, Akaike N 1992 ATP-gated current in dissociated rat nucleus solitarii neurons. J Neurophysiol 69:778–785

Valera S, Hussy N, Evans RJ et al 1994 A new class of ligand-gated ion channel defined by P_{2X} receptor for extracellular ATP. Nature 371:516–519

Van Belle H 1995 Adenosine uptake blockers for cardioprotection. In: Belardinelli L, Pelleg A (eds) Adenosine and adenine nucleotides: from molecular biology to integrative physiology. Kluwer Acad, Norwell, MA, p 373–378

van Rhee AM, Van der Heijed MPA, Beukers MW, IJzerman AP, Soudijn W, Nickel P 1994 A novel competitive antagonist for P2 purinoceptors. Eur J Pharmacol 268:1–7

Venter JC, Di Porzio U, Robinson DA et al 1988 Evolution of neurotransmitter receptor systems. Prog Neurobiol 30:105–169

Vidal M, Hicks PE, Langer SZ 1986 Differential effects of α,β-methylene ATP on responses to nerve stimulation in SHR and WKY tail arteries. Naunyn-Schmiedebergs Arch Pharmacol 332:384–390

von Kügelgen I, Kurz K, Starke K 1994 P_2-purinoceptor-mediated autoinhibition of sympathetic transmitter release in mouse and rat vas deferens. Naunyn-Schmiedebergs Arch Pharmacol 349:125–132

Voogd TE, Vansterkenburg ELM, Wilting J, Janssen LH 1993 Recent research on the biological activity of suramin. Pharmacol Rev 45:177–204

Wang D, Huang N-N, Heppel LA 1990 Extracellular ATP shows synergistic enhancement of DNA synthesis when combined with agents that are active in wound healing or as neurotransmitters. Biochem Biophys Res Commun 166:251–258

Webb TE, Simon J, Krishek BJ et al 1993 Cloning and functional expression of a brain G-protein-coupled ATP receptor. FEBS Lett 324:219–225

Webb TE, Henderson D, King BF et al 1996a A novel G protein-coupled P_2 purinoceptor (P_{2Y3}) activates preferentially by nucleoside diphosphates. Submitted Mol Pharmacol

Webb TE, Kaplan MG, Barnard EA 1996b Identification of 6H1 as a P_{2Y} purinoceptor: $P2Y_5$. Biochem Biophys Res Commun 219:105–110

Westfall DP, Fedan JS, Colby J, Hogaboom GK, O'Donnell JP 1983 Evidence for a contribution by purines to the neurogenic response of the guinea-pig urinary bladder. Eur J Pharmacol 87:415–422

Wieraszko A, Ehrlich YH 1994 On the role of extracellular ATP in the induction of long-term potentiation in the hippocampus. J Neurochem 63:1731–1738

Wieraszko A, Goldsmith G, Seyfried TN 1989 Stimulation-dependent release of adenosine triphosphate from hippocampal slices. Brain Res 485:244–250

Wiley JS, Chen R, Jamieson GR 1993 The ATP^{4-} receptor-operated channel (P2Z class) of human lymphocytes allows Ba^{2+} and $ethidium^{+}$ uptake: inhibition of fluxes by suramin. Arch Biochem Biophys 305:54–60

Williams M 1993 Purinergic drugs: opportunities in the 1990s. Drug Dev Res 28:438–444

Woolridge S, van Helden DF 1990 Enhanced excitatory junction potentials in (mesenteric) arteries of the spontaneously hypertensive rat. Proc Austr Physiol Pharmacol Soc 21:60P

Yatani A, Tsuda Y, Akaike N, Brown AM 1982 Nanomolar concentrations of extracellular ATP activate membrane Ca channels in snail neurones. Nature 296:169–171

Young JZ 1988 Sympathetic innervation of the rectum and bladder of the skate and parallel effects of ATP and adrenalin. Comp Biochem Physiol 89:101C–107C

Yu H, Ferrier J 1993 ATP induces an intracellular pulse in osteoclasts. Biochem Biophys Res Commun 191:357–363

Zheng LM, Zychlinsky A, Liu C-C, Ojcius, DM, Young JD-E 1991 Extracellular ATP as a trigger for apoptosis or programmed cell death. J Cell Biol 112:279–288

Zheng J-S, Boluyt MO, O'Neill L, Crow MT, Lakatta EG 1994 Extracellular ATP induces immediate-early gene expression but not cellular hypertrophy in neonatal cardiac myocytes. Circ Res 74:1034–1041

Ziganshin AU, Hoyle CHV, Bo X et al 1993 PPADS selectively antagonizes P_{2X}-purinoceptor-mediated responses in the rabbit urinary bladder. Br J Pharmacol 110:1491–1495

Ziganshin, AU, Hoyle CHV, Lambrecht G, Mutschler E, Bäumert HG, Burnstock G 1994 Selective antagonism by PPADS at P_{2X}-purinoceptors in rabbit isolated blood vessels. Br J Pharmacol 111:923–929

Zimmer-Faust RK, Gleeson RA, Carr WES 1988 The behavioural response of spiny lobsters to ATP: evidence for mediation by P2-like chemosensory receptors. Biol Bull 175:167–174

Zimmerman AM 1962 Action of ATP on amoeba. J Cell Comp Physiol 60:271–280

DISCUSSION

Leff: You mentioned the $P2X_3$ receptor. Is α,β-methylene ATP acting as an agonist on this receptor, as with other subtypes, or, conversely, does it act as an antagonist?

Burnstock: It does act as an agonist, but it is not the most active. 2-MethylthioATP is the most potent, with α,β-methylene ATP is in about third or fourth place. We haven't looked yet to see whether it antagonizes the response, but we know that suramin does.

Kennedy: In neuronal cells of the dorsal root ganglion α,β-methylene ATP is active in the low micromolar concentration range (Robertson et al 1995). In this respect these cells are similar to smooth muscle cells of the vas deferens.

Burnstock: There have been a couple of nice papers on that *in situ* (Bean & Friel 1990, Robertson et al 1995). The results tie in well with those we have obtained with the $P2X_3$ receptor.

North: α,β-Methylene ATP clearly is a good agonist at some P2X receptors but not at others. I'm more concerned about its other actions: in your experiments with autoradiography, are you detecting binding to P2X receptors only or binding to the host of other ATP-binding proteins that are present in brain?

Burnstock: That is a fair question. The binding can only be selective for those P2X subclasses that are sensitive to α,β-methylene ATP.

North: That is not my concern. Obviously, the radiolabelled ATP is going to bind only to a subclass of P2X receptors, but how do you know that it's not binding to many other nucleotide-binding proteins, such as kinases and ATP synthases?

Barnard: I can throw a little light on that. On rat brain membranes we have studied the binding of ATP derivatives (Simon et al 1995) and Michel & Humphrey (1993) have looked at the binding of α,β-methylene ATP. The binding of ATP to enzymes of brain membranes and so forth is definitely at a very much lower affinity and is also absolutely dependent on divalent cations (Bo et al 1992, Humphrey et al 1995). When binding studies are conducted in the absence of divalent cations and of Na^+, then what Geoff Burnstock says is perfectly correct: there is a definite component binding to α,β-methylene ATP which corresponds to a class of P2X receptors. The limitation to this for autoradiography is that this component is in overall number approximately 2% of the binding sites of $P2Y_1$-like receptors (as measured by the ATPαS labelled with ^{35}S). The pharmacology of that latter binding is unmistakable: it shows the profile of the $P2Y_1$ receptor. So overall in the brain there are about 50 times more $P2Y_1$ and related P2Y receptors than that class of P2X receptors. There is much less P2U receptor binding. There is a set of binding sites for α,β-methylene ATP in the brain (which we can measure by binding to brain membranes and which Geoff Burnstock and co-workers have seen by autoradiography) that shows the pharmacology of certain subtypes of the P2X receptor such as $P2X_1$ (Michel & Humphrey 1993). But, of course, what would not be seen are the other P2X subtypes ($P2X_2$ etc.) which do not respond to α,β-methylene ATP. The distribution of those within the brain as discerned by ligand-binding autoradiography is not presently available, but at the mRNA level the $P2X_1$ and $P2X_2$ receptors have recently been shown *in situ* to be distributed at a very low level throughout adult rat brain, while in the neonate rat brain they are somewhat more abundant and the two types are at different locations (Kidd et al 1995). What is unmistakeable is that there is in the brain, in contrast, a very large population of P2Y receptors which have a pharmacology equal or very similar to that of $P2Y_1$. This is true in both the chick and the rat brain. By *in situ* hybridization of $P2Y_1$ mRNA (Webb et al 1994) and by ligand-specific autoradiography (J. Simon & E. A. Barnard, unpublished results), this receptor does not occur everywhere in the brain, but it is present in most areas. Although it is such a large amount overall, it is not universal but occurs in anatomically discrete loci. This obviously must be an important component in the brain, which is full of P2 purinergic pathways about which we know very little.

Harden: Has anyone expressed P2X receptors and done α,β-methylene ATP binding studies with them?

North: I do not know of binding studies with α,β-methylene ATP that have been published, but some of the expressed receptors are sensitive to α,β-methylene ATP and some are not.

Harden: Can you carry out radioligand binding assays in cells in which the receptor has been expressed and find a site that wasn't there before?

Surprenant: Our colleague Anton Michel at Cambridge has been doing that with an HEK cell line (Michel et al 1996). He's not been able to make it work because he has found too high a background. He has made it work with viral-infected CHO cells, but he's happier using ATPγS in divalent-ion-free solutions, because he feels that this has been giving him a much better profile for the cloned receptors.

IJzerman: Radioligand binding studies on P2 purinergic receptors are performed with radiolabelled agonists, i.e. nucleotides, only. This is a serious drawback. First of all, labelling receptors with agonists will certainly change the system: receptors are prone to desensitization, internalization, etc. Second, the nucleotides are liable to degradation. I find it increasingly hard to explain to others that we still use these to see what happens with our receptors. There's a real need for radiolabelled antagonists that are stable.

Burnstock: You're right; there really is a need for such compounds.

Jacobson: We've seen that P2Y–radioligand binding reaches a peak within five minutes or so, and then beyond that we have diminishing returns. Have you seen that also in your studies with deoxyATPαS?

Barnard: No. Was that in divalent-cation free medium?

Jacobson: Divalent cations were not added to the incubation medium.

Barnard: Because if Mg^{2+} or even Ca^{2+} ions are present, the ATPases will affect the result.

Jacobson: Are you able to calculate the affinity from the kinetics?

Barnard: K_d for dATPαS binding was 6.6 ± 0.3 nM at 25 °C, pH 7.4; the k_i values for the other nucleotides are given in our chapter (Barnard et al 1996, this volume). I believe that degradation need not be a risk in the binding experiments on P2Y receptors, because we have done controls in which we have exposed these labelled compounds to the membranes in our divalent-cation-free medium and then run a chromatogram to find the percentage degraded. There is no degradation if you omit Mg^{2+}, Ca^{2+} and Na^+.

Edwards: Shouldn't the binding be blocked by suramin? Presumably this is a way one might get around this problem of degradation. I know that suramin is a very 'dirty' drug, but I don't think it is particularly dirty for other ATP-binding proteins.

North: Suramin doesn't seem to behave as a strictly competitive antagonist in any of the functional studies. You could come up with all kinds of models in which suramin would not block the binding of the agonist.

Barnard: All of the high binding of [^{35}S]ATPαS to brain membranes is sensitive to suramin (K_i 0.79 μM rat brain and 1.05 μM chick brain). The same is true for the expressed $P2Y_1$ receptor (Barnard et al 1996, this volume).

Edwards: Another comment I had relates to desensitization with α,β-methylene ATP. It's clear that the early measurements of potency ratios must be reassessed because they did not take degradation due to nucleotidases into account. However, potency ratios will also be affected by desensitization. Depending on the method of application used with α,β-methylene ATP, for example, you may see no response or a substantial response. In contrast, another agonist which is less prone to desensitization will cause a similar response under both conditions.

Burnstock: You do have to be well aware of that, I agree.

North: This has all been gone over in the nicotinic field. The high-affinity binding sites for most of the agonists correspond to a desensitized form of the receptor.

Edwards: It's not a problem for comparisons within binding studies, because everything is desensitized under those conditions.

North: No, it's a problem for correlating the numbers you get in binding assays with the numbers that you get for agonists which cause marked desensitization.

Kennedy: In quite a few of the single-cell studies where α,β-methylene ATP has no effect, it has been applied in millisecond timescales. This would rule out desensitization as an explanation for its inactivity.

Boeynaems: You mention that the recognition of ATP^{4-} is a unique characteristic of the P2Z receptor. I'm a little concerned by that, in relation to what Dr Barnard said about binding in the absence of Ca^{2+} and Mg^{2+}. There's now evidence that for most if not all subtypes of P2 receptor, the active ligand is not complexed to Ca^{2+} or Mg^{2+}. I don't know if anybody knows of a subtype which would recognize a divalent cation-complexed form of ATP.

Wiley: The P2Z purinoceptor shows a much higher selectivity for ATP^{4-} compared with $MgATP^{2-}$—so much so, that the latter divalent species is probably inactive as an agonist (Wiley et al 1993). However, with the other purinoceptor subtypes, the difference in selectivity between these two ionized forms of ATP may be much less.

Leff: I disagree with that as regards P2U receptors. I believe there are going to be subtypes of the P2U receptor as well. The one that people are most familiar with has a strong preference for the ATP^{4-} species.

Just to pick up on another point that Frances Edwards made, where we are thinking about classification in terms of potency orders, differential chelating ability of the agonists must be taken into account in this sort of experiment. I think the tendency is for people to believe that all triphosphates chelate Ca^{2+} similarly. In some structure–activity relationships that's true, but not in all. The definitive classification experiments should be done, taking into account desensitization and the ionic species. We may have recognized and got rid of ectoATPases as a problem, although we have to admit that the compound we're using to block ectoATPases really isn't potent enough to exclude their influence completely. Nevertheless, we're making some headway there. Now we have to take on the issue of ionic species and receptor desensitization when we are quoting potency orders—that's another necessary level of refinement in functional studies.

Hourani: I have some results which show that the ionic species recognized by the P2T receptor is ATP^{4-} (Hall et al 1994). We are fortunate with platelets in that we can look at functional studies in the absence of Ca^{2+} and Mg^{2+}, because we can look at effects that are not dependent on extracellular cations. With the P2X receptors it is very difficult to look at a response in the absence of Ca^{2+}.

Wiley: Could you clarify that? I thought that the P2T receptor responded to ADP^{3-}, and was actually antagonized by ATP^{4-}.

Hourani: Yes, but the ligand in each case is the unchelated form.

Barnard: As I've said, the binding series found for recombinant P2Y receptors, with the pharmacology predicted from agonist activities, is that measured in the absence of divalent cations. If you add Mg^{2+} but keep Na^{+} out, you are taking a small risk of some degradation by ATPase activity. The activity never goes up with Mg^{2+}: it goes down a little, as Paul Leff also indicated for other cases. The $P2Y_1$ receptor, at least, therefore recognizes both the Mg^{2+}-complexed and the uncomplexed forms of ATP. The same has been shown to hold in the binding of α,β-methylene ATP to the P2X receptor type in vas deferens membranes (Michel & Humphrey 1994).

Surprenant: Our discussions concerning neural release of purines all assume that it is ATP rather than adenosine that is released. But what is the direct evidence that this is the case?

Burnstock: It's clear enough in the periphery; adenosine is taken up by neurons and converted largely to ATP. Breakdown enzymes are not present inside the nerve cells, but rather as ectoenzymes on the membranes. In the brain it's more contentious. On the basis of studies of the localization of adenosine deaminase it has been suggested that adenosine might be a neurotransmitter in the brain (Nagy et al 1984). I'm doubtful myself: adenosine is so avidly taken up by most cells (ATP is not) and is converted to and stored as ATP. It seems to be released as ATP and then it's broken down extracellularly; both 5′-nucleotidase and Ca^{2+}/Mg^{2+} ATPase have been shown to be present in the membranes of cells in the CNS (Yoshioka & Tanaka 1989, Zenker et al 1992, Schoen et al 1993). But it's a good point for debate (see Phillis 1990).

Edwards: It just depends on us being able to use this Astra compound (ARL67085). If you can block adenosine effects by blocking the breakdown of ATP then you've answered that question.

Westfall: We follow the release of ATP, ADP, AMP and adenosine with our HPLC–fluorescence ATP assay. From chromaffin cells and PC12 cells very little adenosine is released and very little is formed by extracellular metabolism. From sympathetic neurons it's a little more difficult to determine whether adenosine is released, because in these there is a more rapid metabolism of nucleotides to adenosine. But if you do the experiments carefully you can follow the emergence of ATP, followed by ADP, AMP and adenosine. The production of these can be interfered with by various drugs. So the evidence is pretty strong that it's mostly ATP that is released from sympathetic nerves.

Hourani: Going back to the brain, Peter Richardson looked at cholinergic synapses in the brain and blocked the prejunctional inhibitory effects with an antibody against 5′ nucleotidase (Richardson et al 1987). This is quite good evidence that it is not adenosine that is released but a nucleotide.

References

Barnard EA, Webb TE, Simon J, Kunapuli SP 1996 The diverse series of recombinant P2Y purinoceptors. In: P2 purinoceptors: localization, function and transduction mechanisms. Wiley, Chichester (Ciba Found Symp 198) p 166–188

Bean BP, Friel DD 1990 ATP-activated channels in excitable cells. In: Naharashi (ed) Ion channels, vol 2. Plenum, New York, p 169–203

Bo X, Simon J, Burnstock G, Barnard EA 1992 Solubilization and molecular size determination of the P_{2X} purinoceptor from rat vas deferens. J Biol Chem 267:17581–17587

Hall DA, Frost V, Hourani SMO 1994 Effects of extracellular divalent cations on responses of human blood platelets to adenosine 5′-diphosphate. Biochem Pharmacol 48:1319–1326

Humphrey PPA, Buell G, Kennedy I et al 1995 New insights on P_{2X} purinoceptors. Naunyn-Schmiedebergs Arch Pharmacol 351:1–12

Kidd EJ, Grahames CBA, Simon J, Michel AD, Barnard EA, Humphrey PPA 1995 Localization of P_{2X} purinoceptor transcripts in the rat nervous system. Mol Pharmacol 48:569–573

Michel AD, Humphrey PPA 1993 Distribution and characterization of $[^3H]\alpha,\beta$-methylene ATP binding sites in the rat. Naunyn-Schmeidebergs Arch Pharmacol 348:608–617

Michel AD, Humphrey PPA 1994 Effects of metal cations on $[^3H]\alpha,\beta$-methylene ATP binding in rat vas deferens. Naunyn-Schmeidebergs Arch Pharmacol 350:113–122

Michel AD et al 1996 $[^3H]\alpha,\beta$-meATP, $[^{35}S]ATP\gamma S$ and $[^{33}P]ATP$ label the human bladder recombinant P_{2X} receptor. Br J Pharmacol 117:1254–1260

Nagy JI, LaBella LA, Buss M 1984 Immunohistochemistry of adenosine deaminase: implications for adenosine neurotransmission. Science 224:166–168

North RA 1996 P2X receptors: a third major class of ligand-gated ion channels. In: P2 purinoceptors: localization, function and transduction mechanisms. Wiley, Chichester (Ciba Found Symp 198) p 91–109

Phillis JW 1990 Adenosine deaminase and the identification of purinergic neurons. Neurochem Int 16:223–226

Richardson PJ, Brown SJ, Bailyes EM, Luzio 1987 Ectoenzymes control adenosine modulation of immunoisolated cholinergic synapses. Nature 327:232–234

Robertson SJ, Rowan ER, Kennedy CK 1995 Characterisation of the electrophysiological actions of P_2 purinoceptor agonists in cultured neurones of the rat dorsal root ganglia. Br J Pharmacol 116:54P

Schoen SW, Kreutzberg GW, Singer W 1993 Cytochemical redistribution of 5′-nucleotidase in the developing cat visual cortex. Eur J Neurosci 5:210–222

Simon J, Webb TE, Barnard EA 1995 Characterization of a P_{2Y} purinoceptor in the brain. Pharmacol Toxicol 76:302–307

Webb TE, Simon J, Bateson AN, Barnard EA 1994 Transient expression of the recombinant chick brain $P2Y_1$ receptor and localization of the corresponding mRNA. Mol Cell Biol 40:437–442

Wiley JS, Chen R, Jamieson GP 1993 The ATP^{4-} receptor-operated channel (P_{2Z} class) of human lymphocytes allows Ba^{2+} and ethidium$^+$ uptake: inhibition of fluxes by suramin. Arch Biochem Biophys 305:54–60

Yoshioka T, Tanaka O 1989 Histochemical localization of Ca^{2+}/Mg^{2+} ATPase of the rat cerebellar cortex during postnatal development. Int J Dev Neurosci 7:181–193

Zenker W, Rinne B, Bankoul S, Lehir M, Kaissling B 1992 5′-Nucleotidase in spinal meningeal compartments in the rat: an immuno and enzyme histochemical study. Histochemistry 98:135–139

The diadenosine polyphosphate receptors: P2D purinoceptors

M. Teresa Miras-Portugal, Enrique Castro, Jesus Mateo and Jesus Pintor

Departamento de Bioquimica, Facultad de Veterinaria, Universidad Complutense, E-28040 Madrid, Spain

Abstract. Diadenosine polyphosphates—Ap_4A, Ap_5A and Ap_6A—are co-stored in neurosecretory vesicles together with ATP and aminergic compounds. They are released from neural cells and synaptic terminals in a Ca^{2+}-dependent process. Ligand binding and displacement experiments carried out with [^{3}H]Ap_4A on isolated chromaffin cells and synaptosomal preparations result in curvilinear Scatchard plots with K_d values close to 0.1 nM for the high-affinity binding sites. Displacement curves with two steps are obtained for homologous and heterologous nucleotide ligands; the lowest-affinity step exhibits K_i values in the micromolar range for Ap_nA compounds. The high-affinity binding sites were named P2D purinoceptors on the basis of their binding characteristics. Single-cell studies in neurochromaffin cells indicate the presence of P2X purinoceptors in noradrenergic cells that do not respond to Ap_4A and in which noradrenaline secretion can be induced by influx of extracellular Ca^{2+}. P2Y receptors that respond to ATP analogues and Ap_nAs are present in endothelial cells from adrenal medulla. Those cells that express P2U purinoceptors are unresponsive to Ap_nAs. Ectodiadenosine polyphosphate hydrolases with K_m values of 0.3 to 2 μM are present in both neural and endothelial cells from adrenal medulla. In midbrain synaptic terminals diadenosine polyphosphates induce Ca^{2+} entry from the extracellular medium. The fact that the synaptic response is not cross-desensitized by ATP and its non-hydrolysable analogues, the non-blocking effect of suramin, and the differential effect of Ca^{2+} channel blockers, together suggest that there are different receptors for nucleotides and dinucleotides in rat brain synaptosomes, which we have called P4 purinoceptors on the basis of functional studies.

1996 P2 purinoceptors: localization, function and transduction mechanisms. Wiley, Chichester (Ciba Foundation Symposium 198) p 35–52

The occurrence of dinucleoside polyphosphates, mainly guanine dinucleotides, in crustacea and the formation of adenine dinucleoside polyphosphate (Ap_4A) as a by-product of reactions with an adenilyl intermediate, caused widespread interest in their intracellular biological actions, including the initiation of DNA replication (for review see McLennan 1992). Nevertheless, this interest was slowly disappearing, until the possibility of extracellular activity was suspected when the diadenosine polyphosphates Ap_3A and Ap_4A were found co-stored with adenine nucleotides and serotonin in the dense granules of platelets (Flodgaard & Klenow 1982, Lüthje & Ogilvie 1983).

The presence of diadenosine polyphosphates—Ap_4A, Ap_5A and Ap_6A—in neurosecretory vesicles together with adenine nucleotides, acetylcholine or catecholamines, strengthened the search for their plasma membrane receptors and functions not only in neural tissues, but in the adjacent muscular and vascular structures (Rodriguez del Castillo et al 1988, Pintor et al 1992a).

The structural analogies between diadenosine polyphosphates (Ap_nAs) and adenine nucleotides suggest the possibility of overlapping actions on the nucleotide receptors. Today, the field of nucleotide P2 purinoceptors is one of the most productive areas in cellular communication research. The existence of two families has been demonstrated, one of ionotropic receptors (P2X) and the other including a large variety of metabotropic receptors (P2Y, P2Z, P2T, P2U, P2D, etc., also called generically $P2Y_{1\ldots n}$; for review see Abbracchio & Burnstock 1994, Fredholm et al 1994, Zimmermann 1994). However, the main question concerning Ap_nAs still remains: do the Ap_nAs have specific receptors and, if not, to what extent do they share the P2 purinoceptors with other nucleotides (Pintor et al 1991b, 1993)? The fact that the stacked form is the most probable conformation for Ap_nAs in physiological solutions (Fig. 1A), indicates that a careful analysis not only of the pharmacology of binding sites, but also of the cellular responses will be required in order to define what is unique to Ap_nAs and what they have in common with other nucleotides.

Experimental procedures

Chromaffin and endothelial cells from bovine adrenal medulla were obtained and cultured as previously reported (Castro et al 1994). Rat brain synaptosomal preparations for ligand binding and functional studies were obtained as described by Herrero et al (1992). We used high performance liquid chromatography (HPLC) to quantify and characterize the diadenosine polyphosphates and the nucleotides present in cellular and granular preparations and extracellular fluids, as described previously (Rodriguez del Castillo et al 1988, Pintor et al 1991a, 1992b). An HPLC control of all the nucleotides and dinucleotides employed was also carried out to ensure the purity of the effectors. We have carried out binding experiments and displacement studies with $[^3H]Ap_4A$, as described by Pintor et al (1991b, 1993). The study of binding sites by autoradiography required the synthesis of $[^3H]Ap_4A$ with a highly specific activity. In short, $[^3H]$ATP was condensed with non-labelled AMP in the presence of carbodiimide, according to the procedure of Ng & Orgel (1987).

Intracellular Ca^{2+} was measured using fura 2 as a fluorescent dye as described by Castro et al (1992). In synaptic terminals some modifications were required; these are described elsewhere (Pintor & Miras-Portugal 1995b).

Microfluorescence techniques were used to record the $[Ca^{2+}]_i$ from single adrenal endothelial or chromaffin cells as described by Castro et al (1994). Ectonucleotidase activities for nucleotides and dinucleotides were studied by HPLC as described by Rotllan et al (1991) and Ramos et al (1995).

Results and discussions

Diadenosine polyphosphates in adrenal medulla

The adrenomedullary tissue constituted by neurosecretory and endothelial cells represents the best known model for studying the action of diadenosine polyphosphates. Ap_4A, Ap_5A and Ap_6A stored in secretory granules are released from chromaffin cells in a Ca^{2+}-dependent manner after acetylcholine stimulation (Pintor et al 1991a, 1992b). Chromaffin cells are considered to be differentiated sympathetic neurons. After exocytosis, the extracellular levels of each Ap_nA can reach 15–30 μM and that of ATP about 0.5–1 mM. These values were estimated by assuming an extracellular volume of distribution similar to that of the secretory cells. Much lower concentrations can be assumed when they dilute in the surrounding media and reach the vascular endothelial cells. High-affinity binding sites for Ap_4A at the plasma membrane were first demonstrated in cultured chromaffin cells (Pintor et al 1991b). Two high-affinity binding sites were obtained from the ligand saturation studies (Fig. 1).

Other neural preparations, such as rat brain and Torpedo synaptic terminals, and endothelial cells from adrenal medulla exhibited similar Scatchard curves and affinity values in the same range (Pintor et al 1993, Miras-Portugal et al 1994). Autoradiographic studies performed at a concentration of ligand 10 times the value of K_d for the first binding site, with labelled $[^3H]Ap_4A$, supported the existence of high-affinity binding sites in adrenomedullary tissue and their scarcity in the adrenal cortex (Fig. 1B).

Studies of Ap_nA signalling through plasma membrane receptors in populations of neurochromaffin cells demonstrated a Ca^{2+} mobilization from internal stores and activation of protein kinase C (PKC) (Castro et al 1992).

Although chromaffin cells have been used in neural studies largely because of their claimed homogeneity and non-tumoral characteristics, new techniques permitting single-cell studies have shown that they are clearly a heterogeneous population. Concerning secretion, the chromaffin cells containing adrenaline are quite different from the ones containing noradrenaline, and the membrane receptors for most signals are not equally distributed. For this reason, the effects of purinergic agents (mainly on the Ca^{2+} signal) were studied at single-cell level.

Figure 2 shows that there are two types of response to ATP and its analogues distinguishable in terms of mechanism and pharmacology. A population of chromaffin cells displays ATP-evoked $[Ca^{2+}]_i$ transients originated exclusively

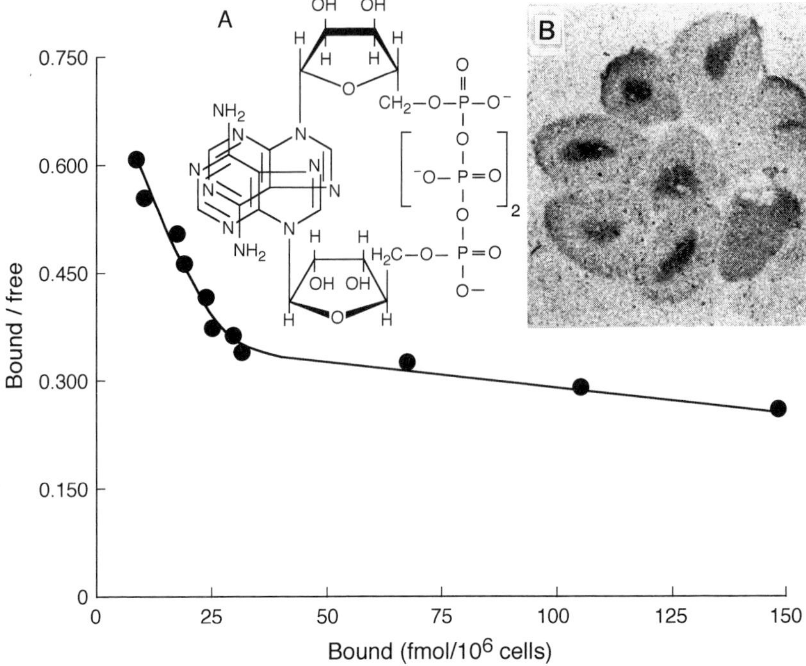

FIG. 1. Diadenosine tetraphosphate binding to adrenal preparations. Scatchard analysis of equilibrium of [^{3}H]Ap$_4$A binding to chromaffin cells in culture. The K_d and V_{max} values were $8 \pm 0.65 \times 10^{-11}$ M and 5420 ± 450 sites per cell (90 ± 8 fmol/mg protein equivalent to 9 fmol/mg of tissue) for the high-affinity binding site and $5.6 \pm 0.53 \times 10^{-9}$ M and 70 000 sites per cell (1.15 ± 0.01 pmol/mg protein) for the second binding site, respectively. (Inset A) Stacked representation of Ap$_4$A structure. (Inset B) Autoradiography of consecutive slices of adrenal gland showing [^{3}H]Ap$_4$A binding at 1 nM concentration (25.7 Ci/mmol). Binding site densities were 11 fmol/mg tissue and 3 fmol/mg tissue for the adrenal gland medulla and cortex, respectively.

by Ca^{2+} influx and blocked by suramin (Castro et al 1995). This purinoceptor was also activated with a similar potency by 2-methylthioATP (2-MeSATP) but was insensitive to UTP, ADPβS, Ap$_4$A and α,β-methylene ATP (α,β-meATP) (Castro et al 1994). The $[Ca^{2+}]_i$ transients were associated with Na^+ entry (measured with the sodium-binding fluorescent indicator probe). Thus, we classify this ionotropic receptor as a P2X purinoceptor. The pharmacological profile of this receptor does not agree with the P2X subtype from smooth muscle (Bean 1992, Valera et al 1994). However, it fits smoothly in the known pharmacology of other neuronal ATP-gated channels (Ueno et al 1992) represented by the P2X subtype cloned from the PC12 cell line (Brake et al 1994). It is important to remember that this line is derived from a rat phaeochromocytoma: a tumour of chromaffin cells.

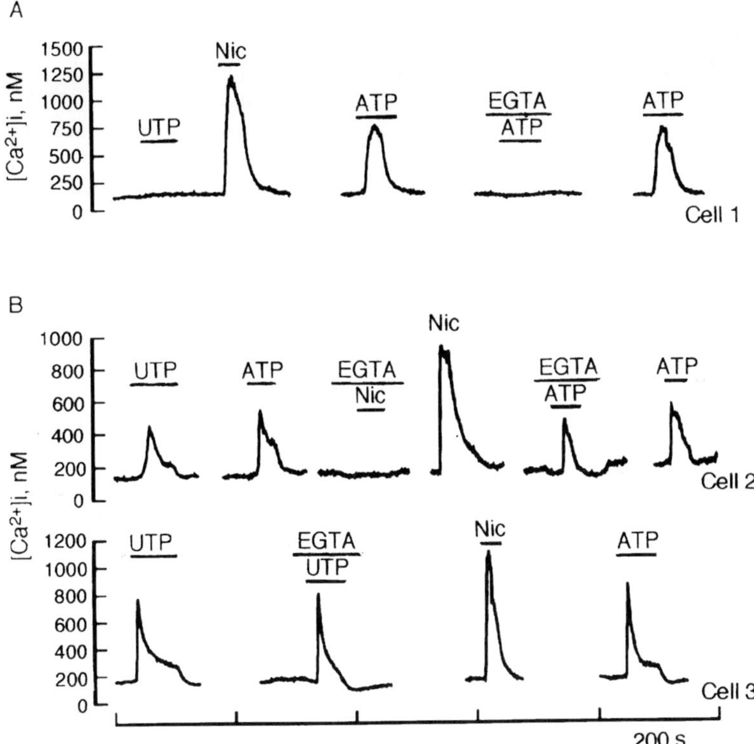

FIG. 2. $[Ca^{2+}]_i$ responses of adrenal chromaffin cells to ATP and analogues reveal two separate mechanisms of action. The traces shown were obtained from morphologically identified chromaffin cells. This identification was confirmed functionally by positive responses to nicotine $10\,\mu M$ (Nic). Successive drug challenges are indicated by the horizontal bars. To avoid receptor desensitization or cell damage by photobleaching, cells were allowed to rest for 5 min (trace breaks) between stimulations. (A) Example of a cell unresponsive to UTP and displaying ATP-evoked $[Ca^{2+}]_i$ transients absolutely dependent on extracellular Ca^{2+}. ATP and UTP were applied at $100\,\mu M$. 'EGTA' means replacement of normal extracellular medium with a low-Ca^{2+} solution buffered with EGTA to produce 100 nM $[Ca^{2+}]_o$. (B) A different population of cells displayed $[Ca^{2+}]_i$ transients arising from mobilization of internal stores and was UTP sensitive.

A separate population of chromaffin cells is equipped with an ATP receptor coupled to intracellular Ca^{2+} mobilization (Fig. 2B). This action was also evoked by UTP, indicating the involvement of a P2U purinoceptor. However, suramin was unable to block ATP- or UTP-evoked $[Ca^{2+}]_i$ transients in these cells. This suggests that a novel P2U subtype may be involved (see below). P2X and P2U purinoceptors were segregated to different chromaffin cell types. The ATP-gated channel was localized exclusively to noradrenaline-secreting cells, while the P2U purinoceptor was present in both adrenaline- and noradrenaline-secreting cells. Furthermore, ATP evoked catecholamine secretion from noradrenergic cells only (Castro et al 1995), which agrees with the known ineffectiveness of $[Ca^{2+}]_i$ rises coming from internal stores in eliciting secretion from chromaffin cells (Augustine & Neher 1992). This emphasizes the functional importance of purinoceptors present in each subtype of chromaffin cells: the pharmacological differences serve to translate a qualitatively different biological message in response to ATP by adrenergic and noradrenergic chromaffin cells.

Purinoceptors are also present in endothelial cells from blood capillaries in the adrenal gland. In primary cultures, these cells seem to be equipped with both P2U and P2Y purinoceptor subtypes, as indicated by the dose–response curves to ATP, UTP and 2-MeSATP (Castro et al 1994). Thus, primary endothelial cells displayed clear, although somewhat smaller, responses to 2-MeSATP and Ap_4A (see Fig. 3A). However, P2Y purinoceptors are lost when these cells are subcultured: endothelial cells at third or higher passage display a homogeneous population of a single P2U purinoceptor subtype (Fig. 3B). As in the case of chromaffin cells, this purinoceptor was not sensitive to the antagonists suramin or pyridoxalphosphate-6-azophenyl-2′,4′-disulfonic acid (PPADS), or the agonists 2-MeSATP and Ap_4A (Fig. 3A; J. Mateo, unpublished results). Thus, this P2U purinoceptor may represent a novel subtype, different from cloned P2U purinoceptors (Lustig et al 1993) which are activated by Ap_4A and antagonized by suramin (Lazarowsky & Harden 1994).

It is important to take into account that most of the responses to Ap_4A have been obtained at high ligand concentration. They are still active at very low concentrations (in the nanomolar range) due to the presence of many high-affinity binding sites of Ap_4A in adrenomedullary cells (Fig. 1B). The low density of high-affinity binding sites for Ap_4A in the adrenal cortex is also noteworthy, perhaps related to the blood circulation that goes from the cortex to the medulla and from there to the extra-adrenal vessels. Thus the secretory components of chromaffin cells do not reach the cortical cells.

The presence of ecto-diadenosine polyphosphate hydrolase, with K_m values for Ap_nAs ranging from 0.3 to 2 μM in adrenal medulla endothelial and neural cells, accounts for the extracellular hydrolysis of these compounds and increases their physiological relevance (Rotllan et al 1991, Ramos et al 1995).

FIG. 3. $[Ca^{2+}]_i$ responses of adrenal endothelial cells to ATP and other purinergic agonists. (A) Purinergic pharmacology in primary cultures. $[Ca^{2+}]_i$ recordings were obtained from single, identified, cells exposed successively with ATP and several analogues for the period indicated by the horizontal bars. Cells were allowed to rest for 5 min after each challenge (spaces between individual traces). In cells 1 and 2, all drugs were applied at 50 μM. In cell 3, ATP and UTP were 50 μM, suramin 100 μM. 'EGTA' means same as in Fig. 2. (B) Results from endothelial cells at the 15th passage. All agonists were used at 100 μM, except ADP (1 mM).

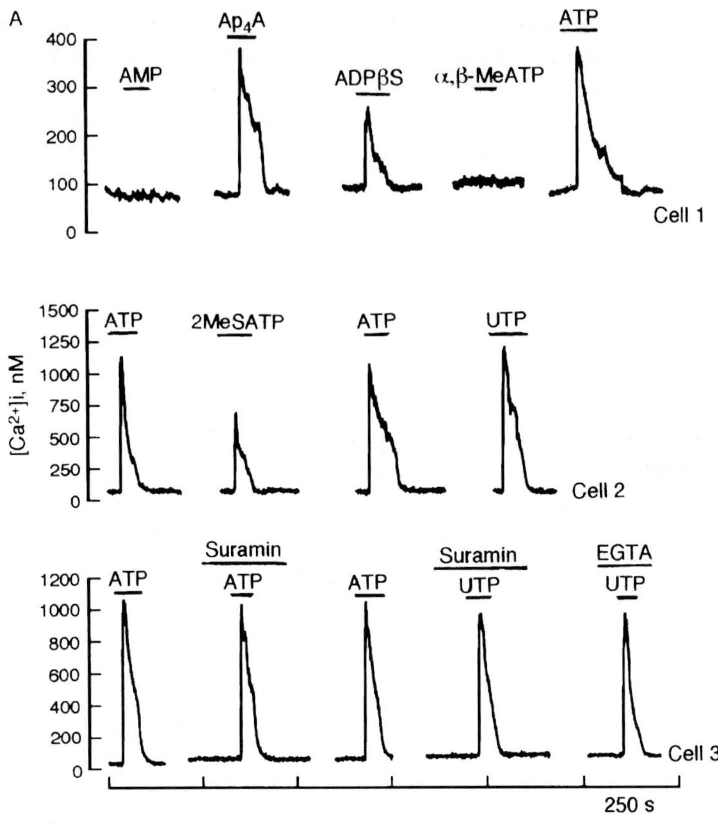
A
400
300
200
100
0
AMP
Ap4A
ADPβS
α,β-MeATP
ATP
Cell 1
[Ca2+]i, nM
1500
1250
1000
750
500
250
0
ATP
2MeSATP
ATP
UTP
Cell 2
1200
1000
800
600
400
200
0
ATP
Suramin
ATP
ATP
Suramin
UTP
EGTA
UTP
Cell 3
250 s

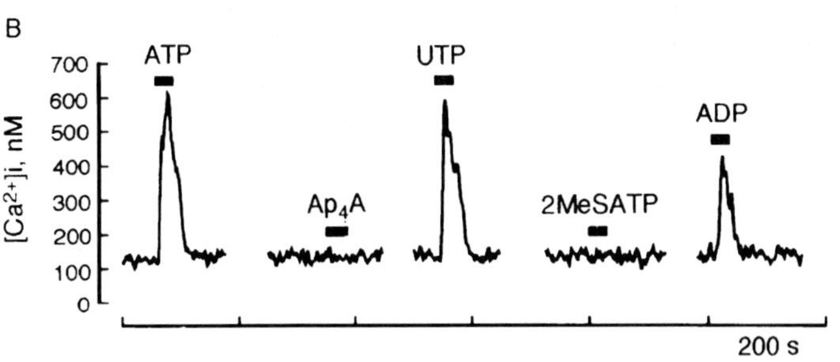
B
700
600
500
400
300
200
100
0
[Ca2+]i, nM
ATP
Ap4A
UTP
2MeSATP
ADP
200 s

The V_{max} of these enzymes is very small when compared with ecto-ATPase activity from the same cells, providing a longer life as messengers for the extracellular Ap_nAs (Miras-Portugal et al 1994). This finding is analysed in the CNS later in this paper. It is also noteworthy that the asymmetrical hydrolysis carried out by these enzymes always produces AMP and the corresponding $(n-1)$ nucleotide, which in the case of Ap_5A is the adenosine tetraphosphate (Ap_4). This compound is only destroyed with difficulty by the ecto-ATPases and persists longer outside the cell (Ramos et al 1995). Recently, Bailey & Hourani (1995) have reported a highly significant effect of Ap compared with ATP on the contractions of the guinea-pig vas deferens. These data are important for us to understand the actions of Ap_nAs and also the possible *in situ* generation of active compounds from the Ap_nAs.

Diadenosine polyphosphates in the CNS

The presence of diadenosine polyphosphates (Ap_4A, Ap_5A and Ap_6A) has been reported in rat brain synaptosomes, all three at similar levels (100–170 pmol/mg of protein) (Pintor et al 1992a). Synaptosomal depolarization in the presence of Ca^{2+} induces a 10% release of the total content. Due to the heterogeneity of neurotransmitters in brain synaptosomes, it is almost impossible to estimate the Ap_nA concentration in the synaptic cleft and surrounding areas from the values of the Ca^{2+}-dependent release. Nevertheless, brain perfusion studies in conscious rats with a push–pull technique indicate that the control values in the perfusion media are under 2 nM. These levels are significantly increased after amphetamine stimulation in a dose-dependent manner, reaching maximal values, about 70 nM, in the neostriatum 30 min after amphetamine administration (Pintor et al 1995). Concentrations about two orders of magnitude higher can be assumed for the synaptic terminals. Moreover, the high nanomolar and low affinity (micromolar) binding sites found in rat brain could be associated with different functional receptors (Pintor et al 1993). The long life of Ap_4A and Ap_5A in perfused brain after stimulation, together with the absence of ATP that is supposedly co-released, confirms that the extracellular degradation of these compounds is also much slower in the CNS (Pintor et al 1995).

Diadenosine polyphosphate receptors in the CNS

The receptor types and actions of diadenosine polyphosphates in the CNS are far from clear. Binding studies with $[^3H]Ap_4A$ have demonstrated the presence of a P2 purinoceptor with specificity for adenine dinucleotides. Displacement studies for some diadenosine polyphosphates at this receptor showed the following pattern: $Ap_4A > Ap_5A > Ap_6A$. In addition, the potency order of the different ATP analogues did not display any of the classical pharmacological profiles. Consequently, a new name—P2D purinoceptor—was given to this receptor (Pintor et al 1993).

Additional information can be obtained from cellular responses. In this way, the interactions with purinergic P1 non-nucleotidic receptors have been described, as it is the modulation of the firing rate of cortical neurons (Stone & Perkins 1981). The interaction with A_1 receptors has also been reported by Klishin et al (1994) on hippocampal slices, where inputs of pyramidal CA1 neurons were stimulated via the Schaffer collateral–commisural pathway. Ap_4A or Ap_5A, at very low concentrations, inhibited both the field potential and the excitatory postsynaptic current measured by means of *in situ* whole-cell patch-clamp. The actions were imitated by adenosine analogues, but the Ap_5A effects were reversed by protein kinase C (PKC) inhibitors, which was not the case for adenosine. Nevertheless, recent data in freshly isolated hippocampal CA3 neurons using a whole-cell patch-clamp technique demonstrated the selective potentiation of N-type Ca^{2+} channels by Ap_5A. It is relevant that ATP even at higher concentrations had no effect, as was also the case for adenosine analogues (Panchenko et al 1996). It is not clear which receptors mediate the specific responses to diadenosine polyphosphates in CA1 and CA3 cells from hippocampus.

The action of Ap_4A on P2X purinoceptors in the CNS can be deduced from the autoradiographic studies with $[^3H]\alpha,\beta$-meATP. Whereas most purinergic agonists are unable to displace the ligand binding, Ap_4A is very effective (Balkar et al 1995). However, agonist or antagonist action cannot be excluded in these studies.

Synaptosomal preparations have proved to be a useful model for studying the actions of diadenosine polyphosphates in rat brain, in spite of the heterogeneity of their content and function. In previous work in our laboratory, we discovered the presence of a presynaptic metabotropic glutamate receptor involved in the positive feedback of glutamate exocytosis using such preparations (Herrero et al 1992). Our previous experience suggested we should study the action of diadenosine polyphosphates, ATP and its non-hydrolysable analogues, on Ca^{2+} levels and membrane potential in synaptosomal preparations. We carried this work out systematically with rat mid-brain synaptic terminals. All nucleotides and dinucleotides were able to induce an increase in intrasynaptosomal Ca^{2+} levels, measured with fura 2 dye, and as expected, this was lower than that obtained by non-selective depolarization with 30 mM K^+ (Pintor & Miras-Portugal 1995a). Although it could be considered that both agonists, Ap_nA and ATP, could activate the same receptor, pharmacological studies demonstrated the activation of different purinoceptor populations. Whereas the Ca^{2+} increase induced by ATP was blocked by suramin, the effect by diadenosine polyphosphates was not. Another important point was the non-desensitization effect when both agonists were applied consecutively (Pintor & Miras-Portugal 1995b).

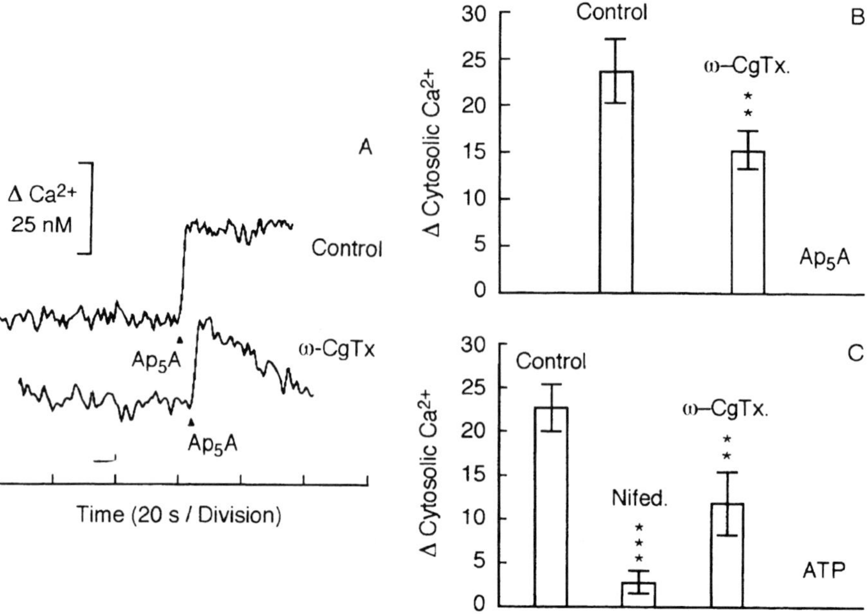

FIG. 4. Effect of Ca^{2+} antagonists on Ap_5A and ATP responses in midbrain synaptic terminals. (A) Effect of ω-conotoxin G-VI-A on the Ca^{2+} transient induced by diadenosine pentaphosphate. The N-type Ca^{2+} channel blocker was able to inhibit only the sustained phase of the cytosolic cation increment, indicating the voltage independency of the initial transient. (B) Quantification of the effect of ω-conotoxin G-VI-A on the total Ca^{2+} entry induced by Ap_5A. (C) Effect of ω-conotoxin G-VI-A and nifedipine on the Ca^{2+} increase induced by ATP. Both blockers significantly inhibited the cation entry, suggesting the involvement of N- and L-voltage-dependent Ca^{2+} channels in the presynaptic Ca^{2+} increase induced by ATP.

The study of the Ca^{2+} entry mechanisms with specific voltage-dependent Ca^{2+} channel (VDCC) inhibitors showed significant differences between both compounds. Both ATP and diadenosine polyphosphates allowed a voltage-independent cation entry, but as a secondary step the dinucleotides opened an N-type VDCC, whereas ATP opened up both N- and L-type VDCCs, as deduced from the effects of nifedipine and ω-conotoxin G-VI-A (Fig. 4). All these differences suggested the existence of a new purinoceptor exclusively for adenine dinucleotides, which we tentatively termed a P4 purinoceptor (Pintor & Miras-Portugal 1995a,b). These results agree with the activation of the N-type Ca^{2+} channel specifically by diadenosine polyphosphates, and not by ATP analogues or A_1, A_2 agonists in hippocampal neurons already discussed (Panchenko et al 1996).

The mechanism by which through specific receptors diadenosine polyphosphates induce Ca^{2+} entry and activate the N-type VDCC remains unclear.

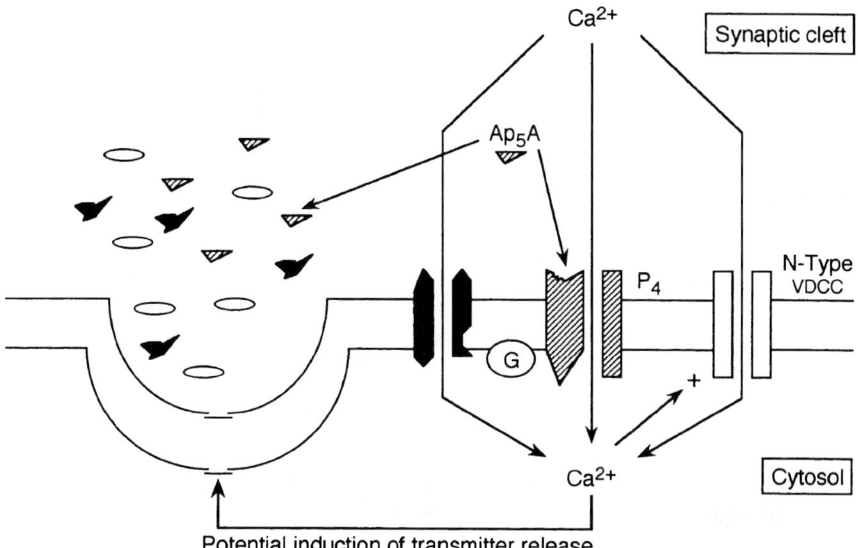

FIG. 5. Molecular mechanisms and putative role of diadenosine polyphosphates in the CNS acting through P4 purinoceptors. The release of diadenosine polyphosphates Ap_4A and Ap_5A from the synaptic terminal, induced by the membrane depolarization, allows them to reach extracellular concentrations sufficient to stimulate the P4 purinoceptor. This receptor could be an intrinsic membrane protein containing a Ca^{2+} ion pore or a receptor coupled to a Ca^{2+} channel via G protein. The stimulation of the receptor by extracellular Ap_5A induces a voltage-independent Ca^{2+} entry, which produces a membrane depolarization responsible for the activation of the N-type voltage-dependent Ca^{2+} channel (VDCC). The entry of the cation would contribute significantly to the potentiation of the transmitter release.

Two possibilities are that the receptor either is at the same time a channel (as occurs with the P2X purinoceptor) or it is coupled to a voltage-independent Ca^{2+} channel. It is clear that the initial Ca^{2+} entry is not abolished neither by any of the Ca^{2+} channel blockers currently known, nor by depolarization prior to the dinucleotide application. All the evidence suggests that the initial cation entry induces the opening of the N-type VDCC. The mechanism of action proposed for the P4 purinoceptor is described in Fig. 5. The presence of these receptors in presynaptic terminals suggests they have an important role in the regulation of neurotransmitter release. However, much work needs to be done to identify the specific synapses where diadenosine polyphosphates are present, and if they are co-released with more classical transmitters. The understanding of their actions as fast transmitters or modulators in neural tissues will help to clarify the relevance of these compounds in the functioning of nerve cells.

Acknowledgements

This work has been supported by the research grant PB 92/230 from the Spanish Ministry of Education and Sciences, DGICYT, and the Areces Foundation Neurosciences Programme 1994. J. M. and J. P. are respectively, predoctoral and postdoctoral students from the Spanish Ministry of Education and Sciences and the Areces Foundation. The adrenal medulla $[^3H]Ap_4A$ autoradiography was carried out by F. Rodriguez-Pascual and R. Cortes. We thank E. Lundin for his help in the preparation of the manuscript.

References

Abbracchio MP, Burnstock G 1994 Purinoceptors: are there families of P_{2X} and P_{2Y} purinoceptors? Pharmacol Ther 64:445–475

Augustine GJ, Neher E 1992 Calcium requirements for secretion in bovine chromaffin cells. J Physiol 450:247–271

Bailey SJ, Hourani SMO 1995 Effects of suramin on contractions of the guinea-pig vas deferens induced by analogues of adenosine 5′-triphosphate. Br J Pharmacol 114:1125–1132

Balcar VJ, Li Y, Killinger S, Bennett MR 1995 Autoradiography of P_{2X}ATP receptors in the rat brain. Br J Pharmacol 115:302–306

Bean BP 1992 Pharmacology and electrophysiology of ATP-activated ion channels. Trends Pharmacol Sci 13:87–90

Brake AJ, Wagenbach MJ, Julius D 1994 New structural motif for ligand-gated ion channels defined by an ionotropic ATP receptor. Nature 371:519–523

Castro E, Pintor J, Miras-Portugal MT 1992 Ca^{2+} stores mobilization by diadenosine tetraphosphate, Ap_4A, through a putative P_{2Y} purinoceptor in adrenal chromaffin cells. Br J Pharmacol 106:833–837

Castro E, Tomé AR, Miras-Portugal MT, Rosario LM 1994 Single-cell fura-2 microfluorometry reveals different purinoceptor subtypes coupled to Ca^{2+} influx and intracellular Ca^{2+} release in bovine adrenal chromaffin and endothelial cells. Pflügers Arch Eur J Physiol 426:524–533

Castro E, Mateo J, Tomé AR, Barbosa RM, Miras-Portugal MT, Rosário LM 1995 Cell-specific purinergic receptors coupled to Ca^{2+} entry and Ca^{2+} release from internal stores in adrenal chromaffin cells. J Biol Chem 270:5098–5106

Flodgaard H, Klenow H 1982 Abundant amounts of diadenosine 5′,5‴-p1,p4-tetraphosphate are present and releaseable, but metabolically inactive in human platelets. Biochem J 208:737–742

Fredholm BB, Abbracchio MP, Burnstock G et al 1994 Nomenclature and classification of purinoceptors. Pharmacol Rev 46:143–156

Herrero H, Miras-Portugal MT, Sanchez-Prieto J 1992 Positive feedback of glutamate exocytosis by metabotropic presynaptic receptor stimulation. Nature 360:163–166

Klishin A, Lozovaya N, Pintor J, Miras-Portugal MT, Krishtal O 1994 Possible functional role of diadenosine polyphosphates: negative feedback for excitation in hippocampus. Neuroscience 58:235–236

Lazarowski ER, Harden TK 1994 Identification of a uridine nucleotide-selective G-protein-linked receptor that activates phospholipase C. J Biol Chem 269:11830–11836

Lustig KD, Shiau AK, Brake AJ, Julius D 1993 Expression cloning of an ATP receptor from mouse neuroblastoma cells. Proc Natl Acad Sci USA 90:5113–5117

Lüthje J, Ogilvie A 1983 The presence of diadenosine 5′,5‴-p^1,p^4-triphosphate (Ap_3A) in human platelets. Biochem Biophys Res Commun 115:253–260

McLennan AG 1992 Ap_4A and other dinucleoside polyphosphates. CRC Press, Boca Raton, FL

Miras-Portugal MT, Pintor J, Castro E, Rodriguez-Pascual F, Torres M 1994 Diadenosine polyphosphates from neuro-secretory granules: the search for receptors, signals and function. In: Municio AM, Miras-Portugal MT (eds) Cell signal transduction, second messengers and protein phosphorylation in health and disease. Plenum, New York, p 169–186

Ng KME, Orgel LE 1987 The action of a water-soluble carbodiimide on adenosine 5′-polyphosphates. Nucleic Acids Res 15:3573–3580

Panchenko VA, Pintor J, Tsydrenko AY, Miras-Portugal MT, Krishtal OA 1996 Diadenosine polyphosphates selectively potentiate N-type Ca^{2+} channels in the rat central neurons. Neuroscience 70:353–360

Pintor J, Miras-Portugal MT 1995a P_2 purinergic receptors for diadenosine polyphosphates in the nervous system. Gen Pharmacol 26:229–235

Pintor J, Miras-Portugal MT 1995b A novel receptor for diadenosine polyphosphates coupled to calcium increase in rat midbrain synaptosomes. Br J Pharmacol 115:895–902

Pintor J, Torres M, Miras-Portugal MT 1991a Carbachol induced release of diadenosine polyphosphates -Ap_4A and Ap_5A- from perfused bovine adrenal medulla and isolated chromaffin cells. Life Sci 48:2317–2324

Pintor J, Torres M, Castro E, Miras-Portugal MT 1991b Characterization of diadenosine tetraphosphate (Ap_4A) binding sites in cultured chromaffin cells: evidence for a P_{2Y} site. Br J Pharmacol 103:1980–1984

Pintor J, Diaz-Rey MA, Torres M, Miras-Portugal MT 1992a Presence of diadenosine polyphosphates -Ap_4A and Ap_5A- in rat brain synaptic terminals: Ca^{2+} dependent release evoked by 4-aminopyridine and veratridine. Neurosci Lett 136:141–144

Pintor J, Rotllan P, Torres M, Miras-Portugal MT 1992b Characterization and quantification of diadenosine hexaphosphate in chromaffin cells: granular storage and secretagogue-induced release. Anal Biochem 200:296–300

Pintor J, Diaz-Rey MA, Miras-Portugal MT 1993 Ap_4A and ADP-β-S binding to P2 purinoceptors present on rat brain synaptic terminals. Br J Pharmacol 108:1094–1099

Pintor J, Porras A, Mora F, Miras-Portugal MT 1995 Dopamine receptor blockade inhibits the amphetamine-induced release of diadenosine polyphosphates, diadenosine tetraphosphate and diadenosine pentaphosphate, from neostriatum of the conscious rat. J Neurochem 64:670–676

Ramos A, Pintor J, Miras-Portugal MT, Rotllán P 1995 Use of fluorogenic substrates for detection and investigation of ectoenzymatic hydrolysis of diadenosine polyphosphates: a fluorometric study on chromaffin cell. Anal Biochem 228:74–82

Rodriguez del Castillo A, Torres M, Delicado EG, Miras-Portugal MT 1988 Subcellular distribution studies of diadenosine polyphosphates—Ap_4A and Ap_5A—in bovine adrenal medulla: presence in chromaffin granules. J Neurochem 51:1696–1703

Rotllan P, Ramos A, Pintor J, Torres M, Miras-Portugal MT 1991 Di(1,N6-ethenoadenosine)5′,5‴-P1,P4-tetraphosphate, a fluorescent enzymatically active derivative of Ap_4A. FEBS Lett 280:371–374

Stone TW, Perkins MN 1981 Adenine dinucleotide effect on rat corical neurones. Brain Res 229:241–245

Ueno S, Harata N, Inoue K, Akaike N 1992 ATP-gate current in dissociated rat nucleus solitarii neurons. J Neurophysiol 68:778–786

Valera S, Hussy N, Evans RJ et al 1994 A new class of ligand-gated ion channel defined by a P_{2X} receptor for extracellular ATP. Nature 371:516–519

Zimmermann H 1994 Signalling via ATP in the nervous system. Trends Neurosci 17:420–426

DISCUSSION

Burnstock: To clarify: you are saying that the diadenosine polyphosphates can act through P2X and P2Y receptors. In fact, Ap_4A, Ap_5A and Ap_6A seem to work through P2X purinoceptors in some experimental models (see Ralevic et al 1995). But are you also saying that there is a unique receptor for these diadenosine polyphosphates in parts of the brain and that you are calling it a P4 purinoceptor?

Miras-Portugal: The receptor that we have called a P4 purinoceptor is a specific receptor for diadenosine polyphosphates that is able to induce Ca^{2+} transients in synaptic terminals. The main reason for differentiating it from the ATP receptors is that it is not cross-desensitized by ATP or its non-hydrolysable analogues, neither is it blocked by suramin. But its final name and acceptance in the classification of nucleotide receptors is a matter for this audience to decide.

Leff: The separate identity of what you're calling a P2D or P4 receptor has obviously been the subject of considerable controversy over the last few years. What do you consider to be the definitive data which give this receptor a separate identity?

Miras-Portugal: I've already outlined the reasons for postulating a P4 purinoceptor; they correspond to a functional role for diadenosine polyphosphates in synaptosomal Ca^{2+} influx. In the case of P2D, the evidence for separate receptors is the existence of very high affinity binding sites (K_d 0.1 nM), together with displacement pharmacological profiles different from those of P2X and, to a lesser extent, P2Y. For that reason, the P2D receptor has also been called $P2Y_7$ in recent reviews. Moreover, autoradiographic studies of $[^3H]Ap_4A$ binding sites in rat brain showed a different distribution from those reported for P2X and P2Y.

Burnstock: You will need to clone it as well if you want to define a new subtype.

Starke: With relation to the chromaffin cells, you showed the nicotinic receptor and the P2 purinoceptor located side-by-side. Have you tested whether suramin will influence the effect of a nicotinic agonist and vice versa? I'm asking this because in cultured sympathetic neurons, Allgaier et al (1995) have shown that the noradrenaline release response elicited by ATP is partly blocked by hexamethonium and the release response to nicotine is partially blocked by suramin. Have you found anything like this happening?

Miras-Portugal: We did not do those studies in single cells. We only studied the nicotinic response to make sure that we were working with a neural cell, not an endothelial cell or some other cell type. Suramin blocks the P2X receptor from adrenal noradrenergic cells completely, but it has no effect on Ca^{2+} mobilization from internal stores of adrenergic cells. However, we haven't studied the effects of suramin and hexamethonium on the same receptors in the same cells.

Illes: You have looked at the effects of diadenosine polyphosphates in chromaffin cells of the adrenal medulla. These cells derive from the neuronal crest and secrete catecholamines. We are ourselves interested in a group of noradrenergic central neurons situated in the nucleus locus coeruleus. Both ATP analogues (Tschöpl et al 1992) and diadenosine polyphosphates (Illes et al 1996, this volume) depolarize rat locus coeruleus neurons and increase their spontaneous firing rate. In our experiments the diadenosine tetraphosphate-induced stimulation was, in contrast to the α,β-methylene ATP-induced stimulation, only slightly antagonized by 100 μM suramin. I was pleased to see that your diadenosine polyphosphate effects were also rather insensitive to suramin.

Did you find any binding of diadenosine polyphosphates in the area of the locus coeruleus?

Miras-Portugal: Yes, although the high affinity binding was very poor. Perhaps this is because in the locus coeruleus you have the bodies of the noradrenergic cells but not their synaptic terminals. The Ca^{2+} transients diadenosine polyphosphates induce in synaptic terminals from rat midbrain are suramin-insensitive. This is not the case with the ATP effect in synaptic terminals or in noradrenergic cells from bovine adrenal medulla, where suramin effectively blocks extracellular Ca^{2+} entry to the cells.

Barnard: Can you say something more about the requirement for amphetamine for stimulating diadenosine polyphosphate release in the CNS? You have to give it 30 min before maximal values are reached. There was no release of these polyphosphates unless you go through an aminergic system. Can you explain this?

Miras-Portugal: Our studies on brain perfusion are limited and have only been carried out in the neostriatum of conscious rats. This brain area is particularly rich in dopaminergic terminals and amphetamine induces the non-exocytotic release of this compound to the extracellular space. Since diadenosine polyphosphates are co-stored with ATP and aminergic compounds in synaptic vesicles, it can be assumed that they are also co-stored with dopamine. The released dopamine is essential, in this specific area, for the further release of diadenosine polyphosphates, because the blockade of dopaminergic receptors prevents their release. Nevertheless, we do not know the specific requirements in other brain areas. It is relevant that ATP and ADP do not appear in the perfusion samples, but Ap_4A and Ap_5A are present at concentrations of roughly 10 nM 30 min after amphetamine injection and undergo a slow decrease with time.

Barnard: But that could not explain the effects in other brain areas that you see. For example, there is no dopamine in the cerebellar cells that were indicated.

Miras-Portugal: It is possible that each area of CNS where diadenosine polyphosphates are present needs a specific set of effectors to stimulate their release. With regard to their storage vesicles, acetylcholine, 5-HT and all catecholamines seem to be co-stored with Ap_nA and ATP, but these results do not exclude other vesicles with different transmitters.

North: What is the proportion of PNMT-containing cells in the bovine chromaffin cell cultures?

Miras-Portugal: It depends on the methodology one uses. Usually, in bovine adrenal glands, you can reach about 80% of adrenergic cells.

North: How does this compare with the rat?

Miras-Portugal: I don't know exactly.

North: The reason I ask the question is because Bob Elde (University of Minnesota) has raised antibodies against two of the P2X receptors; they both stain uniformly almost all of the cells in the adrenal medulla in the rat. This suggests that most cells in the adrenal medulla are expressing both the P2X receptor from the vas deferens and the one from the PC12 cells. So if there were many PNMT-positive cells in the rat, perhaps we wouldn't expect them to be staining.

Miras-Portugal: In rat, most cells are PNMT-positive. Nevertheless, the expression of receptors in bovine and rat adrenomedullary cells is very different. For instance, rat cells lack nicotinic receptors in adrenomedullary cells and the muscarinic receptors present there induce the secretion through a G protein able to open a Ca^{2+} channel (Felder 1995).

Westfall: I was curious about how you separate these cells. We separate them by passing them over columns and so forth. What we see in bovine adrenal chromaffin cells is that you get some that are enriched for the secretion of adrenaline and some are enriched for secretion of noradrenaline, but even those enriched for secretion of one catecholamine also release the other. So the separation of cells is not always so clear-cut.

Miras-Portugal: I do not need to separate adrenergic or noradrenergic cells, because the studies of microfluorometry have been carried out in single cells that we can identify by further immunostaining. But we can obtain enriched populations of cells if we need them by the renografin gradient method (Moro et al 1990).

Harden: Eduardo Lazarowski, who works with Richard Boucher and myself, has been looking at the relative effects of diadenosine polyphosphates on cloned P2 receptors. He finds that Ap_4A is almost as potent as UTP and ATP at the P2U receptor, whereas with the phospholipase C-activating P2Y receptor that we've cloned from turkey, none of these compounds are very good agonists. It seems that at least Ap_4A is a very good P2U agonist. Ap_5A is also a fairly good agonist at the P2U receptor.

Miras-Portugal: I agree with you that there is some heterogeneity of P2U receptors. We have also found that P2U receptors on adrenal endothelial cells were not sensitive to Ap_4A after the second passage of culture. The response to Ap_4A completely disappeared in cells that had been cultured for a long time.

Harden: You didn't show the relative potencies in any of those figures.

Miras-Portugal: With regard to the Ca^{2+} signal, Ap_4A and UTP had similar potency in the primary cultures of endothelial cells and also in adrenergic, but not in noradrenergic cells.

Williams: There were a couple of papers published in the early 1990s by the South Carolina group (Hilderman et al 1991, Walker et al 1993) who showed that there was an Ap_4A receptor in the heart that was inducible. Has anyone been looking at these receptors for inducibility, and are we focusing too much on the sensitization *vis-à-vis* inducibility? Are there any thoughts on those two papers, because I've seen no follow-up to them at all. In terms of pathophysiology, should we be looking at inducibility? This brings up the whole issue of transfected receptors. What is actually being measured and what is the relevance, if any, to the physiological situation? Paul Leff was a co-author on an interesting paper (Bond et al 1995) showing that in a mouse model overexpressing β-adrenoceptors, compounds that had previously been shown to be 'neutral' antagonists were actually inverse agonists. The Glaxo group (Stables et al 1995) have shown in a CHO cell line transfected with the adenosine A_{2a} receptor that the non-xanthine adenosine antagonist, CGS15943, is an inverse agonist rather than a neutral antagonist while the efficacy of the agonist, 5′-*N*-ethylcarboxamidoadenosine (NECA) is proportional to receptor density. Thus, ligand pharmacology can change markedly as a function of expressed receptor density and functional coupling. This raises issues as to their relevance in reflecting the native, 'wild-type' receptor, its pharmacology and function and the physiological importance of transfected receptors systems.

Burnstock: I'm glad you raised this issue. I have a hunch that we are going to find a lot of changes in pathological situations that will surprise us, both in the expression of receptors and co-transmitters. It is easy just to look at controls and conclude that a receptor or transmitter is not present, and miss their appearance in pathological circumstances where they might be very important.

Illes: In bovine chromaffin cells, nicotinic agonists open a cationic channel and cause adrenaline and noradrenaline secretion in consequence to Ca^{2+} entry via this channel and also via voltage-dependent Ca^{2+} channels. Muscarine doesn't have much effect on secretion, although it increases intracellular Ca^{2+}. What is the physiological significance of the increase of intracellular Ca^{2+} that you find in some cases in your chromaffin cells?

Miras-Portugal: Only Ca^{2+} influx through the plasma membrane by ionotropic receptors or Ca^{2+} channels is able efficiently to elicit secretion in neural cells. The Ca^{2+} released from internal stores by the phospholipase C/inositol 1,4,5-trisphosphate pathway has a modulatory role by phosphorylation mediated by protein kinase C or Ca^{2+}/calmodulin kinase. A large set of proteins—channels, receptors, cytoskeletal components and metabolic enzymes—can be involved.

References

Allgaier C, Wellmann H, Schobert A, Kurz G, von Kügelgen I 1995 Cultured chick sympathetic neurons: ATP-induced noradrenaline release and its blockade by nicotinic receptor antagonists. Naunyn-Schmiedebergs Arch Pharmacol 352:25–30

Bond RA, Leff P, Johnson TD et al 1995 Physiological effects of inverse agonists in transgenic mice with overexpression of the β_2-adrenoceptor. Nature 374:272–276

Felder CC 1995 Muscarinic acetylcholine receptors: signal transduction through multiple effectors. FASEB J 9:619–625

Hilderman RH, Martin M, Zimmerman JK, Pivorun EB 1991 Identification of a unique membrane receptor for adenosine 5′,5‴-P_1,P_4 tetraphosphate. J Biol Chem 266:6915–6918

Illes P, Nieber K, Fröhlich R, Nörenberg W 1996 P2 purinoceptors and pyrimidinoceptors of catecholamine-producing cells and immunocytes. In: P2 purinoceptors: localization, function and transduction mechanisms. Wiley, Chichester (Ciba Found Symp 198) p 110–129

Moro MA, Lopez MG, Gandia L, Michelena P, Garcia AG 1990 Separation and culture of living adrenaline and noradrenaline-containing cells from bovine adrenal medulla. Anal Biochem 185:243–248

Ralevic V, Hoyle CHV, Burnstock G 1995 Pivotal role of phosphate chain length in vasoconstrictor versus vasodilator actions of adenine nucleotides in rat mesenteric arteries. J Physiol 483:703–713

Stables JM, Rees ES, Sheehan MJ et al 1995 Coexpression of the adenosine A2a receptor and the stimulatory G-protein alpha subunit in a stable cell line using a novel polycistronic expression vector. Pharmacol Res 31:201S

Tschöpl M, Harms L, Nörenberg W, Illes P 1992 Excitatory effects of adenosine 5′-triphosphate on rat locus coeruleus. Eur J Pharmacol 213:71–77

Walker J, Lewis TE, Pivorun EB, Hilderman RH 1993 Activation of the mouse heart adenosine 5′,5‴-P_1,P_4 tetraphosphate receptor. Biochemistry 32:1264–1269

P2T purinoceptors: ADP receptors on platelets

Susanna M. O. Hourani and David A. Hall*

Receptors and Cellular Regulation Research Group, School of Biological Sciences, University of Surrey, Guildford, Surrey GU2 5XH, UK

Abstract. ADP acts on platelets via the P2T purinoceptor to cause aggregation, but the way in which it does so is not fully understood. Most aggregating agents act via G protein-coupled receptors to stimulate phospholipase C (PLC) and so mobilize Ca^{2+} via inositol trisphosphate, whereas ADP clearly causes the mobilization of Ca^{2+} from internal stores but is only a weak activator of PLC. ADP also inhibits adenylate cyclase and it has been suggested that this effect is mediated by a different receptor, although evidence from antagonist studies argues against this. Studies of Ca^{2+} influx have shown that ADP is unique in causing a rapid influx of Ca^{2+}, and patch-clamp studies have confirmed the activation by ADP of non-selective cation channels. This would imply the existence of two ADP receptors on platelets, a receptor-operated channel responsible for the rapid Ca^{2+} influx and a G protein-coupled receptor possibly linked to both inhibition of adenylate cyclase and mobilization of Ca^{2+}. In this review the structure–activity relationships for aggregation, inhibition of adenylate cyclase and increases in cytoplasmic Ca^{2+} are summarized, and the relationship between these effects discussed.

1996 P2 purinoceptors: localization, function and transduction mechanisms. Wiley, Chichester (Ciba Foundation Symposium 198) p 53–70

Whereas until recently there has been some resistance to the concept of receptors for extracellular ATP, the existence of receptors for ADP on platelets has been widely accepted for over 30 years, since the seminal discovery that ADP could cause platelets to clump together ('aggregate') *in vitro* and that this aggregation could be antagonized by ATP (Born 1962).

Although the platelet ADP receptor is clearly of the P2 purinoceptor family and has been called the P2T purinoceptor (Gordon 1986), it is unique in that whereas ADP is an agonist, ATP is a competitive antagonist. However, it may

*Present address: The Biological Laboratory, The University, Canterbury, Kent CT2 7NJ, UK.

be an oversimplification to refer to the P2T purinoceptor, as it is not at all clear that there is only one ADP receptor on platelets; indeed there have been a number of suggestions that the effects of ADP cannot be described simply by activity at a single receptor.

Although platelets are small, anucleate cells and are technically easy to isolate and work with, they show a complexity and rapidity of responses that can complicate the interpretation of results. Typically, the response to an aggregating agent consists of shape change (from a biconvex disc to a spiny sphere), aggregation, and the release of mediators and the contents of platelet dense granules and α granules. As a general rule, these events occur in that order and require increasingly intense stimuli. In particular, most aggregating agents can cause the release from platelets of highly potent aggregating agents—notably thromboxane A_2 (TXA_2) and its precursors prostaglandins G_2 and H_2 (PGG_2 and PGH_2), which are rapidly synthesized, as well as ADP and 5-hydroxytryptamine (5-HT) which are stored in platelet dense granules. Adhesive proteins such as thrombospondin may also be released from the α granules and make the aggregation irreversible, and acidic phospholipids ('platelet factor 3') are exposed at the platelet surface and enhance the coagulation cascade. The act of aggregation itself, which is due to the exposure on the platelet surface of receptors to which fibrinogen binds in a Ca^{2+}-dependent manner and so cross-links the platelets, can also induce the release of mediators (Fig. 1). Platelets are therefore very responsive cells, and stimulation results in a rapid amplification of the initial stimulus. Substances

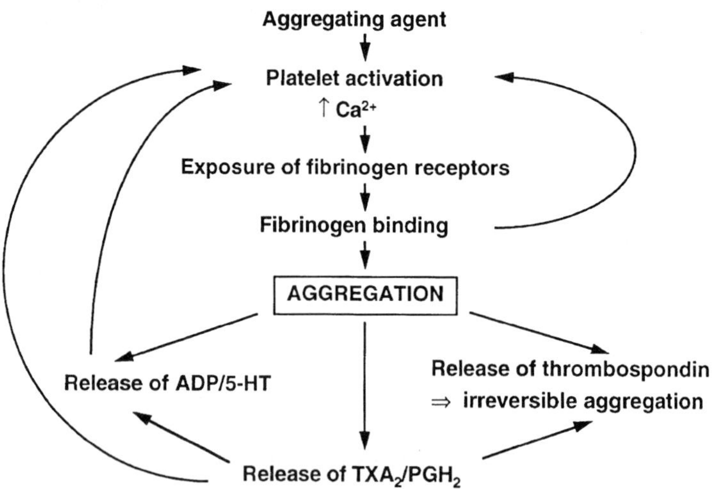

FIG. 1. Schematic diagram showing some of the processes occurring as a result of stimulation of platelets by an aggregating agent, resulting in release of mediators and amplification of the initial stimulus.

released from platelets are also crucial to the interaction between platelets and blood vessels, as in general they cause vasodilation via the release of NO and prostacyclin (PGI_2) if an intact endothelium is present, but vasoconstriction via receptors on the smooth muscle if the endothelium is absent or damaged. NO and PGI_2 are also potent inhibitors of platelet aggregation, as is adenosine which may be formed from released ADP and ATP by the action of endothelial ectonucleotidases. The result is that if platelets start to aggregate in an intact blood vessel, the vessel will dilate and the aggregation will be limited (Fig. 2), whereas if they aggregate in a cut or damaged blood vessel, the vessel will constrict and the aggregation will be enhanced and will spread (Fig. 3). This, of course, is the physiological process of haemostasis, but if these events should be triggered inappropriately the same processes will result in thrombosis. Because ADP plays such a central role in the amplification mechanisms of platelets, antagonists at the P2T receptor have great potential as antithrombotic drugs. Indeed, one such drug, ticlopidine, is in clinical use already, although its mechanism of action is not entirely clear and, as it requires hepatic metabolism, it is inactive *in vitro* (Schrör 1993). Stable analogues of ATP and of P_1,P_4-diadenosine tetraphosphate (Ap_4A) have also shown promise in animal models of thrombosis (Kim et al 1992, Humphries et al 1994a). Because of the potential clinical importance of the platelet P2T purinoceptor and the popularity of platelets in studies of signal transduction, the responses of

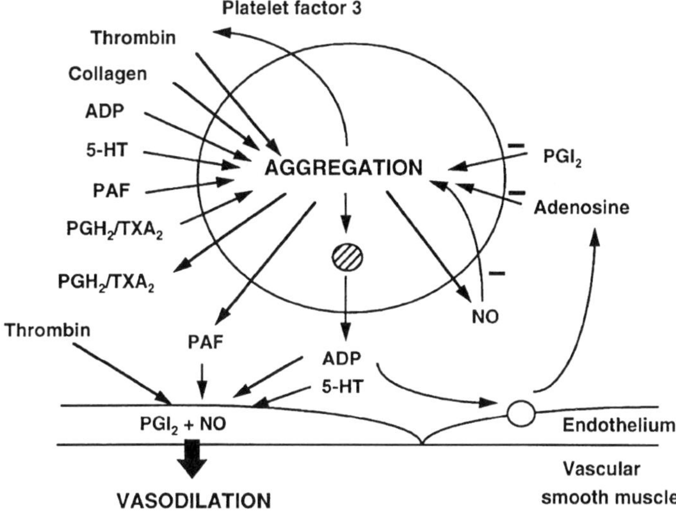

FIG. 2. Schematic diagram showing interaction between platelets and blood vessels with an intact endothelium. Release of mediators results in vasodilation and inhibition of platelet aggregation, due to the release of PGI_2 and NO as well as the formation of adenosine by ectonucleotidases. PAF, platelet-activating factor; other abbreviations as in text.

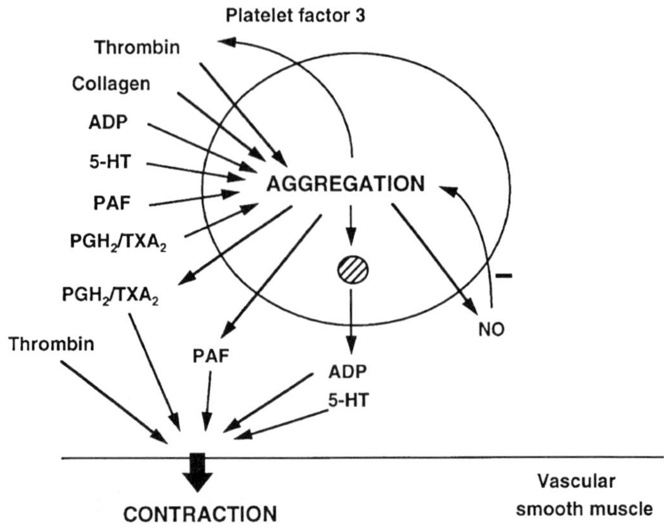

FIG. 3. Schematic diagram showing interaction between platelets and blood vessels if the endothelium is absent, where release of mediators results in vasoconstriction and enhancement of aggregation. PAF, platelet-activating factor; other abbreviations as in text.

platelets to ADP have been extensively studied, but there are still a number of unanswered questions.

Structure–activity relationships

The structure–activity relationships of the induction of platelet aggregation by P2T purinoceptors have been extensively investigated (for reviews see Haslam & Cusack 1981, Hourani & Cusack 1991, Hourani & Hall 1994; see also Table 1). Indeed, this is the only P2 receptor for which a range of truly competitive antagonists exists, in the form of ATP and its analogues. It should be noted here that the effect of ectonucleotidase activity on the potency of nucleotides is unlikely to have a significant influence on these observed structure–activity relationships, although prolonged incubation, especially in plasma, may lead to misleading results due to the formation of adenosine, which is a non-competitive inhibitor of aggregation. For ADP analogues, 2-substitution increases potency, with 2-chloroADP and 2-methylthioADP both being more potent than ADP itself, while substitution of a methylene group for the linking oxygen in the diphosphate chain (α,β-methylene ADP) reduces potency. Substitution of a sulfur for an oxygen on either the α or the β phosphate

(ADPαS or ADPβS) results in a lower efficacy, and these compounds are partial agonists, capable of inhibiting the effect of ADP as well as inducing aggregation themselves. For the ATP analogues (antagonists), again 2-substitution (2-chloroATP) increases affinity while methylene substitution in the phosphate chain (β,γ-methylene ATP) reduces it. ATPαS and ATPβS are both more potent antagonists than ATP itself, while P_1,P_5-diadenosine pentaphosphate (Ap_5A) has roughly the same affinity as ATP. The platelet P2T receptor is stereospecific, as the enantiomers of ADP and ATP, L-ADP and L-ATP, are completely inactive. ADPαS and ATPαS both exist as a pair of diastereoisomers, and in each case the S_P isomer is roughly five times more potent than the R_P as an agonist or an antagonist, showing that some stereoselectivity is also displayed towards the phosphate chain. The P2 antagonist suramin is also an antagonist at the P2T receptor, roughly equipotent with ATP, but appears to have a non-specific component to its action (Hourani et al 1992). Interestingly, the 2-alkylthio analogues of ATP and AMP, including 2-methylthioATP and 2-methylthio-β,γ-methylene ATP, act as specific but non-competitive inhibitors of ADP-induced aggregation in plasma, and their mechanisms of action have not been elucidated (Cusack & Hourani 1982a). However, the potential antithrombotic drug ARL66096 (2-propylthio-β,γ-difluoromethylene ATP; formerly known as FPL66096), which has a very similar structure, has recently been reported to be a highly potent and apparently competitive antagonist of ADP-induced aggregation in platelets in buffer, so it may be that the environment of the platelets influences the observed activity of these compounds in some as yet unknown way (Humphries et al 1994b).

While the P2Z receptor has always been defined as a receptor for the uncomplexed form of the ligand, ATP^{4-} (Gordon 1986), evidence has been presented that this is also the case for the P2Y and P2U receptors, for example on aortic endothelial cells (Motte et al 1993). It has also been suggested that the P2X receptor on the guinea-pig vas deferens is a receptor for ATP^{4-} (Fedan et al 1990), although this conclusion is not so clear cut because of the influence of ectonucleotidases, which are dependent on the presence of divalent cations, and the possible effect of divalent cations on the conductance of the ligand-gated channel (see Kennedy & Leff 1995). Although it is known that divalent cations are not required for the binding of [β-^{32}P]2-methylthioADP to platelets (Macfarlane et al 1983), the form of ADP and ATP preferred by the receptor had not been investigated. We therefore decided to examine the effects of extracellular Ca^{2+} and Mg^{2+} on two responses which are not (unlike aggregation) dependent on extracellular divalent cations—shape change and the inhibition by ADP of adenylate cyclase. We found that removal of divalent cations increased the potency of ADP roughly threefold for both effects, and increased the apparent affinity of ATP roughly 20-fold (Hall et al 1994). This is consistent with the preferred form of the agonist and antagonist being ADP^{3-} and ATP^{4-}, respectively, as suggested for the other P2 receptor subtypes.

Signal transduction

Most aggregating agents act largely via G protein-coupled receptors to stimulate phospholipase C (PLC), with the resultant stimulation of protein kinase C (PKC) via diacylglycerol (DAG) and mobilization of intracellular Ca^{2+} via inositol 1,4,5-trisphosphate ($InsP_3$) (Siess 1989, Hourani & Cusack 1991). Such aggregating agents also cause an influx of Ca^{2+} which, as in other cells, is thought to be triggered by the discharge of intracellular Ca^{2+} stores (Sage et al 1992, Heemskerk & Sage 1994). Many aggregating agents also inhibit adenylate cyclase via G_i, but this inhibition is not thought to be responsible for aggregation. However, as agonists such as adenosine and PGI_2 inhibit platelet aggregation by stimulating adenylate cyclase, its inhibition by aggregating agents may enhance aggregation *in vivo*, where, as discussed above, platelets may be exposed to a mixture of agonists with opposing effects (Siess 1989, Hourani & Cusack 1991). ADP and 2-methylthioADP have been shown to activate a G protein in human platelet membranes (Gachet et al 1992a), an effect blocked by the ticlopidine analogue clopidogrel (Gachet et al 1992b). Recent work suggests that this G protein may be G_{i2}, which would be consistent with an inhibition of adenylate cyclase (C. Gachet, P. Ohlmann, K. Laugwitz, B. Nürnberg, K. Spicher, G. Schultz & J.-P. Cazenave, unpublished results 1995).

ADP itself is roughly equipotent for aggregation and the inhibition of adenylate cyclase, but this is not true for some of its analogues. In particular, 2-methylthioATP is rather more potent as an inhibitor of adenylate cyclase than as an aggregating agent, and this has led to suggestions that these two effects are mediated by two different receptors (Macfarlane et al 1983). This suggestion has been supported by the use of an affinity reagent, 5′-fluorosulfonylbenzoyladenosine (FSBA), which inhibits ADP-induced aggregation but not the effect of ADP on adenylate cyclase. FSBA labels a 100 kDa protein which has been called 'aggregin', and is claimed to be the ADP receptor mediating shape change and aggregation, but not inhibition, of adenylate cyclase (see Colman 1992 for review, and Gachet & Cazenave 1991, Hourani & Hall 1994 for discussion). However, the different potencies of the agonists need not reflect different affinities for two separate receptors. Indeed, in the case of ADPαS there is some evidence that it is the efficacy that differs rather than the affinity: although both diastereoisomers of this compound act as partial agonists for aggregation, they act as competitive antagonists for the effects of ADP on adenylate cyclase, and in each case the S_P isomer was five times more potent than the R_P (Cusack & Hourani 1981a). Also, using a series of antagonists shown to be competitive by Schild analysis we showed a good correlation between their ability to inhibit ADP-induced aggregation and the effect of ADP on adenylate cyclase, suggesting that these two effects are mediated by the same receptor (Cusack & Hourani 1982b). A criticism of that study has been that these compounds were all nucleotide analogues, and so may not be able to discriminate between the two putative receptors, but more recently we have shown that suramin also inhibits both effects of ADP in a similar manner (Hourani et al 1992, Hall & Hourani 1994). Ticlopidine also inhibits both effects of ADP (Gachet et al 1990, Cattaneo et al 1991, Schrör 1993), although the effects of this compound are not so easy to interpret. In addition, our recent demonstration that for both shape change and the inhibition of adenylate cyclase the uncomplexed forms of the ligands seem to be the preferred agonist is consistent with these effects being mediated by the same receptor (Hall et al 1994).

There are some doubts as to the site of action of FSBA. It is possible that it acts on some process occurring after the initial receptor interaction. It has been reported not to inhibit the ability of ADP to increase intracellular Ca^{2+} levels, and the suggestion has been made that this effect too may be mediated by a different receptor from that inducing shape change and aggregation (Rao & Kowalska 1987). Because of this, we decided to look in detail at the structure–activity relationships for agonists and antagonists in increasing cytoplasmic Ca^{2+} concentrations, and compare these with their known effects on aggregation. We found that 2-methylthioADP was about 20 times more potent than ADP, whereas ADPαS and ADPβS were both partial agonists, and ATP inhibited all of these compounds in an apparently competitive manner. 2-ChloroATP, ATPαS, Ap_5A and β,γ-methylene ATP were also antagonists of the effect of ATP, while α,β-methylene ATP and UTP were only very weakly active and 2-methylthioATP was apparently non-competitive (Hall & Hourani 1993). Suramin also inhibited ADP-induced increases in cytoplasmic Ca^{2+}, in a very similar manner to its inhibition of ADP-induced aggregation (Hourani et al 1992, Hall & Hourani 1994). Overall, a clear correlation was observed for both agonists and antagonists for their effects on two parameters, both qualitatively and quantitatively (see Hourani & Hall 1994), suggesting that they are, as expected, causally related and mediated by the same receptor. It follows that FSBA cannot be acting at the ADP binding site, as it does not inhibit the ability of ADP to raise cytoplasmic Ca^{2+} levels. Indeed, if one takes into account all the analogues for which data on the effect on Ca^{2+} are available, an excellent correlation is obtained between all three parameters—aggregation, inhibition of adenylate cyclase and increases in Ca^{+2}—which does not support suggestions that they are mediated by different receptors (Table 1, Fig. 4).

However, although these overall responses may be mediated by one receptor, detailed investigations of the Ca^{2+} response to ADP suggests another complication. ADP clearly causes an increase in cytoplasmic Ca^{2+} in platelets, but its effect on PLC is inconsistent, some workers being unable to detect any

TABLE 1 pA_2 values for antagonists and pD_2 values for agonists at the human platelet ADP receptor for aggregation, inhibition of adenylate cyclase and increases in cytoplasmic Ca^{2+} levels

	Aggregation	*Adenylate cyclase*	*Ca^{2+}*
Antagonists			
ATP	4.64	5.21	5.01
2-chloroATP	4.13	4.54	5.60
Sp-ATPαS	5.44	5.34	5.92
Ap_5A	4.45	4.79	5.10
α,β-methylene ATP	4.08	4.22	4.27
Suramin	4.6	5.09	4.63
Sp-ATPβS	5.46	Not determined	5.92
Agonists			
ADP	5.5	5.5	6.10
2-methylthioADP	6.5	7.5	7.39
Sp-ADPαS	5.4*	5.13†	6.30*
ADPβS	4.7*	5.3*	5.40*

*These analogues are partial agonists for these effects.
†Sp-ADPαS is an antagonist for this effect of ADP, so this value is an apparent pA_2. (Data are taken from Cusack & Hourani 1981a,b, 1982b,c, Hall et al 1994, Hall & Hourani 1993, 1994, Hourani et al 1992, Hall 1993.)

stimulation at all (for reviews see Gachet & Cazenave 1991, Hourani & Hall 1994). These inconsistent reports may reflect the difficulty of ensuring that only the primary responses to ADP are being measured, not responses due to other mediators. In any case, it seems likely that ADP, is, if anything, a rather weak activator of PLC. This could imply that some other pathway not dependent on PLC mediates mobilization of Ca^{2+} from intracellular stores in response to ADP, but no such pathway has clearly been shown to exist. However, ADP is unique among platelet agonists in that, as well as causing mobilization of Ca^{2+} from intracellular stores and an associated delayed influx, it also causes a very rapid influx of Ca^{2+}. Patch-clamp studies on platelets have shown the activation by ADP of non-selective cation channels, suggestive of ligand-gated channels (Sage et al 1992). This Ca^{2+} entry may account for the apparent mismatch between the ability of ADP to activate PLC and its ability to raise cytoplasmic Ca^{2+}. In addition, because small increases in cytoplasmic Ca^{2+} may be enough to trigger shape change but not aggregation, this may account for the reported resistance of shape change to inhibitors of ADP-induced aggregation such as ticlopidine (Gachet & Cazenave 1991). It appears possible, therefore, that there are indeed two receptors for ADP on platelets, although

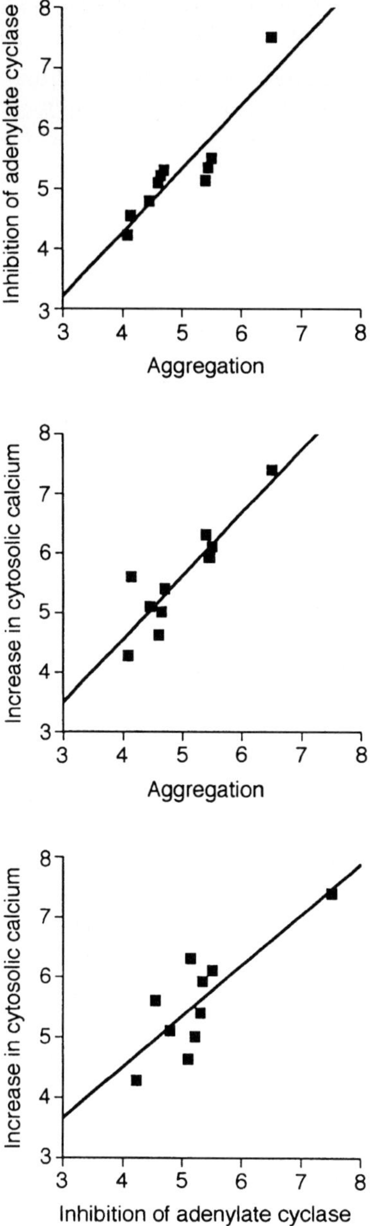

FIG. 4. Correlations between the effect of the agonists and antagonists on aggregation, inhibition of adenylate cyclase and increases in cytoplasmic Ca^{2+} levels. Data are taken from Table 1, and in each case there is a statistically significant correlation, $P < 0.01$.

defined differently from the two receptors proposed by Macfarlane et al (1983) and reviewed by Colman (1992). There may be a G protein-coupled receptor linked to inhibition of adenylate cyclase and possibly to stimulation of PLC, which is normally responsible for the observable responses to ADP. In addition, there may be a ligand-gated cation channel operated by ADP, which may be functionally of less importance although its effects may be observed when the G protein-linked receptor is blocked, for example by ticlopidine. Both G protein-coupled P2 receptors and ligand-gated cation channels responsive to ATP have now been cloned (for reviews see Barnard et al 1994, Kennedy & Leff 1995, Surprenant et al 1995), so the existence of both types of receptor on platelets is plausible.

Acknowledgements

D. A. H. was supported by an MRC studentship in collaboration with Ciba-Geigy Pharmaceuticals.

References

Barnard EA, Burnstock G, Webb TE 1994 G protein-coupled receptors for ATP and other nucleotides: a new receptor family. Trends Pharmacol Sci 15:67–70

Born GVR 1962 Aggregation of blood platelets by adenosine diphosphate and its reversal. Nature 194:927–929

Cattaneo M, Akkawat B, Lecchi A, Cimminiello C, Capitanio AM, Mannucci PB 1991 Ticlopidine selectively inhibits human platelet responses to adenosine diphosphate. Thromb Haemostasis 66:694–699

Colman RJ 1992 Platelet ADP receptors stimulating shape change and inhibiting adenylate cyclase. News Physiol Sci 7:274–278

Cusack NJ, Hourani SMO 1981a Effects of R_p and S_p diastereoisomers of adenosine 5′-O-(1-thiodiphosphate) on human platelets. Br J Pharmacol 73:409–412

Cusack NJ, Hourani SMO 1981b Partial agonist behaviour of adenosine 5′-O-(2-thiodiphosphate) on human platelets. Br J Pharmacol 73:405–408

Cusack NJ, Hourani SMO 1982a Specific but noncompetitive inhibition by 2-alkylthio analogues of adenosine 5′-monophosphate and adenosine 5′-triphosphate of human platelet aggregation induced by adenosine 5′-diphosphate. Br J Pharmacol 75:397–400

Cusack NJ, Hourani SMO 1982b Adenosine 5′-diphosphate antagonists and human platelets: no evidence that aggregation and inhibition of stimulated adenylate cyclase are mediated by different receptors. Br J Pharmacol 76:221–227

Cusack NJ, Hourani SMO 1982c Competitive inhibition by adenosine 5′-triphosphate of the actions on human platelets of 2-chloroadenosine 5′-diphosphate, 2-azidoadenosine 5′-diphosphate and 2-methylthioadenosine 5′-diphosphate. Br J Pharmacol 77:329–333

Fedan JS, Dagirmanjian JP, Attfield MD, Chideckel EW 1990 Evidence that the P_{2X} purinoceptor of the smooth muscle of the guinea-pig vas deferens is an ATP^{4-} receptor. J Pharmacol Exp Ther 255:46–51

Gachet C, Cazenave JP 1991 ADP induced blood platelet activation: a review. Nouv Rev Fr Hematol 33:347–358

Gachet C, Cazenave JP, Ohlmann et al 1990 The thienopyridine ticlopidine selectively prevents the inhibitory effects of ADP but not of adrenaline on cAMP levels raised by stimulation of the adenylate cyclase of human platelets by PGE_1. Biochem Pharmacol 40:2683–2687

Gachet C, Cazenave JP, Ohlmann P, Hilf G, Wieland T, Jakobs KH 1992a ADP receptor-induced activation of guanine-nucleotide-binding proteins in human platelet membranes. Eur J Biochem 207:259–263

Gachet C, Savi P, Ohlmann P, Maffrand JP, Jakobs KH, Cazenave JP 1992b ADP receptor-induced activation of guanine nucleotide binding proteins in rat platelet membranes—an effect selectively blocked by the thienopyridine clopidogrel. Thromb Haemostasis 68:79–83

Gordon JL 1986 Extracellular ATP: effects, sources and fate. Biochem J 233:309–319

Hall DA 1993 The biochemical pharmacology of the human platelet ADP receptor. PhD thesis, University of Surrey, Guildford, Surrey, UK

Hall DA, Hourani SMO 1993 Effects of analogues of adenine nucleotides on increases in intracellular calcium mediated by P_{2T} purinoceptors on human blood platelets. Br J Pharmacol 108:728–733

Hall DA, Hourani SMO 1994 Effects of suramin on increases in cytosolic calcium and on inhibition of adenylate cyclase induced by adenosine 5′-diphosphate in human platelets. Biochem Pharmacol 47:1013–1018

Hall DA, Frost V, Hourani SMO 1994 Effects of extracellular divalent cations on responses of human blood platelets to adenosine 5′-diphosphate. Biochem Pharmacol 48:1319–1326

Haslam RJ, Cusack NJ 1981 Blood platelet receptors for ADP and adenosine. In: Burnstock G (ed) Receptors & recognition series B, vol 12: Purinergic receptors. Chapman & Hall, London, p 221–285

Heemskerk JWM, Sage SO 1994 Calcium signalling in platelets and other cells. Platelets 5:295–316

Hourani SMO, Cusack NJ 1991 Pharmacological receptors on blood platelets. Pharmacol Rev 43:243–298

Hourani SMO, Hall DA 1994 ADP receptors on human blood platelets. Trends Pharmacol Sci 15:103–108

Hourani SMO, Hall DA, Nieman CJ 1992 Effects of the P_2-purinoceptor antagonist suramin on human platelet aggregation induced by adenosine 5′-diphosphate. Br J Pharmacol 105:453–457

Humphries RG, Tomlinson W, Leff P, Ingall AH, Kindon ND 1994a The P_{2T}-purinoceptor antagonist, FPL 67085, is a potent, efficacious and selective inhibitor of dynamic arterial thrombosis in the phenobarbitone-anaesthetized dog. Br J Pharmacol 113:63P(abstr)

Humphries RG, Tomlinson W, Ingall AH, Cage PA, Leff P 1994b FPL 66096: a novel, highly potent and selective antagonist at human platelet P_{2T}-purinoceptors. Br J Pharmacol 113:1057–1063

Kennedy C, Leff P 1995 How should P_{2X} purinoceptors be classified pharmacologically? Trends Pharmacol Sci 16:168–174

Kim BK, Zamecnik P, Taylor G, Guo MJ, Blackburn GM 1992 Antithrombotic effect of β,β'-monochloromethylene diadenosine 5′,5‴-p1,p4-tetraphosphate. Proc Natl Acad Sci USA 90:5113–5117

Macfarlane DE, Srivastava PC, Mills DCB 1983 2-Methylthioadenosine [β-^{32}P]diphosphate, an agonist and radioligand for the receptor that inhibits the accumulation of cyclic AMP in intact blood platelets. J Clin Invest 71:420–428

Motte S, Pirotton S, Boeynaems J-M 1993 Evidence that a form of ATP uncomplexed with divalent cations is the ligand of P_{2Y} and nucleotide/P_{2U} receptors on aortic endothelial cells. Br J Pharmacol 109:967–971

Rao AK, Kowalska MA 1987 ADP-induced platelet shape change and mobilization of cytoplasmic ionised calcium are mediated by distinct binding sites on platelets: 5′-p-fluorosulphonylbenzoyladenosine is a weak platelet agonist. Blood 70:751–756

Sage SO, Mahaut-Smith MP, Rink TJ 1992 Calcium entry in nonexcitable cells: lessons from human platelets. News Physiol Sci 7:108–113

Schrör K 1993 The basic pharmacology of ticlopidine and clopidogrel. Platelets 4:252–261

Siess W 1989 Molecular mechanisms of platelet activation. Physiol Rev 69:58–178

Surprenant A, Buell G, North RA 1995 P_{2X} receptors bring new structure to ligand-gated ion channels. Trends Neurosci 18:224–229

DISCUSSION

Burnstock: Paul Leff, you have a very exciting antagonist for the P2T purinoceptor. Have you any comments?

Leff: Regarding surmountable or insurmountable kinetics, we are unable to distinguish the kinetics of ARL67085 (2-propylthio-D-β,γ-dichloromethylene ATP) from simple competition, but amongst this family of compounds there is a variable tendency to produce saturable depression of curves. It also appears that under different experimental conditions the same compound can produce different extents of curve depression. I'm not sure why; we've speculated about a two-state receptor model which does allow for specific saturable depression of dose–response curves accompanied by a right shift. At least you don't have to invoke something too esoteric to explain the findings. The specificity of the compound is the important thing as far as we are concerned. You're right to point out there is only a trivial difference between ARL67085 and the compounds that Noel Cusack made. If you had stuck another methylene in, going from ethyl to propyl, you would have got that compound. Having said that, the compounds we have made since ARL67085 are far more chemically interesting. The features that we have introduced will hopefully make compounds that are orally active—ARL67085 requires intravenous delivery. It is in phase II trials at the moment, and the animal studies look fantastic in terms of its antithrombotic efficacy. There is an increased window between antithrombotic activity and bleeding time compared with fibrinogen antagonists. In the same sort of animal models we've tested thrombin inhibitors and found that these are actually unable to produce an antithrombotic effect, yet they still produce a bleeding time increase, which is the complete converse of what you want. Perhaps more by luck than judgement, we've chosen a mechanism which is poised between too much and too little efficacy in the antithrombotic sense, which itself translates to only a modest increase in bleeding time. Having discovered this, we can draw very nice biochemical cascade pictures and explain why.

Kennedy: My question concerns how the rise in internal Ca^{2+} induced by ADP comes about. Two mechanisms come to mind. On the one hand there is the possibility of Ca^{2+} influx through a ligand-gated cation channel. Could that Ca^{2+} influx produce Ca^{2+}-induced Ca^{2+} release? Is the Ca^{2+} release totally independent of extracellular Ca^{2+}?

Hourani: Ca^{2+} mobilization does occur in the absence of extracellular Ca^{2+}. But Sage et al (1992) have suggested that the Ca^{2+} influx enhances the Ca^{2+} mobilization and also speeds it up. If you're suggesting that Ca^{2+}-induced Ca^{2+} release occurs via ryanodine receptors, platelets have been reported to lack ryanodine receptors (Heemskerk et al 1993), which presumably rules out cADP ribose as a possible mechanism for Ca^{2+} release.

Kennedy: The second possible mechanism is release of Ca^{2+} from internal stores via a G protein-coupled receptor. You know that the ADP receptor is linked to adenylate cyclase by G_i. In theory, $\beta\gamma$ subunits released from G_i could activate PLC, and that way you would have the one G protein-coupled receptor linked to both adenylate cyclase and PLC.

Hourani: I have no problem with that idea. The problem is that many people have been unable to detect $InsP_3$ or diacylglycerol formed in response to ADP.

Kennedy: But this $\beta\gamma$ effect is best seen at levels of overexpression of the G protein. It is possible simply that in cells where people have seen an increase in $InsP_3$ there is already a high level of G_i.

Hourani: My feeling is that there probably is an $InsP_3$ effect, but that it's very weak, and there may be synergism between the $InsP_3$ pathway and the Ca^{2+} influx pathway. ADP is a much better Ca^{2+} mobilizer than you would expect from looking at the $InsP_3$ levels formed in response to it.

Jacobson: The dual control of these two second messengers by a single receptor you proposed is reminiscent of a receptor in the P1 field. The adenosine A_3 receptor, which has been shown to be coupled to inhibition of adenylate cyclase, also stimulates a rise in intracellular Ca^{2+}. This occurs in brain slices. You might keep an eye out for that for comparison.

Hourani: There are many cases where receptors which couple to adenylate cyclase via G_i have been shown also to couple to PLC in different cells. Platelets are an example of a cell where it's obvious that these two things are going on in response to one agonist. This is true not just for ADP but for a number of other agonists too. So if you think these are two receptors for ADP, you need to propose two receptors for each of these other agonists too, such as thrombin, thromboxane A_2 and platelet-activating factor (PAF), which seems excessively complex.

Wiley: You mentioned that 2-methylthioADP was more potent than ADP as an agonist. I think we're all familiar with the action of ectoATPase in modifying agonist potency sequences, but plasma also contains adenylate kinase (Haslam & Mills 1967), which equilibrates ADP with AMP and ATP. Is 2-methylthioADP equilibrated the same way and might the actions of adenylate kinase interfere with comparisons of the potency of ADP with 2-methylthioADP?

Hourani: The major enzymic effect on platelets is phosphorylation of ADP. But the time courses of the responses are generally very rapid, and I don't think that enzymic modification of the agonists or the antagonists is really a problem.

Wiley: The second point I wanted to make is that we have tried to confirm the presence of a cation channel operated by ADP as ligand (R_2; Mahaut-Smith et al 1990), using Ba^{2+} as the permeant and fura-2-loaded platelets in a fluorescence assay. However, we were not able to detect a clear-cut Ba^{2+} influx through that channel. This doesn't necessarily mean that the ion channel receptor is absent; it may simply mean that the Ba^{2+} influx through the ion channel receptor is much less than the amount of Ba^{2+} which enters through the refilling (capacitative) channel which, of course, is activated when you discharge internal Ca^{2+} stores through the G protein-coupled (R_1) receptor.

Hourani: In the patch-clamp study, Mahaut-Smith et al (1992) calculated that per platelet there were between 13 and 35 receptor operated channels, whereas binding studies using [^{32}P]methylthioADP had shown around 500 binding sites per platelet (Cristalli & Mills 1993). Certainly, if we believe both of those calculations, there are many more G protein-coupled receptors than there are ion channel receptors.

Di Virgilio: Could you tell us more about the ion selectivity of the ion channel receptor?

Hourani: The channel is permeable to Na^+, K^+ and Ba^{2+} and Mn^{2+}, as well as presumably to Ca^{2+}, but is blocked by Ni^{2+} (Mahaut-Smith et al 1992, Sage et al 1992).

Starke: In your experiments, why didn't you see anything of the ion channel receptor? You saw no change in Ca^{2+}, no effect on aggregation, only pharmacological evidence for the G protein-coupled receptor-mediated effect. Did you work in the absence of extracellular Ca^{2+}?

Hourani: No, and one of the things I would like to do would be to look at the structure–activity relationships of these two possible receptors. If you're looking at the overall responses of platelets, then it is likely that the responses of the G protein-coupled receptors dominate, because there are so many more of them.

IJzerman: You did aggregation studies on washed platelets and on platelet-rich plasma. We have used whole blood to measure aggregation by the impedance method. With 10 μM ADP one observes an immediate aggregation in whole blood. If you do dose–response curves with, say 1, 3, 5 and 10 μM ADP, you don't see any aggregation at all. This could imply that the P2 receptor here is very rapidly desensitized.

Hourani: It is known to be desensitized. If you incubate platelets with ADP without stirring them, so they can't aggregate, they are then unresponsive to further challenge with ADP.

IJzerman: We thought if you gave ATP that, with all the ectoATPase present in the blood, perhaps ATP by being degraded to ADP would aggregate platelets, but it doesn't.

Hourani: ATP has been reported to cause aggregation of rat platelets in plasma; this is thought to be due to high levels of creatine kinase in rat plasma (Haskel et al 1979).

Harden: José Boyer has seen an ADP receptor that activates PLC in a pertussis toxin-insensitive way on a megakaryocytic cell line and which has no effect on adenylate cyclase.

Miras-Portugal: Ap_3A, Ap_4A, Ap_5A and Ap_6A are present in secretory granules and they are co-released with ATP and ADP. Zamecnik et al (1992) have published a paper about anti-aggregational compounds derived from the diadenosine polyphosphates. Have you any comments about this?

Hourani: Antithrombotic effects have been demonstrated in animal models for an Ap_4A analogue with a chloromethylene linkage (Kim et al 1992).

Has anyone attempted to clone the P2T receptor?

Barnard: Firstly, of course, platelets do not have normal mRNA. They do not have a nucleus and they only have fragments of mRNA; I do not think platelets can be used as a source for receptor cDNA cloning. So one is thrown back on the precursors, the megakaryocytes, and the cell lines derived from them. Several megakaryocytic cell lines have an ADP receptor which has a somewhat different pharmacology to that which you have described on the platelets. Obviously, as you inferred yourself, before they form platelets the gene expression in the megakaryocytes may not be representative of specific products in the platelet population. David Mills and co-workers in Philadelphia, to whose protein-labelling studies on platelets you made reference, have employed a human erythroid leukaemic cell line (HEL cells), also derived from megakaryocytes. From this they have photoaffinity-labelled a protein band (Akbar et al 1995) which seems to have the same properties (with sensitivity to ADP) as the band they observed from platelets (Cristalli & Mills 1993).

Hourani: Is that the aggregin band or the 43 kDa band?

Barnard: It is not the aggregin band but a 55 kDa band. This could indicate that some cell types in the megakaryocyte lineage might be a source from which to clone P2T receptor cDNA. However, the precise pharmacology of the platelet P2T receptor so far does not appear to have been demonstrated in such precursor cell lines and it would be necessary to determine if a cell line can be cultured where the differentiation involved has gone fully to the point of the P2T receptor expression, which may be terminal.

Surprenant: I would like to make two points here. Firstly, it is of primary importance to develop antibodies to recognize this receptor (as there is obviously no DNA or RNA in platelets). Such an antibody may indeed demonstrate the presence of this receptor in platelets. Secondly, I think one must be very careful in experiments testing the functionality of desensitizing purinoceptors. For example, no one has ever described a P2X-like response in immune cells, but we found very high levels of mRNA for $P2X_1$ purinoceptors in a number of immune-like cells (HL-60 and RBL cells) as well as high levels of protein expression based on antibody immunofluorescence. However, we never detected an ATP-gated ionic current in these cells until we washed them with apyrase to metabolize endogenously released ATP. Our conclusion is that the continuous release of ATP from these immune cells (and perhaps platelets) acts chronically to desensitize the $P2X_1$ receptor, thus masking its functional presence under these *in vitro* conditions.

Di Virgilio: Bretschneider et al (1995) describe a non-selective cationic channel activated by extracellular ATP in human B lymphocytes.

Illes: Platelets and immunocytes are non-excitable cells and therefore it may be hard to find the P2X receptor on them in its classic form. On the other hand, many non-excitable cells exhibit the so-called capacitative Ca^{2+} current in their plasma membrane (Putney & Bird 1993). Such a current is activated by the depletion of intracellular Ca^{2+} stores. Do platelets have this type of Ca^{2+} channel which opens after the activation of P2T receptors and the subsequent depletion of the $InsP_3$-sensitive Ca^{2+} pool?

Di Virgilio: I don't think so. The reason I asked about the permeability is because the capacitative Ca^{2+} channel is very Ca^{2+} selective.

Illes: I wouldn't completely exclude this type of conductance. The capacitative Ca^{2+} channel becomes a non-selective cation channel if you leave out divalent cations.

Hourani: Platelets certainly also have the capacitative entry, but all the platelet agonists induce that entry, whereas only ADP induces the much more rapid entry consistent with the presence of a receptor-operated channel. There's a time lag for the capacitative entry, whereas the Ca^{2+} influx in response to ADP has been reported to occur immediately (Sage et al 1992).

Can I ask a question of the P2Z group? ATP^{4-} has been used to permeabilize platelets to get aequorin in (e.g. Salzman et al 1985), so does this mean that platelets have P2Z receptors on them as well?

Wiley: Yes, the observation is correct. The permeabilization technique has been around for many years, but it usually employs ATP in conjunction with very high concentrations of EDTA in a KCl medium for periods of 20–30 min. It has always been unclear to me whether it's the ATP^{4-} by itself that produces this permeabilization effect or whether this also requires the removal of Ca^{2+} from cell membranes. Certainly, the results are rather suggestive that permeabilization may be produced at least partially through P2Z receptors, but I think the jury is still out on that.

Di Virgilio: I disagree: there is pretty clear evidence that permeabilization is really a receptor-dependent mechanism. I don't think that the experiments Susanna Hourani was referring to really implicate the P2Z receptor in permeabilization. One of the key features of this procedure is reversibility and I'm not sure that in those experiments these criteria were observed. Aequorin is a large molecule of about 20 kDa—we know that P2Z-dependent permeabilization has a fixed molecular cut off point of about 900 Da.

Hourani: The permeabilization was reversible and they proceeded to use the platelets for other studies, so the platelets were functioning normally.

Lustig: There are a few other fibroblast cell lines in which exposure to high levels of ATP will activate the P2Z receptor and cause this channel to open. Molecules as large as 10–20 kDa can enter the permeabilized cells, and the cells can then be re-sealed by addition of Ca^{2+} or Mg^{2+}. At least some of the cells that have incorporated these large molecules are viable (Saribas et al 1993).

Di Virgilio: That is the only paper in which this has been fully described. However, the incubation medium in which Saribas et al (1993) performed their permeabilization was rather peculiar in that it contained 5 mg/ml of T-500 dextran.

Lustig: No, fluorescently labelled dextrans were the large molecules that were sealed inside the cells. Otherwise, they used the standard permeabilization buffer that has been used for many years.

Barnard: I wanted to say something about the various discrepancies which you recorded for platelets. I wonder if some of these are due to the fact that platelet aggregation is a very complicated mechanism, which is not the same as pure transduction of a signal into the cell; it may be too much to expect simple correlations with one of these metabolic pathways. For example, in an established technique, as you will know, the fibrinogen activation of platelet aggregation will proceed in its first stage on formalin-fixed platelets (Greco et al 1991). There cannot be many second messenger systems operating there. There must be difficulties in looking for correlations with aggregation.

Hourani: I think you're right. The problem does arise from the fact that platelets are complex cells. But, not withstanding that, we have shown correlations between these intracellular events and the overall functional response of aggregation.

References

Akbar GVM, Dasari VR, Sheth S, Mills DCB, Kunapuli SP 1995 Identification of two calcium-signalling P_{2T} and P_{2U} purinergic receptors in human erythroleukaemia cells. FASEB J 9:117A

Bretschneider F, Klapperstück M, Löhn M, Markwardt F 1995 Non-selective cation currents elicited by extracellular ATP in human B lymphocytes. Pflügers Arch 429:691–698

Cristalli G, Mills DCB 1993 Identification of a receptor for ADP on blood platelets by photoaffinity labelling. Biochem J 291:875–881

Greco NJ, Yamamoto N, Jackson B, Tandon NN, Moos Mjr, Jamieson GA 1991 Identification of a nucleotide-binding site on glycoprotein IIb. Relationship to ADP-induced platelet activation. J Biol Chem 266:13627–13633

Haskel EJ, Agarwal KC, Parks RE 1979 ATP- and ADP-induced rat platelet aggregation: significance of plasma in ATP-induced aggregation. Thromb Haemostasis 42:1580–1588

Haslam RJ, Mills DCB 1967 The adenylate kinase of human plasma, erythrocytes and platelets in relation to the degradation of adenosine diphosphate in plasma. Biochem J 103:773–784

Heemskerk JWM, Vis P, Feige MAH, Hoyland J, Mason WT, Sage SO 1993 Roles of phospholipase C and Ca^{2+}-ATPase in calcium responses of single, fibrinogen bound platelets. J Biol Chem 268:356–363

Kim BK, Zamecnik P, Taylor G, Guo MJ, Blackburn GM 1992 Antithrombotic effect of β,β'-monochloromethylene diadenosine 5′,5‴-P_1,P_4-tetraphosphate. Proc Natl Acad Sci USA 89:1056–1058

Mahaut-Smith MP, Sage SO, Rink TJ 1990 Receptor-activated single channels in intact human platelets. J Biol Chem 265:10479–10483

Mahaut-Smith MP, Sage SO, Rink TJ 1992 Rapid ADP-evoked currents in human platelets recorded with the nystatin permeabilized patch technique. J Biol Chem 267:3060–3065

Putney JW, Bird GSJ 1993 The signal for capacitative calcium entry. Cell 75:199–201

Sage SO, Mahaut-Smith MP, Rink TJ 1992 Calcium entry in nonexcitable cells: lessons from human platelets. News Physiol Sci 7:108–113

Salzman EW, Johnson PC, Ware JA 1985 Measurement of intracellular platelet calcium with aequorin and quin 2. Adv Exp Med Biol 192:163–170

Saribas AS, Lustig KD, Zhang X, Weisman GA 1993 Extracellular ATP reversibly increases the plasma membrane permeability of transformed mouse fibroblasts to large macromolecules. Anal Biochem 209:45–52

Zamecnik PC, Kim B, Gao MJ, Taylor G, Blackburn M 1992 Analogues of diadenosine 5′,5‴-P_1,P_4-tetraphosphate (Ap_4A) as potential anti-platelet aggregation agents. Proc Natl Acad Sci USA 89:2370–2373

P2Z purinoceptors

S. E. Hickman, C. E. Semrad and S. C. Silverstein

Departments of Physiology and Medicine, Columbia University, College of Physicians and Surgeons, 630 West 168th Street, New York, NY 10032, USA

Abstract. In response to tetra-anionic ATP^{4-}, P2Z receptors signal opening of a non-selective plasma membrane pore which permits passage across cell membranes of ions, nucleotides and other small molecules that are usually membrane impermeant. P2Z receptor-induced pores on murine macrophages, macrophage-like cell lines and human culture-matured macrophages are permeable to molecules of up to 831 Da. The function of P2Z receptors is unknown. Also unknown is whether the binding site for ATP^{4-} and the transmembrane pore, the properties that characterize P2Z receptors, reside on a single protein or reflect the activities of two or more proteins. That ATP^{4-}-unresponsive cell lines do not express connexin 43 has led Beyer and Steinberg to suggest that opening or surface expression of this gap junction protein is induced by P2Z receptors. *Xenopus* oocytes, injected with cRNA transcribed from a pool of 100 cDNA clones isolated from a murine macrophage-derived cDNA library, and treated with ATP^{4-}, express a non-selective membrane conductance characteristic of P2Z receptors. The conductance induced with cRNA is smaller than that induced by mRNA from macrophages, suggesting the presence of a dominant subunit of a multicomponent receptor in this pool of 100 cDNA clones.

1996 P2 purinoceptors: localization, function and transduction mechanisms. Wiley, Chichester (Ciba Foundation Symposium 198) p 71–90

Extracellular ATP mediates a variety of responses in many cell types. These responses include altered contraction in smooth muscle cells (Gordon 1986), neurotransmission (Burnstock 1971, Edwards et al 1992), release of endothelium-derived relaxing factor (EDRF) (Gordon 1990), cell proliferation in fibroblasts (Huang et al 1989), apoptosis in thymocytes (Zanovello et al 1990), degranulation in rat mast cells (Gomperts 1983) and chloride secretion in human airway epithelial cells (Mason et al 1991). A distinct effect of the free acid form of ATP (ATP^{4-}) is permeabilization of the plasma membranes of several types of cells to molecules <900 Da (Steinberg et al 1987).

Cellular responses to extracellular ATP are mediated by purinergic receptors. Pharmacological studies indicate that there are at least three classes of ATP receptors. P2U/P2Y-type receptors are G protein-coupled receptors which trigger release of intracellular Ca^{2+} through activation of phospholipase C (Dubyak & El-Moatassim 1993). P2X receptors directly activate ATP-ligand gated cation channels (Fredholm et al 1994). P2Z receptors mediate the opening of large 'pores' in plasma membranes making cells permeable to membrane-impermeant molecules of up to ~900 Da (Steinberg et al 1987). P2Z receptors are present on rat mast cells (Gomperts 1983), murine peritoneal macrophages (Steinberg et al 1987) and several murine transformed cell lines including J774 and BAC1.2F5 macrophage-like cell lines (Steinberg et al 1987, El-Moatassim & Dubyak 1992). Uptake of membrane-impermeant dyes such as Lucifer yellow or ethidium bromide in response to ATP has been used to confirm P2Z receptor activity in murine macrophage-like cell lines (J774 and BAC1), mouse peritoneal macrophages and cultured human monocyte-derived macrophages (Steinberg et al 1987, Hickman et al 1994). Prolonged exposure of P2Z-bearing cells to ATP results in lysis of cells.

Why do cells express a membrane receptor which, in the presence of extracellular ATP, opens a non-selective pore that allows relatively large solutes to enter and leave the cytoplasm? While several roles for P2Z receptors have been proposed—intercellular communication (Beyer & Steinberg 1991), macrophage polykarion (giant cell) formation (Falzoni et al 1995), inhibition of phagocytosis and cell movement (Sung et al 1985)—we have insufficient information to provide definitive answers. To gain insight into this question we have used expression cloning to define the molecular identity of the P2Z receptor of mouse macrophages.

Pharmacology of the P2Z receptors

P2Z receptors can be distinguished from the other P2 receptor classes pharmacologically. Tetra-anionic ATP, ATP^{4-}, is the species of ATP that induces permeabilization of P2Z receptor-expressing cells. Divalent cations such as Mg^{2+} and Ca^{2+} block the P2Z-mediated responses by converting ATP^{4-} to $MgATP^{2-}$ (Cusack 1993, Fredholm et al 1994). In general, the rank order of potency of ATP and ATP analogues as P2Z receptor agonists is 3′-*O*-(4-benzoyl)benzoyl ATP > ATP > ATPγS (Cusack 1993). Hydrolysis of ATP is not required for P2Z receptor activation since non-hydrolysable ATP analogues such as AMP-PNP (5′-adenylylimidodiphosphate) and AMP-PCP (β,γ-methylene ATP) function as agonists (Steinberg et al 1987, Heppel et al 1985), albeit at higher concentrations than ATP.

There is some variation in the relative efficacies of ATP analogues in activating P2Z receptors expressed by different cell types. For example, ATPγS is a much less potent P2Z agonist in cultured human monocyte-derived macrophages than in mouse macrophage-like cell lines (Hickman et al 1994). This may reflect small differences in a family of closely related molecules.

Electrophysiological studies of P2Z receptors

In mast cells (Tatham & Lindau 1990) and macrophages (Buisman et al 1988, Albuquerque et al 1993) voltage clamped at -20 and $-100\,mV$, ATP^{4-} induces a fast inward current with a reversal potential (E_{rev}) of 0 mV. So long as ATP was maintained in the medium this current was sustained. Brief application of small amounts of ATP (100 μM) onto voltage-clamped mouse peritoneal macrophages and macrophage polykaryons resulted in several different responses, even among cells of the same population. Some cells demonstrated only a fast inward current characteristic of the P2Z receptor-activated pore. Others exhibited this fast inward current followed by an outward current. This outward current was carried by K^+ and was thought to be due to a Ca^{2+}-activated K^+ channel. In other cells of the same type, only an outward current was seen in response to ATP.

P2Z receptor expression is regulated by development and cytokines in human mononuclear phagocytes

Freshly explanted human blood monocytes express little P2Z receptor activity (Hickman et al 1994). However, monocytes maintained in culture for 3–10 days mature into macrophages, 30–40% of which express P2Z receptor activity (Blanchard et al 1991, Hickman et al 1994, Falzoni et al 1995). Gamma interferon up-regulates the rate and extent of P2Z receptor expression by these cells (Blanchard et al 1991, Falzoni et al 1995).

Extracellular ATP promotes lysis of cells that express P2Z receptors

Prolonged (>15 min at 37 °C) exposure to extracellular ATP of human monocyte-derived macrophages treated with γ-interferon (Blanchard et al 1991), L929R cells (a tumour necrosis factor-resistant mouse fibroblast line) (Pizzo et al 1992), and J774 macrophage-like cells (Steinberg et al 1987) results in lysis of these cells by a colloid-osmotic mechanism secondary to opening of the cells' P2Z receptors. This early lysis is not accompanied by the DNA fragmentation pattern which is a hallmark of apoptosis.

Is connexin 43 involved in the response of P2Z receptor-expressing cells to ATP?

Steinberg et al (1987) isolated an ATP-resistant J774 cell line by repeated incubation of ATP-sensitive J774 cells with ATP. The resulting ATP-resistant J774 (ATPR J774) cells were neither permeabilized to fluorescent dyes nor lysed by prolonged treatment with millimolar concentrations of ATP. These

ATPR J774 cells retained P2U activity as shown by their release of Ca^{2+} from intracellular stores in response to ATP or UTP (Greenberg et al 1988).

Beyer & Steinberg (1991) noted that J774 cells express connexin 43, the cardiac muscle gap junction-forming protein. Gap junctions formed by connexin 43 are permeable to molecules up to 1000 Da. They found that ATPR J774 cells do not express connexin 43 at the protein (Western blots) or RNA level (PCR and Northern blot). These findings led them to suggest that connexin 43 is itself the P2Z receptor or is inserted into the plasma membrane in response to signals transmitted by ATP-activated P2Z receptors.

To examine this interesting idea we obtained a baby hamster kidney (BHK) cell line stably transfected with connexin 43 cDNA from Drs Kumar and Gilula (Scripps Research Institute). These cells stain strongly with antiserum against a connexin 43 peptide, especially in the areas of intercellular contact (Fig. 1A and B). High concentrations of ATP did not permeabilize these cells to Lucifer yellow (Fig. 1C and D). Figure 1 (E and F) shows uptake of Lucifer yellow by J774 cells tested in the same experiment. These findings suggest that connexin 43 is not itself an ATP-activated membrane channel. They do not rule out the possibility that activation of P2Z receptors by ATP signals surface recruitment and/or assembly of connexins.

Expression of J774 cell P2Z receptors in *Xenopus* oocytes

As noted above, J774 cells express two different purinergic receptors: (1) P2U/P2Y type G protein-coupled receptors which respond to ATP or UTP by activating phospholipase C, thereby triggering the release of intracellular Ca^{2+} (Greenberg et al 1988); and (2) P2Z receptors which respond to ATP by opening a large non-selective pore in the plasma membrane (Steinberg et al 1987). In contrast, ATP has little or no effect on *Xenopus* oocytes, indicating that they express few if any endogenous P2 receptors. These properties make *Xenopus* oocytes excellent substrates for expression of purinergic receptors. Indeed, they have been used successfully for cloning of P2U (Lustig et al 1993), P2Y (Webb et al 1993) and P2X (Valera et al 1994, Brake et al 1994) receptors and for expression of macrophage P2Z receptors (Nuttle et al 1993, Hickman et al 1993).

When expressed in *Xenopus* oocytes, J774 macrophage P2Y/U and P2Z receptors can be distinguished from one another by the distinctive electrophysiological signatures they produce when they are activated by nucleotides (Hickman et al 1993). Oocytes contain an endogenous Ca^{2+}-activated Cl^- channel (Fournier et al 1990). Thus a rise in oocyte $[Ca^{2+}]_i$ signals opening of a Cl^- channel. This produces an inward current when the oocyte is voltage clamped at $-100\,mV$, an outward current when the oocyte is clamped at $0\,mV$, and little current change when clamped at $-20\,mV$, which is close to the reversal potential for Cl^- (between -20 and $-30\,mV$) under our buffer

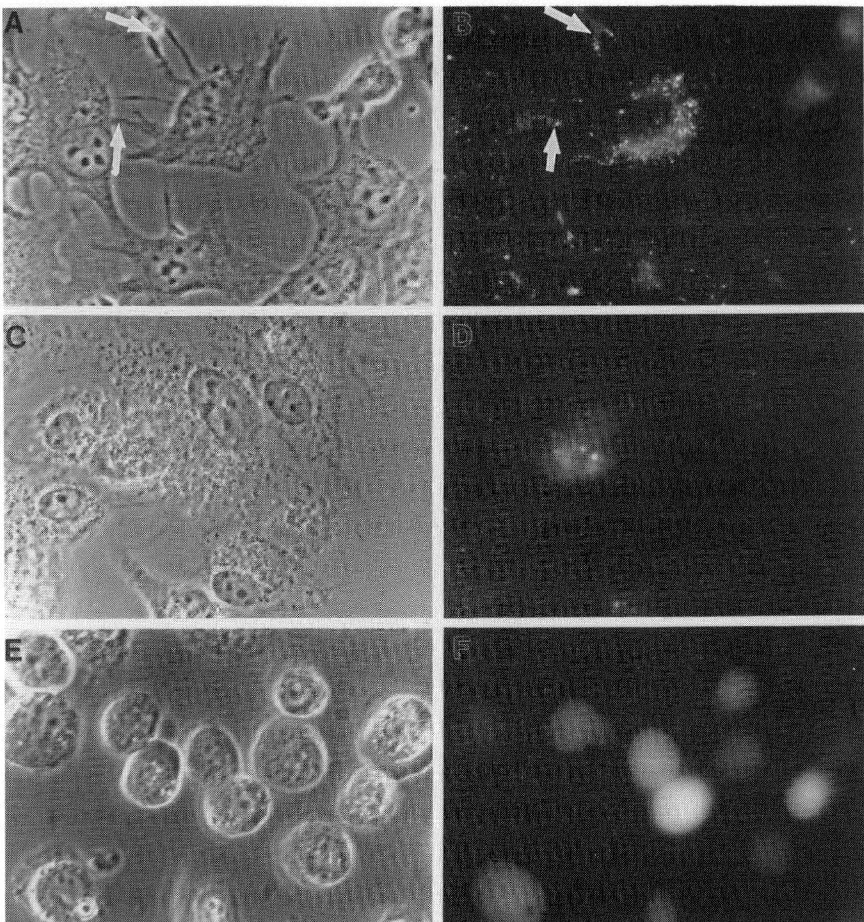

FIG. 1. Phase contrast (A) and fluorescent (B) micrographs of BHK cells transfected with connexin 43 and stained with rabbit anti-connexin 43 IgG (gift from Nalin Kumar). Staining is evident within the BHK cells and where gap junctions form at zones of cell–cell contact (arrows). (C) and (D) show that connexin 43-expressing BHK cells are not permeabilized to Lucifer yellow when incubated with 5 mM ATP. (E) and (F) show J774 cells permeabilized to Lucifer yellow following treatment with 5 mM ATP.

conditions. Therefore, expression of P2Y/U receptors by oocytes renders these cells capable of responding to ATP or UTP with a Cl^- current whose direction is dependent upon the membrane potential.

P2Z receptors do not permeabilize cells in response to UTP. In responding to ATP^{4-}, P2Z receptors signal opening of a membrane pore that is freely permeable to cations and anions. This pore mediates an inward current when the cell's membrane potential is naturally or artifically held at a negative value, and an outward current when the cell's membrane potential is clamped at a positive value. There is no net flux of ions across the pore when the cell's membrane potential is 0 mV.

Thus, at appropriate membrane potentials the movement of ions through the ion non-selective pore formed by P2Z receptors will induce different current profiles from those induced by P2U receptors. Using a pClamp software program (Axon Instruments), oocytes were voltage clamped at −100 mV (8 s), −20 mV (2 s) and 0 mV (2 s). This three-step voltage clamp protocol was used to distinguish a Cl^- conductance (reversal potential between −20 and −30 mV), from the non-selective conductance of the ATP-activated pore (reversal potential near 0 mV). We used these differences in current flow, as well as differences in agonist specificity, to identify expression of P2Z receptors in *Xenopus* oocytes.

For our initial studies, oocytes were injected with either 25–40 ng J774 mRNA or 10–20 ng cRNA synthesized from a reporter gene (secretory alkaline phosphatase; SEAP) (Tate et al 1990). Oocytes injected with SEAP mRNA were used as negative controls. The oocytes were incubated for 2–9 days at 20 °C in modified Barth's solution (MBS: 88 mM NaCl, 1 mM KCl, 0.33 mM $Ca(NO_3)_2$, 0.41 mM $CaCl_2$, 0.82 mM $MgSO_4$, 2.4 mM $NaCO_3$, 10 mM Hepes, pH 7.4) and monitored electrophysiologically for expression of ATP- and UTP-induced currents.

For measurements, individual oocytes were placed in a continuous perfusion chamber, impaled with two electrodes containing 3 M KCl (0.5–2.0 megaΩ resistance), and perfused with MBS. Outward currents are reported as positive. The resting membrane potential of injected oocytes ranged between −30 and −60 mV. After recording a baseline current, we perfused each oocyte with Mg^{2+}-free MBS containing UTP (500 μM), or ATP (500 μM), and current responses were monitored continuously for 2–5 min using the three-step voltage clamp protocol described above. When different nucleotides were added sequentially to an individual oocyte, the oocyte was perfused with Mg^{2+} containing MBS between nucleotide treatments until current readings returned to baseline.

After establishing that oocytes injected with a sample of total J774 mRNA responded to ATP or UTP, and that these responses could be distinguished from one another, mRNA was size-selected by centrifugation through a 5–25% sucrose gradient. Oocytes injected with mRNAs from two different gradient fractions gave two distinct electrophysiological responses to nucleotides. Oocytes injected with mRNA from gradient fraction II (enriched in 3–6 kb mRNA) responded predominantly to ATP, whereas oocytes injected with mRNA from gradient fraction III (1.5–4 kb mRNA), responded to both ATP and UTP.

UTP activates an endogenous Cl^- conductance

Individual oocytes injected with 25 ng mRNA from sucrose gradient fraction III were initially perfused with Mg^{2+}-containing MBS solution to obtain a baseline current. Perfusion with 500 μM UTP resulted in an inward oscillating current (-70 to -160 nA below baseline) at -100 mV clamping potential, a very small outwardly directed current at -20 mV, and an outward current at 0 mV (130 nA over baseline) ($n > 20$) (Fig. 2A). We believe the current oscillations observed at -100 mV are due to oscillations in $[Ca^{2+}]_i$. Such $[Ca^{2+}]_i$ oscillations have been observed in oocytes injected with inositol 1,4,5-trisphosphate (Fournier et al 1990). The magnitude of current responses to UTP was dependent on the concentration of UTP used. We obtained a minimal conductance using 1 μM UTP and a maximal effect at 300 μM UTP. Although UTP was the most effective activator of this current, ATP and ATPγS also acted as agonists. UTP activated a current that reversed at -25 mV ($n = 5$) (Fig. 2B), which is close to the predicted equilibrium potential for Cl^- with 90 mM Cl^- in the bath. Reduction of Cl^- in the medium to 2 mM by replacement with methanesulfonate changed the reversal potential (E_{rev}) of the UTP-activated Cl^- conductance to a positive value (data not shown). To test whether UTP's effect was mediated through increases in $[Ca^{2+}]_i$ we buffered oocyte cell Ca^{2+} with the intracellular Ca^{2+} chelator MAPTAM. Oocytes that responded to UTP at three days post-injection were allowed to reseal their cell membranes overnight and were used again the following day. Some of these oocytes were treated with MAPTAM. Others were incubated with buffer alone. Oocytes incubated in buffer alone responded to UTP on day 4 with currents that were similar in magnitude to their responses of the previous day. Figure 2B shows a current–voltage (I–V) plot of a single oocyte that had responded to UTP on day 3 after mRNA injection, but that no longer responded to UTP on day 4 after MAPTAM treatment. MAPTAM also blocked the current response normally induced in oocytes by the Ca^{2+} ionophore ionomycin, indicating that MAPTAM effectively buffered $[Ca^{2+}]_i$. Taken together, these results show that oocytes injected with mRNA from fraction III express a P2U-like purinergic receptor which when bound by UTP opens an endogenous oocyte Cl^- channel by increasing $[Ca^{2+}]_i$.

During the course of these experiments Lustig et al (1993) reported the cloning of a P2U receptor from mouse brain. They generously provided us with the clone encoding this receptor. The response to UTP of oocytes injected with 5 ng of RNA transcribed from this cDNA was of a shorter duration than that seen with J774 mRNA. Both responses were self-limited despite the continued presence of UTP in the bath. This suggests that, like other heterotrimeric G protein-coupled receptors, this receptor becomes desensitized by prolonged agonist application.

ATP activates a non-selective membrane conductance characteristic of a P2Z receptor

Baseline currents were measured in oocytes injected with 25 ng mRNA from sucrose gradient fraction II (3–6 kb). Perfusion with 500 μM ATP in Mg^{2+}-free

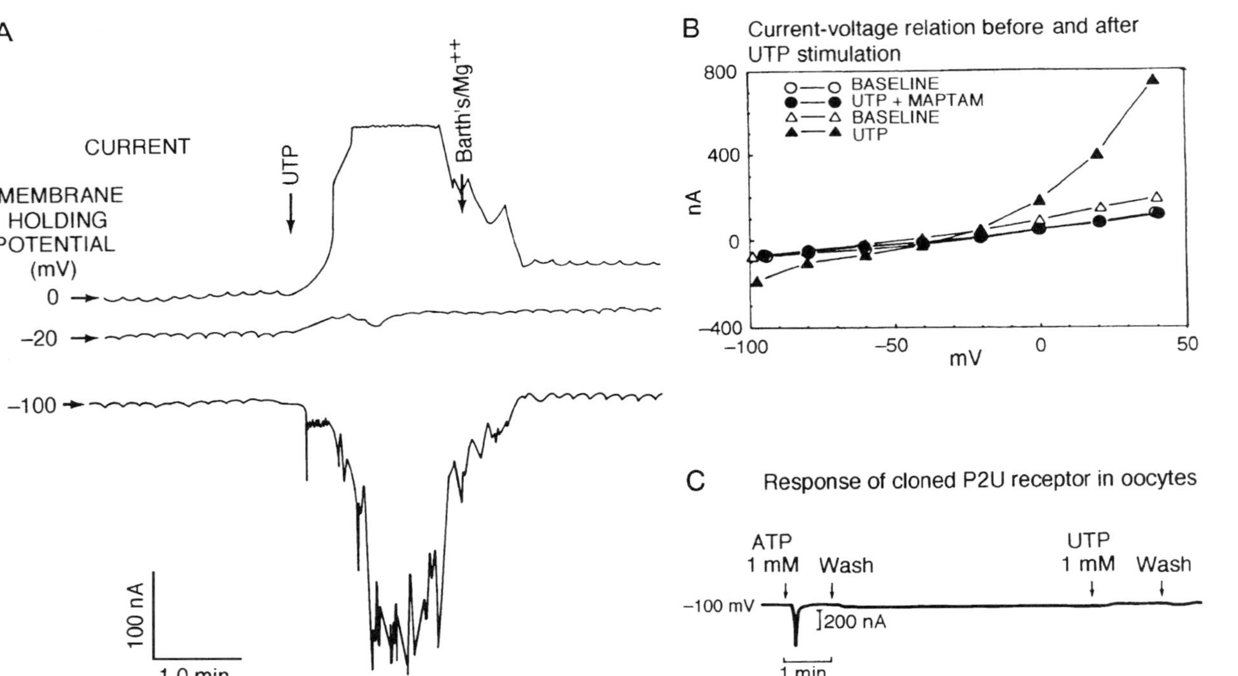

FIG. 2. (A) Response to UTP of an oocyte injected with mRNA from J774 cells. Measurements were taken with the membrane clamped at −100, −20 and 0 mV using the three-step voltage clamp protocol. (B) Comparison of I–V plots of an oocyte before perfusion with UTP (open triangles), and after exposure to UTP (closed triangles). The same oocyte was allowed to reseal overnight in MBS and then incubated for two hours with MAPTAM before restimulation with UTP. Baseline currents measured before UTP application are represented by open circles. MAPTAM chelates intracellular Ca^{2+} and inhibits the oocyte's UTP response (closed circles). I–V plots were obtained using ramp-command voltages with slopes of about 25 mV/s. (C) Response of an oocyte injected with 5 ng cRNA synthesized from the cloned P2U receptor. The oocyte was perfused with 1 mM ATP and the current response measured at −100 mV. The oocyte was washed for 6 min and then perfused with 1 mM UTP.

MBS resulted in the rapid appearance of a large inward current (up to 1.2 μA over baseline), in oocytes clamped at -100 mV ($n > 20$). In addition, a small inward current of about 20 nA was observed at -20 mV, and a small or no outward current at 0 mV (Fig. 3A). Addition of 1 mM Mg^{2+} to the perfusate reduced the conductances to baseline. I–V plots showed an E_{rev} of 0 mV, which corresponds to the E_{rev} determined for the ATP-induced 'pore' in J774 cells (Buisman et al 1988). In contrast to its effect on P2U/Y receptor function, MAPTAM did not inhibit the ATP-activated conductance (Fig. 3B). Similarly MAPTAM buffering of $[Ca^{2+}]_i$ in J774 cells did not block ATP activation of Lucifer yellow-permeant pores in these cells. Thus, fraction II mRNA obtained from J774 cells encodes proteins that are themselves ATP activated 'pores', with the electrophysiological characteristics and divalent cation sensitivity of P2Z receptors.

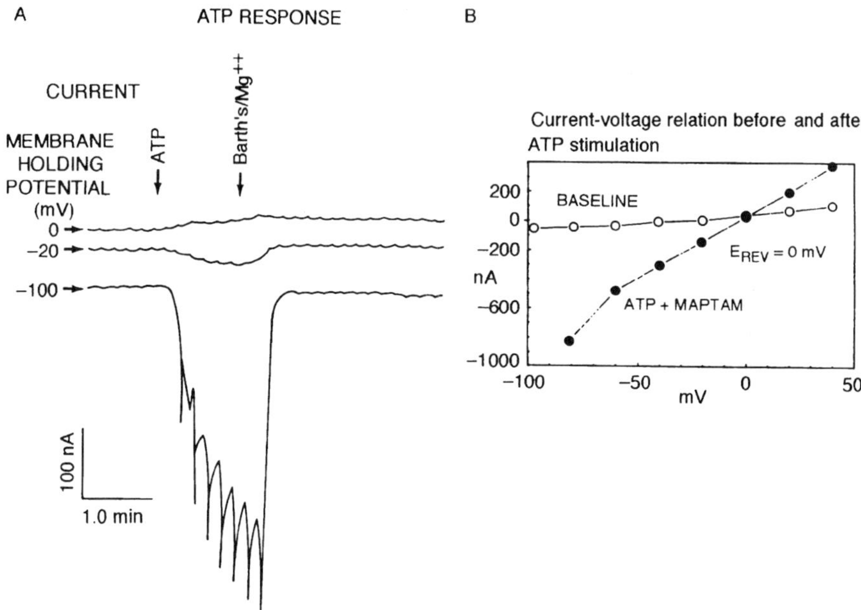

FIG. 3. (A) Electrophysiological response to ATP of an oocyte injected three days earlier with mRNA from J774 cells. Current changes in response to ATP were measured at -100, -20 and 0 mV as described previously. (B) I–V plot of an oocyte that responded previously to ATP, and was allowed to reseal overnight in MBS. It was then incubated for two hours in MAPTAM and re-stimulated with ATP. MAPTAM does not inhibit the P2Z receptor-mediated current response (closed circles). Reversal potential (E_{rev}) for this response was 0 mV, which is the reversal potential of the P2Z receptor induced current in J774 cells responding to ATP. I–V plots were obtained using ramp-command voltages with slopes of about 25 mV/s.

We next created a cDNA library from the fraction II mRNA and cloned it into the pSV-Sport plasmid vector (Gibco). We used directional cloning to insert the library's cDNAs into the vector at the *Sal*I and *Not*I restriction enzyme sites. This vector carries SP6 and T7 RNA polymerase promoters to allow the synthesis of mRNA under the direction of the SP6 promoter or antisense mRNA under the direction of the T7 promoter. The library was originally divided into pools of 20 000 independent clones. cRNA transcribed *in vitro* from the DNA extracted from one of the pools of 20 000 clones was injected into oocytes. Two to four days postinjection the oocytes were assayed for ATP responses as described above except that 1 mM ATP was used instead of 500 μM. Figure 4A shows the response to ATP of oocytes injected with cRNA transcribed from this pool of 20 000 clones. There was no response to UTP (1 mM) perfused for 2 min. The maximum response to ATP was 250 nA and the response was sustained for 2 min in the continued presence of ATP.

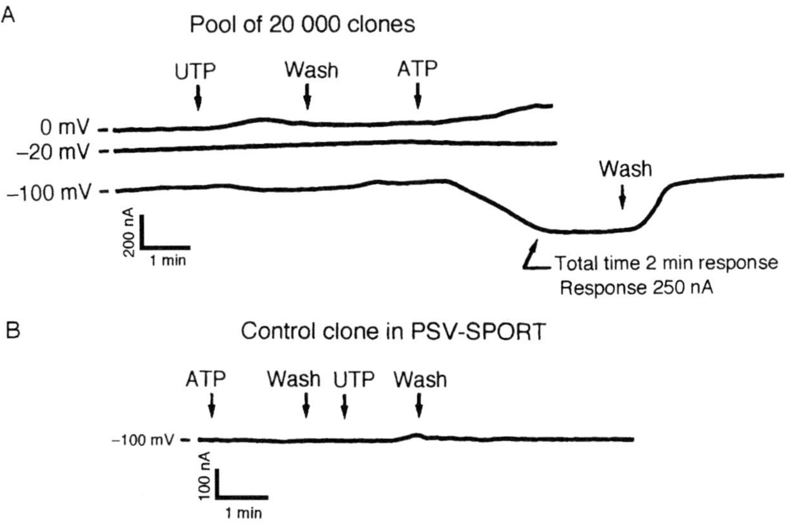

FIG. 4. P2Z receptor expression by oocytes injected with cRNA transcribed from a J774 cDNA library. (A) cRNA was transcribed *in vitro* with SP6 polymerase from DNA isolated from a pool containing 20 000 independent clones from a cDNA library created from the sucrose gradient fraction II J774 mRNA. The cRNA was injected into oocytes and three days later they were tested for expression of P2Z receptor activity. Current measurements were recorded using the three-step voltage clamp protocol. Oocytes showing no response to UTP were washed in MBS and perfused with ATP. Once a P2Z type response was observed, the three-step voltage clamp was stopped and the membrane potential held at −100 mV over two minutes to measure the maximum current response to ATP. (B) The lower tracing shows the lack of response to ATP and UTP of an oocyte injected with cRNA transcribed from the vector alone.

Figure 4B documents the lack of response to ATP of an oocyte injected with cRNA transcribed from the vector alone.

The library was continually subdivided into smaller and smaller pools. We have now identified a pool continuing 100 clones. Oocytes injected with cRNA transcribed from the DNA extracted from these 100 clones responded to ATP as shown in Fig. 5. This oocyte gave a 60 nA response when clamped at −100 mV. This oocyte was washed in MBS, perfused again with ATP and its response measured using the three-step voltage clamp protocol. The chart recorder speed was increased to spread out the tracing lines to distinguish changes between voltage steps. This time the oocyte response was nearly 100 nA at −100 mV, 20 nA at −20 mV, and no change at 0 mV. The oocyte was washed with MBS, voltage clamped at −100 mV, and perfused a third time with ATP. The oocyte responded with a current increase of 110 nA throughout the 3 min that ATP remained in the bath. The perfusate was changed to a solution containing 1 mM ATP in MBS with 10 mM Mg^{2+}. It took 30–40 s from the time the perfusion solution was changed until fresh solution reached the oocyte. Within a minute of changing to a solution containing 10 mM Mg^{2+} the response decreased nearly to baseline, indicating that the presence of Mg^{2+} abrogates the response.

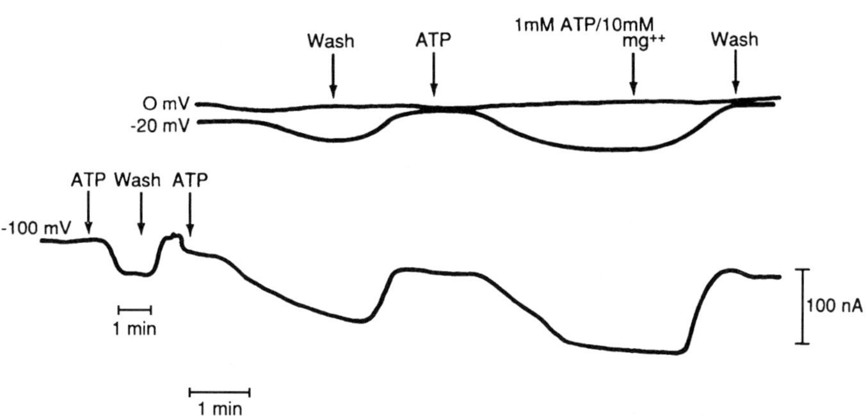

FIG. 5. Response of an oocyte injected with cRNA transcribed from DNA isolated from a pool of 100 clones. Once a response was observed at a membrane holding potential of −100 mV, the oocyte was washed to remove ATP. When the current returned to baseline, the three-step voltage clamp protocol was started and the oocyte was perfused with ATP (1 mM in Mg^{2+} free MBS). After two minutes of ATP perfusion, the oocyte was washed again for one minute and restimulated with ATP. During the peak of the response, the perfusate was changed to ATP/10 mM Mg^{2+}. The Mg^{2+} abrogated the response just as though the ATP had been washed out.

As we divided the pool of 20 000 clones into smaller pools, the signal from the putative P2Z receptor did not increase, but has stayed at a level of 100–160 nA. At present, we have no explanation for this apparent lack of enrichment of P2Z receptor expression, despite a probable 200-fold enrichment in P2Z cDNA. Until we obtain a single clone encoding P2Z receptor activity we will not know whether the macrophage P2Z receptor is formed by a unique polypeptide, or is composed of several different polypeptides. Our current efforts are focused on this goal.

References

Albuquerque C, Oliveira SMC, Countinho-Silva R, Oliveira-Castro GM, Persechini PM 1993 ATP-induced and UTP-induced currents in macrophages and macrophage polykaryons. Am J Physiol 265:1663C–1673C

Beyer EC, Steinberg TH 1991 Evidence that the gap junction protein connexin-43 is the ATP-induced pore of mouse macrophages. J Biol Chem 266:7971–7974

Blanchard DK, McMillen S, Djeu JY 1991 IFN-γ enhances sensitivity of human macrophages to extracellular ATP-mediated lysis. J Immunol 147:2579–2585

Brake AJ, Wagenbach MJ, Julius D 1994 New structural motif for ligand-gated ion channels defined by an ionotropic ATP receptor. Nature 371:519–523

Buisman HP, Steinberg TH, Fischbarg J et al 1988 Extracellular ATP induces a large non-selective conductance in macrophage plasma membranes. Proc Natl Acad Sci USA 85:7988–7992

Burnstock G 1971 Neural nomenclature. Nature 229:282–283

Cusack NJ 1993 P2 receptor: subclassification and structure–activity relationships. Drug Dev Res 28:244–252

Dubyak GR, El-Moatassim C 1993 Signal transduction via P2-purinergic receptors for extracellular ATP and other nucleotides. Am J Physiol 265:577C–606C

Edwards FA, Gibb AJ, Colquhoun D 1992 ATP receptor-mediated synaptic currents in the central nervous system. Nature 359:144–147

El-Moatassim C, Dubyak G 1992 A novel pathway for the activation of phospholipase D by P2Z purinergic receptors in BAC1.2F5 macrophages. J Biol Chem 267: 23664–23673

Falzoni S, Munerati M, Ferrari D, Spisani S, Moretti S, Di Virgilio F 1995 The purinergic P2Z receptor of human macrophage cells, characterization and possible physiological role. J Clin Invest 95:1207–1216

Fournier F, Honore E, Collin T, Guilbault P 1990 Ins(1,4,5)P_3 formation and fluctuating chloride current response induced by external ATP in *Xenopus* oocytes injected with embryonic guinea pig brain mRNA. FEBS Lett 277:205–208

Fredholm BB, Abbracchio MP, Burnstock G et al 1994 Nomenclature and classification of purinoceptors. Pharmacol Rev 46:143–156

Gomperts BD 1983 Mast cell degranulation. Nature 306:64–66

Gordon JL 1986 Extracellular ATP: effects, sources and fate. Biochem J 233:309–319

Gordon JL 1990 The effects of ATP on endothelium. Ann N Y Acad Sci 603:46–52

Greenberg S, Di Virgilio F, Steinberg TH, Silverstein SC 1988 Extracellular nucleotides mediate Ca^{2+} fluxes in J774 macrophages by two distinct mechanisms. J Biol Chem 263:10337-10343

Heppel LA, Weisman GA, Friedberg I 1985 Permeabilization of transformed cells in culture by external ATP. J Membr Biol 86:189–196

Hickman SE, Semrad CE, Field M, Silverstein SC 1993 Expression of macrophage ATP and UTP receptors. FASEB J 7:A712(abstr)

Hickman SE, El Khoury J, Greenberg S, Schieren I, Silverstein SC 1994 P2Z adenosine triphosphate receptor activity in cultured human monocyte-derived macrophages. Blood 84:2452–2456

Huang N, Wang D, Heppel LA 1989 Extracellular ATP is a mitogen for 3T3, 3T6 and A431 cells and acts synergistically with other growth factors. Proc Natl Acad Sci USA 86:7904–7908

Lustig KD, Shiau AK, Brake AJ, Julius D 1993 Expression cloning of an ATP receptor from mouse neuroblastoma cells. Proc Natl Acad Sci USA 90:5113–5117

Mason SJ, Paradiso AM, Boucher RC 1991 Regulation of transepithelial ion transport and intracellular calcium by extracellular ATP in human normal and cystic fibrosis airway epithelium. Br J Pharmacol 103:1649–1656

Nuttle LC, El-Moatassim C, Dubyak GR 1993 Expression of the pore-forming P2Z purinoceptor in *Xenopus* oocytes injected with poly(A^+)RNA from murine macrophages. Mol Pharmacol 44:93–101

Pizzo P, Murgia M, Zambon A et al 1992 Role of P2Z purinergic receptors in ATP-mediated killing of tumor necrosis factor (TNF)-sensitive and TNF-resistant L929 fibroblasts. J Immunol 149:3372–3378

Steinberg TH, Newman AS, Swanson JA, Silverstein SC 1987 ATP^{4-} permeabilizes the plasma membrane of mouse macrophage to fluorescent dyes. J Biol Chem 262: 8884–8888

Sung SJ, Young JDE, Origlio AM, Heiple JM, Kaback HR, Silverstein SC 1985 Extracellular ATP perturbs transmembrane ion fluxes, elevates cytosolic [Ca^{++}], and inhibits phagocytosis in mouse macrophages. J Biol Chem 260:13442–13449

Tate SS, Urade R, Micanovic R, Gerber L, Udenfriend S 1990 Secreted alkaline phosphatase: an internal standard for expression of injected mRNAs in *Xenopus* oocyte. FASEB J 4:227–231

Tatham PER, Lindau M 1990 ATP-induced pore formation in the plasma membrane of rat peritoneal mast cells. J Gen Physiol 95:459–476

Valera S, Hussy N, Evans RJ et al 1994 A new class of ligand-gated ion channel defined by P_{2X} receptor for extracellular ATP. Nature 371:516–519

Webb TE, Simon J, Krishek BJ et al 1993 Cloning and functional expression of a brain G-protein-coupled ATP receptor. FEBS Lett 324:219–225

Zanovello P, Bronte V, Rosato A, Pizzo P, DiVirgilio F 1990 Responses of mouse lymphocytes to extracellular ATP. II. Extracellular ATP causes cell type-dependent lysis and DNA fragmentation. J Immunol 145:1545–1550

DISCUSSION

Hickman: Shortly before this meeting, we isolated a single clone from the J774 library containing a 1.8 kb insert. When this clone was transcribed and the resulting cRNA was injected into oocytes, they responded to ATP with an electrophysiological signature characteristic of the P2Z responses previously described. However, with the cloned message, there was a biphasic response which we had not previously seen with injection of total message from J774 cells. Upon application of ATP to the oocytes injected with the single clone cRNA, there was an initial small inward current (100–150 nA) at −100 mV membrane holding potential followed 1–2 min later by a large inward current (700–1000 nA). The cloned cDNA was transfected into COS-7 cells and they also permeabilized in response to ATP (as assessed by Lucifer yellow uptake). We only have a partial nucleotide sequence on the clone—until we have the full sequence we are unable to provide potential structural models or comparisons with other membrane receptors. We also do not know if there is yet another protein(s) which complements our clone and contributes to the native P2Z response.

Wiley: Brian King, Geoff Burnstock and myself (unpublished work) have seen a similar type of electrophysiological pattern when we have injected P2Z receptor mRNA from human lymphocytes into *Xenopus* oocytes. When ATP is applied there is an inward current, but when the ATP is removed the current decays very slowly and incompletely, and it doesn't come back to the baseline. This is a little atypical for the behaviour of a channel unless there's another modulating feature about this P2Z response. Would you like to speculate on this?

Hickman: That's interesting. My intuitive guess as to why we see a delay between when the initial small ATP response and then the large inward current is that molecules on the cell surface are fusing to form something bigger. But this is just speculation.

Harden: Is there any delay in the response of the COS-7 cells?

Hickman: To detect P2Z expression, I simply incubate them with ATP in PBS with Lucifer yellow for 10 min, then assess the dye uptake. I have not performed a time-course on Lucifer yellow uptake, so I don't know if there is a delay.

Di Virgilio: The activation of the P2Z receptor has been shown to be temperature dependent. Presumably your experiments in *Xenopus* are performed at room temperature.

Hickman: Yes, we have an air conditioned room in which the temperature is 22 °C.

Di Virgilio: At 20 °C there is very little P2Z activity, at least in macrophages. Could it be that the delay and the slow recovery you are seeing is a temperature-dependent phenomenon? If you were operating at 37 °C in a mammalian cell it might not occur.

Hickman: The pore opens at room temperature, certainly. You can't work for very long at 37 °C with oocytes; their membranes rapidly lose integrity. If a protein designed to work at 37 °C is put into an oocyte which, by its nature, works at 18–20 °C, perhaps the nature of the protein changes. I don't know why we're seeing that delay; we do not see this delayed response to ATP in oocytes injected with total message from J774 cells. But now we have all the J774 background messages subtracted out, we're down to a single clone, and this is what we see.

Di Virgilio: Once you shift to the COS cells you can do the experiments at 37 °C.

Hickman: An obstacle with COS cells is that in them the expression of the P2Z receptor is transient and highly variable from experiment to experiment. So, I'm in the process of subcloning the gene encoding our putative P2Z receptor into a vector that will allow stable transfection of BHK cells. Hopefully this will result in a cell line that stably and uniformly expresses our clone. Then I could easily work with these cells at 37 °C.

Wiley: I agree that there's something slightly atypical about the *Xenopus* oocyte expression of the P2Z system. If we look at P2Z in lymphocytes at 37 °C, it opens and closes with agonist addition and removal almost instantaneously, in less than a second, as you would expect for a ligand-gated channel. But in *Xenopus* oocytes, Brian King, Geoff Burnstock and myself find that at room temperature (which at University College is more like 18 °C), closure occurs over a time frame of some minutes, and then closing is sometimes incomplete. This is something we don't fully understand.

Di Virgilio: If George Dubyak is right in saying that at room temperature you have a small channel that grows bigger, or that another channel is inserted in the membrane and is activated when you work at 37 °C, then you might be seeing one channel at room temperature and another channel or pore at physiological temperatures.

Lustig: Why didn't the electrophysiological response get larger as you progressively subdivided the cDNA pools you were screening? Your data show that the response initially gets smaller and then gets much larger at the very last step in the sib selection procedure. How do you account for this?

Hickman: It could just be the quality of the cRNA that we were synthesizing. Each preparation was quite different.

Lustig: Could it be that early in the sib selection procedure the channel is formed by the association of two or more subunits and that you're actually losing one of these as the pool size gets smaller? If that is the case, you could do an experiment where you take the single clone you have isolated and then add it back to earlier cDNA pools and see whether you pick up a different type of response.

Hickman: Sure.

Barnard: Having worked with expression in *Xenopus* oocytes for many years, I know of no expressible single mammalian protein which behaves in the way you describe on expression in the oocyte. If this were the case here, it would be unique. I think the most likely explanation of the behaviour that you have seen is more along the lines that Kevin Lustig was suggesting. Your mRNA is going to code for a protein up to a maximum of about 500 amino acids; that obviously is not enough by itself to form a channel of this huge diameter, so the channel will have a number of those subunits in it. Maybe this homo-oligomer which has formed is unnatural, and therefore has difficulty in assembly. There are precedents for that where you do get this kind of time lag. An explanation of the lag would be simply difficulty in expression, which may be due to the fact that it is missing a partner. For example, we have found that for very large proteins to be expressed this can require a longer incubation time of the messenger in the oocyte. Protein synthesis and processing takes time, and the oocyte is not super-efficient. I wonder if you have tried incubating for longer periods. Do you use the usual two days?

Hickman: We start reading the oocytes at day 2 and we stop when the oocytes die. We have been able to record responses to ATP up to day 6 with recent sets of experiments, but the oocytes aren't in good condition after this time.

Barnard: Perhaps this is an indication that your channel is somewhat lethal for the oocyte. With a channel as large as this it must be like punching big holes in the cells. Presumably in the end it will be lethal. Is this found with P2Z receptors in general? If you activate them for some time in cells which have them in the native state, does this kill the cells?

Hickman: If I inject oocytes with 30 ng of P2Z cRNA, which is a rather large amount of a single message, every single oocyte dies.

Barnard: That is what I thought. What about macrophages: do they die if you leave them in contact with ATP?

Hickman: J774 cells do. That's how the ATP-resistant cell line was isolated.

Barnard: So you probably have a balance between the assembly of this oligomer and the lethality of it. Perhaps this is why the currents go up and down. This will be a difficult system to use for oocyte expression cloning.

Hickman: It is not as clean as we would like it to be.

Di Virgilio: In our experience, the P2Z receptor is lethal to all cells. For it to be lethal it is not necessary for the cell to be continuously exposed to ATP. In many cases a brief pulse of ATP is sufficient if you follow the fate of the cells for a long enough time.

North: It appears that many oocytes secrete ATP all the time. We realized this when we expressed P2X receptors in them, because in some eggs you observe transient inward currents occurring spontaneously which are blocked by suramin. The oocyte is popping out ATP, presumably as a consequence of spontaneous exocytosis.

Barnard: Is this from the oocyte itself or the follicullar cells?

North: These are de-folliculated oocytes. It doesn't kill the oocytes in our experience.

Lustig: So perhaps ATP is being released from the oocyte and is down-regulating or somehow de-sensitizing the channel before you voltage-clamp the oocyte.

Hickman: In our assays with P2Z-expressing oocytes, the P2Z response to ATP does not desensitize. However, if P2Z is overexpressed in oocytes, then the endogenous ATP release from oocytes themselves may result in persistent opening of the channels. This may cause such dramatic ion fluxes over a prolonged period of time that the oocytes cannot maintain their cytoplasmic electrolyte balance and they die.

Wiley: Brian King, Geoff Burnstock and myself (unpublished work) have tried to grapple with this particular problem of oocytes dying after the injection of mRNA from lymphocytes. We put these oocytes into a bath solution containing very high Mg^{2+} concentrations (10 mM) and then we do our electrophysiological readings at 18 h.

Surprenant: That works?

Wiley: Yes, 10 mM Mg^{2+} really reduces the lethality of endogenous ATP release.

Surprenant: Is 3′-*O*-(4-benzoyl)benzoyl ATP (BzATP) really a full agonist of the P2Z receptor? Is it as effective as ATP?

Lustig: In our hands BzATP is about 10-fold more potent than ATP at the P2Z receptor. In fact, Erb et al (1990) have shown that it is possible to cross-link BzATP to the P2Z receptor. Cells permeabilized with cross-linked BzATP can be resealed by addition of Mg^{2+} to the incubation medium. If the cells are re-washed and placed into a divalent cation-free medium, the P2Z channel re-opens and the cells are once again permeable to small molecules.

Hickman: We've done this in J774 cells. We incubated them with BzATP (without Mg^{2+}) in the cold, cross-linked by UV irradiation and rinsed out the BzATP with buffer containing Mg^{2+}. Then we added a solution containing Lucifer yellow (without Mg^{2+}) and warmed the cells to 37 °C and the dye entered the cytoplasm as effectively as if ATP were present. So, the cross-linked BzATP appears effectively to activate the P2Z response.

Di Virgilio: In our hands, too, it is about an order of magnitude more potent than ATP.

Wiley: With the lymphocyte P2Z receptor, the EC_{50} for BzATP is 8 μM. Significantly, BzATP gives a maximum Ba^{2+} flux through the P2Z channel which is twice that observed with the usual agonist, ATP.

Boucher: Are we to assume that connexin 43 is not participating in this? Do you have any more direct experiments, perhaps subtraction experiments using antisense approaches that show this? Is it possible that connexin is still part of this complex?

Hickman: It's possible. However, our clone is not connexin 43.

Lustig: Have you tried to co-express connexin 43 and the putative receptor in the oocyte?

Hickman: No, I've expressed connexin 43 in the oocytes but I haven't co-expressed it with this receptor.

Surprenant: Do you have plans to do that? It sounds like it might lead to something.

Hickman: I have plans to do that, and I also have plans to go back and co-inject the pools, as Kevin Lustig suggested, with the cloned cRNA to see if this results in a different response.

Kennedy: Coming back to the possible involvement of a connexin—whether it is connexin 43 or another connexin—could your receptor be either a large G protein-coupled receptor or a receptor-linked tyrosine kinase, which then generates some message that activates connexin 43, and causes it to open and coalesce?

Hickman: Current thinking is that the P2Z receptor is not G protein-coupled. The response is not blocked by cholera toxin, is not blocked by chelation of intracellular Ca^{2+} and does not desensitize to ATP in oocytes or J774 cells. The theory Sam Silverstein had for a while was that there's an ATP receptor which activates something inside the cell and calls connexin 43 to the surface. I do not know at this point exactly what I have cloned, but it causes a response to ATP in oocytes, and it causes permeabilization to lucifer yellow in transiently transfected COS-7 cells. It may very well be that we were fortunate in that both oocytes and COS cells have a second component that is needed, and if that component were not there, this clone would not be active. This still has to be determined.

Kennedy: Can you rule out the possibility of a tyrosine kinase-linked receptor which, when you activate the receptor, phosphorylates connexin 43?

Lustig: You can probably rule that out by doing a patch-clamp experiment and exposing the extracellular surface to the agonist. In this way you can eliminate the involvement of intracellular second messengers or kinases that require ATP for their action.

Kennedy: Not if the tyrosine kinase activity is intrinsic to the receptor.

Hickman: Some connexins (43 and 32) exist intracellularly as non-phosphorylated proteins until they are inserted into the membrane, at which time they are phosphorylated on serine residues. To form mature gap junctions with other communication-competent cells, connexin 43 is phosphorylated. I do not know if connexin 43 is involved in the P2Z response and I cannot rule out a possible role of a receptor with intrinsic tyrosine kinase activity. With reference to Kevin Lustig's comments, several laboratories have recorded ATP-induced P2Z currents in outside-out patch-clamp experiments using murine mast cells (Tatham & Lindau 1990) and murine peritoneal macrophages (Albuquerque et al 1993). However, for some reason it has been very difficult to measure single channel conductances of the P2Z channels. Both labs used noise analysis to estimate the unitary conductances of the P2Z pore to be less than 10 pS. I am aware of a paper (Coutinho-Silva et al 1996) in which single channel conductances of the P2Z response have been measured at 5–8 pS using outside-out patch clamps from murine peritoneal macrophages. This conductance seems very small for a pore large enough to pass ions and small molecules: it was expected to have a unitary conductance of at least 200 pS (Tatham & Lindau 1990).

Barnard: With regard to the connexins, you talked about the experiments of Beyer & Steinberg (1991) on the ATP-resistant J774 cell line (ATPR J774), where it was proposed that the gap junction protein connexin 43 forms the P2Z pore. I remember in that paper that this was based on the loss of expression of connexin 43 when the resistance was acquired. However, could it not be that in those cells which were selected to give the resistant line there were other proteins whose expression also was changed? It seems to me that this evidence was permissive but not at all conclusive for a gap junction involvement in the P2Z pore.

Hickman: It was more guilt by association, and there may still be some association of connexin 43 with P2Z responses.

Di Virgilio: We have done some experiments with the connexin 43 cDNA in which we screened ATP-sensitive and ATP-resistant cells. Although we confirmed the observation by Tom Steinberg that ATP resistant macrophages do not express connexin 43, we also found ATP-insensitive cells that do express a lot of connexin 43 and vice versa.

Illes: How does 2-methylthioATP act on the P2Z receptor?

Hickman: I haven't used it.

Wiley: I've used it with lymphocytes. It's slightly less potent than ATP, but only slightly, and the maximum flux is about the same as ATP.

Di Virgilio: 2-MethylthioATP is also very potent in microglial cells.

Wiley: Perhaps if I could come back to the unitary channel conductance. In a recent patch-clamp paper, Bretschneider et al (1995) looked at the unitary conductance of the P2Z channel in B lymphocytes. It was only 3 pS. The Hill coefficient for the effective agonist in opening this was 1, which is quite unlike the usual Hill coefficient of 2 which has been obtained in all other studies of P2Z. It's quite possible that with this patch-clamp study they have actually been looking at the monomeric subunit which is just conducting a very small current under conditions preventing the other subunits from coalescing or in which the other subunits are absent.

Hickman: In lymphocyte P2Z channels there seems to be a size selection. Ethidium bromide can go through, which is 394 Da, but propidium iodide, which is 668 Da can't. Whereas in human macrophages, we use a dye called YO-PRO-1, which is 692 Da, and this still passes through.

Jacobson: Is the selection purely on size, or is there some dependence on charge or other chemical features?

Wiley: We've looked at a variety of permeants for the P2Z channel in lymphocytes and, as Suzanne just mentioned, ethidium bromide at 394 Da goes in whereas propidium iodide at 668 Da doesn't go in. We've also looked at organic amines as permeants, using BCECF-loaded lymphocytes and measuring intracellular pH after opening the channel with ATP (Chen et al 1994). The permeability for the mono charged species ammonium$^+$, monomethylammonium$^+$ and dimethylammonium$^+$ is significant, but trimethylammonium$^+$ (69 Da) and triethylammonium$^+$ don't appear to penetrate very well. I think there may be a contribution from the molecular conformation of the permeant as well as permeant size to the exclusion characteristics of the channel. As regards charge, cations seem to be selected over anions.

Kennedy: You have the P2Z purinoceptor, clearly, and also a pore. Are they the same molecule?

Hickman: That point is still open for debate. We may have cloned the P2Z receptor which is itself capable of pore formation. We may have cloned one chain of a multiple subunit receptor or one component of a more complex response pathway. Thorough analysis of our clone at the protein level may give some answers, but I suspect there will still be a haze of ambiguity until we can isolate the native protein from the membranes of J774 cells, for instance, and compare it with the protein product derived from our clone. Until then, I can co-inject oocytes with RNA transcribed from our clone with messages derived from clones of connexin 43, P2X receptors or other pools from our J774 library and determine whether there is another protein involved in the P2Z response.

Wiley: Nuttle & Dubyak's data (1994) show that there is an immediate opening, over a timeframe of milliseconds, of a small channel that conducts Na^+, K^+ and Li^+, but not larger organic cations. I think the channel and the pore could be the same molecule.

Kennedy: So how does Lucifer yellow enter the cell?

Wiley: The larger organic cations become permeants after conversion of the channel to a pore. This pore has larger exclusion limits, which vary between the different cell types that express the P2Z receptor.

References

Albuquerque C, Oliviera SMC, Coutinho-Silva R, Oliveira-Castro GM, Persechini PM 1993 ATP-induced and UTP-induced currents in macrophages and macrophage polykaryons. J Physiol 265:1663C–1673C

Beyer EC, Steinberg TH 1991 Evidence that the gap junction protein connexin-43 is the ATP-induced pore of mouse macrophages. J Biol Chem 266:7971–7974

Bretschneider F, Klapperstück M, Löhn M, Markwardt F 1995 Nonselective cation currents elicited by extracellular ATP in human B lymphocytes. Pflügers Arch 429:691–698

Chen JR, Jamieson GP, Wiley JS 1994 Extracellular ATP increases $NH_4{}^+$ permeability in human lymphocytes by opening a P_{2Z} purinoceptor operated ion channel. Biochem Biophys Res Commun 202:1511–1516

Coutinho-Silva R, Alves LA, Savino W, Persechini PM 1996 A cation nonselective channel induced by extracellular ATP in macrophages and phagocytic cells of the thymic reticulum. Biochem Biophys Acta 1278:125–130

Erb L, Lustig KD, Ahmed AH, Gonzalez FA, Weisman GA 1990 Covalent incorporation of 3′-*O*-(4-benzoyl)benzoyl ATP into a P_2 purinoceptor in transformed fibroblasts. J Biol Chem 265:7424–7431

Nuttle LC, Dubyak GR 1994 Differential activation of cation channels and non-selective pores by macrophage P_{2Z} purinergic receptors expressed in *Xenopus* oocytes. J Biol Chem 269:13988–13996

Tatham PER, Lindau M 1990 ATP-induced pore formation in the plasma-membrane of rat peritoneal mast cells. J Gen Physiol 95:459–476

P2X receptors: a third major class of ligand-gated ion channels

R. Alan North

Glaxo Institute for Molecular Biology, 14 Chemin des Aulx, 1228 Plan-les-Ouates, CH-1211 Geneva, Switzerland

Abstract. Three classes of ligand-gated ion channels are defined by their molecular architecture. The first embraces nicotinic, 5-HT_3, glycine and GABA receptors. The second class contains the glutamate receptors—AMPA, kainate and NMDA types. The third class is the P2X receptors for ATP. Current knowledge of the structure of these channels is reviewed, and set beside what is known of their basic functional properties. The aim of this paper is to consider how our more complete understanding of the first two classes of channels might be helpful in forming a molecular picture of P2X receptor function.

1996 P2 purinoceptors: localization, function and transduction mechanisms. Wiley, Chichester (Ciba Foundation Symposium 198) p 91–109

The past

The initial observations that adenosine 5′-triphosphate (ATP) directly opens membrane ion channels were reported in 1983 by Krishtal et al and by Jahr & Jessel. Those findings, made by recording from dorsal root ganglion cells and spinal cord dorsal horn neurons in culture, have now been extensively replicated on nerve cells, gland cells and smooth muscles (reviewed by Bean 1992, Zimmermann 1994, Surprenant et al 1995). The main characteristics of this response to ATP are that the channels open within a few milliseconds of the agonist application, and that the channels are permeant to cations including Ca^{2+}. Peak responses are related to the ATP concentration applied, with a half-maximal concentration in the low micromolar range and Hill coefficient between 1 and 3.

Two further properties vary among different tissues. First, during a maintained application of ATP for about 1 s, the inward current declines almost completely (desensitization), or almost not at all. Second, the action of ATP is either mimicked by similar concentrations of α,β-methylene ATP (α,β-meATP), or α,β-meATP has little or no action even at hundred-times higher concentrations. These two distinguishing features have been used to define three phenotypes of P2X receptors at the cellular level: (1) desensitizing, α,β-meATP-sensitive (e.g. vas deferens smooth muscle); (2) non-desensitizing,

α,β-meATP-sensitive (e.g. sensory ganglion cells); and (3) non-desensitizing, α,β-meATP-insensitive (e.g. phaeochromocytoma or PC12 cells) (Surprenant et al 1995, Surprenant 1996, this volume).

The main impetus to research on the P2X receptor during the past few years has come from three kinds of development. The first has been the demonstration that suramin (Dunn & Blakeley 1988) and pyridoxalphosphate-6-azophenyl-2′,4′-disulfonic acid (PPADS; Ziganshin et al 1993) are effective antagonists at P2X receptors. The second has been the use of these antagonists to demonstrate that ATP mediates an excitatory junction potential in some smooth muscles (Sneddon & Westfall 1984, Evans & Surprenant 1992), or an excitatory synaptic potential in some neurons (Evans et al 1992, Edwards et al 1992, Galligan & Bertrand 1995). The third has been the isolation of cDNA clones from vas deferens and PC12 cells which code for P2X receptor proteins (Brake et al 1994, Valera et al 1994). When expressed in heterologous cells, these proteins form ion channels with all the basic properties of those in the native tissue (Valera et al 1994, Evans et al 1995, Surprenant 1996, this volume). The protein sequence deduced from the cDNAs indicates that the receptor subunit has a structure unlike that of other ligand-gated ion channels. The remainder of this paper is a commentary on how P2X receptor channels might form in molecular terms. This is approached by reviewing the relation between structure and function of some other, mostly mammalian, channels that are much better understood.

The perspective

Gating by extracellular transmitters

When compared with the rich panoply of molecules that are secreted to act at receptors of the seven-transmembrane domain (7TM) family, only relatively few have been selected to gate channels by binding directly to their extracellular domains (Fig. 1). Those in vertebrates are acetylcholine (ACh), 5-hydroxytryptamine (5-HT), γ-aminobutyric acid (GABA), glycine, glutamate and ATP; these are joined by histamine (Hardie 1989), dopamine (Berry & Cottrell 1979) and some small peptides (Cottrell et al 1990) in invertebrates. The first four transmitters act on receptors that belong to the same class, in overall evolutionary terms (Ortells & Lunt 1995), whereas glutamate and ATP act at two further classes.

The first class of ligand-gated ion channel can be selective for cations (e.g. 5-HT_3 receptors, nicotinic receptors) or anions (GABA and glycine receptors, and a glutamate receptor of *Caenorhabditis elegans* [Cully et al 1994]). Each subunit of these proteins has a long (typically about 200 amino acids) extracellular N-terminal region that includes a disulfided 13-residue loop, four transmembrane domains (TMs) and an intracellular C-terminus. Considerable evidence supports the view that the second transmembrane domain (TM2) is α-helical and contributes to the channel lining (Hucho et al 1986, Leonard et al 1988, Akabas et al 1994, Unwin 1995). In this class of molecule, charge changes in rings of amino acids at key positions near the inner and outer pore mouths can convert from cation to anion selectivity (Galzi et al 1992); in the case of the nicotinic α7 pentameric channel, a glutamate residue near the inner pore mouth is required to confer the relatively high permeability to Ca^{2+} ions (Bertrand et al 1993). A chimaeric receptor in which the N-terminus domain of the nicotinic receptor is joined to the TM domains of the 5-HT_3 receptor has most of the pharmacological properties of the nicotinic receptor, confirming an important role for the N-terminus region in ligand recognition (Eisele et al 1993). Five subunits assemble symmetrically to make the channel; the more evolutionary primordial subunits (Ortells & Lunt 1995) can make channels as homopentamers (e.g. α7 nicotinic, 5-HT_3), but most naturally occurring channels appear to form as heteropentamers.

The second type of ion channel is gated by extracellular glutamate and/or aspartate (Fig. 1); within mammals these channels are found exclusively and abundantly in the CNS, whereas in many invertebrates they also mediate the transmitter action of glutamate at the neuromuscular junction. At least 16 subunits are known, which fall by sequence homology within the broad subfamilies of AMPA (α-amino-3-hydroxy-5-methyl-4-isoxalone propionic acid), kainate and NMDA (*N*-methyl-D-aspartate) receptors (Sprengel & Seeburg 1995). These subunits have very long extracellular N-terminal loops (up to 400 amino acids) and appear to have three TM regions and an intracellular C-terminus (Hollmann et al 1994). The glutamate binding site is formed jointly by a region of the N-terminus domain adjacent to the TM1, and parts of the extracellular loop between TM2 and TM3. A region of the molecule on the cytoplasmic side of the membrane between TM1 and TM2 (which was formerly thought to span the membrane) bears a distant sequence-

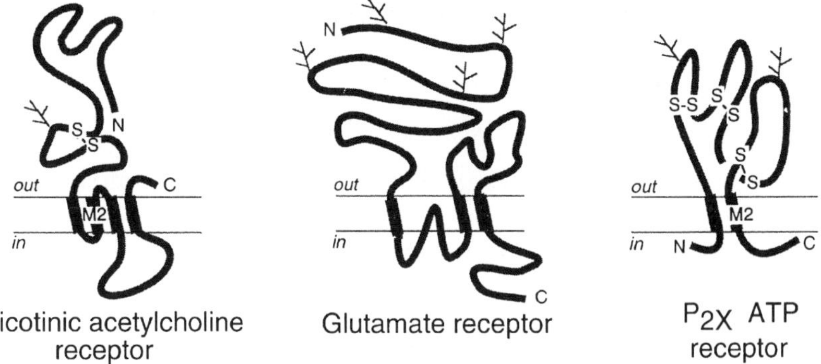

FIG. 1. Schematic representation of the transmembrane topology of individual subunits of three ligand-gated ion channels. Glycosylation sites and disulfide bridges are hypothetical in the case of glutamate and P2X receptors.

relatedness to the pore-forming regions of K^+ channels and cyclic nucleotide-gated cation channels; it is possible that this region loops into the membrane and contributes to the channel pore (Hollmann et al 1994, Wood et al 1995, MacKinnon 1995). As for nicotinic receptors, the Ca^{2+} permeability of glutamate-gated cation channels varies according to the identity of amino acid residues at key positions, which are therefore presumed to lie within the conducting pore (Burnashev et al 1992a,b). For example, in the case of the AMPA receptor subunits the residue is glutamine (all except GluR-B/GluR2) or arginine (GluR-B/GluR2). The arginine actually arises not from a different gene sequence, but by RNA editing (Burnashev et al 1992a). Heteropolymerization occurs between members within the three subfamilies, which have 40–70% identity: channels are not formed by mixing subunits from the different subfamilies, which typically have 30% or less identical amino acids (Sprengel & Seeburg 1995).

P2X receptor subunits comprise the third class of ligand-gated ion channels (Brake et al 1994, Valera et al 1994, Evans et al 1995, see Surprenant et al 1995) (Fig. 1). Each subunit is about 400 amino acids in length, with relatively short intracellular N- and C-termini, and two TM domains separated by a long (almost 300 amino acids) extracellular region. This assignment of topology is supported by mutagenesis experiments on an antagonist binding site (Buell et al 1996). All 10 cysteine residues in this extracellular region are conserved between the receptors, as are most glycine and lysine residues. The subunits from rat vas deferens and human bladder, and from rat PC12 cells reproduce faithfully the properties of the P2X receptors in respective tissues of origin (see Surprenant et al 1995) when they are expressed as individual cDNAs. They are therefore assumed to form as homopolymers, but the number of subunits that comprise a channel is not known. The properties in common between $P2X_1$ and $P2X_2$ include cation-selectivity with a high Ca^{2+} permeability; the properties that differ are α,β-meATP sensitivity and desensitization, which are seen only for the $P2X_1$ clone.

Other channels with two transmembrane domains per subunit

There are other classes of ion channels that have two transmembrane domains per subunit, although these are not known to be gated by extracellular ligands. These include amiloride-sensitive epithelial Na^+ channels (ENaC), inward-rectifier K^+ channels, and the mechanosensitive channel of *Escherichia coli* (Fig. 2). ENaC channels have the most similarity in overall structure to P2X receptors, including a large glycosylated extracellular loop with multiple cysteine residues (Canessa et al 1993). Homologous α, β and γ subunits, which are 34–37% identical, co-assemble to form the ionic channel (Canessa et al 1994); the actual stoichiometry is not known. The pore of this channel is highly

Na^+ selective, but it is not clear which part of the molecule contributes directly to its formation. These mammalian channels are homologous in sequence to the products of the *mec-4* and *mec-10* genes of *C. elegans*, which are therefore also proposed to form ion channels. Those genes were identified from genetic analysis of mutant worms that do not respond to touch; they are expressed in six mechanosensitive neurons of the nematode, where their disruption can lead to death of those neurons (Huang & Chalfie 1994). The genes can be mutagenized so as to give rise to proteins with given amino acid substitutions; correlations with the phenotype (i.e. whether or not the mechanosensitive neurons die) suggests that TM2 and the immediately preceding region might contribute to pore formation (Hong & Driscoll 1994). Despite the overall topological similarity, there is no primary sequence homology between these channels and P2X receptors.

Inward-rectifier K^+ channels also have two transmembrane domains, between which is located the signature pore sequence of K^+-selective channels (typically TMTTVGYGD). In the case of voltage-dependent K^+ channels, this sequence is located between TM5 and TM6, leading to the suggestion that those channels evolved by the addition of four further membrane-spanning domains to a primordial inward rectifier (Jan & Jan 1994). Inward rectifier channels have no primary sequence homology with P2X receptors, and they do not have the large extracellular loop. In some of them, the intracellular N- and C-termini are involved in interactions with $\beta\gamma$ subunits of heterotrimeric GTP-binding proteins; in others, intracellular ATP reduces the probability of channel opening, but the molecular basis of this action has not yet been worked out.

A mechanosensitive channel (mscL) has been cloned from *E. coli* that may represent the most primordial subunit with two transmembrane domains. The protein is predicted to be 136 amino acids long, forms a pore that has a high unit conductance (3 nS) and does not select between cations and anions (Sukharev et al 1994).

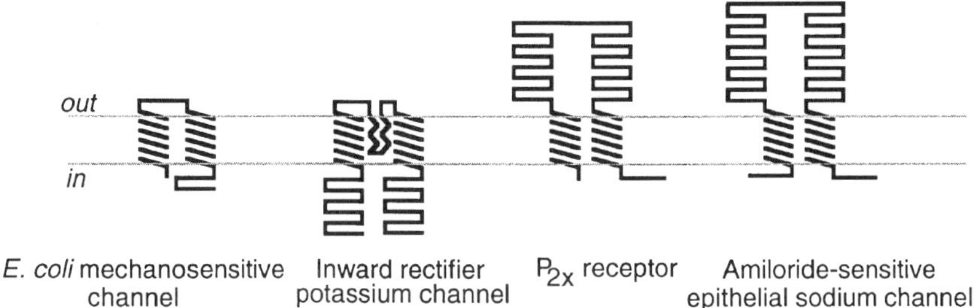

FIG. 2. Schematic representation of channel subunits thought to have two transmembrane domains.

Gating by intracellular nucleotides

Can we learn anything about the operation of P2X receptors from other channels that are gated by nucleotides acting from within the cell? There are three main types of channel for which such actions have been studied. First, a subset of the inward rectifier K^+ channels (K_{ATP} channels) are closed when intracellular ATP rises; this action is relatively specific for ATP over other nucleotides, but it is not understood in molecular terms. Second, a large family of cyclic nucleotide-gated cation channels (CNGCs) has been characterized; these play critical roles in sensory transduction in photoreceptors and olfactory epithelium, and seem likely to be important in other tissues (Firestein & Zufall 1994, Yau & Chen 1995). Third, the cystic fibrosis transmembrane conductance regulator (CFTR) binds ATP.

CNGCs are structurally related to voltage-dependent K^+ channels. Each subunit has six transmembrane domains, and a pore-forming region (P) between TM5 and TM6 that is homologous to the P region of K^+ channels. Relatively minor modifications to the P region of K^+ channels to make them resemble more closely the CNGCs (deletion of the two amino acids YG) results in the K^+ channel becoming cation-non-selective (Heginbotham et al 1992). The nucleotide-binding domain of CNGCs has been identified by its homology to that of the catabolite gene activator protein (CAP) of *E. coli*, for which the crystal structure has been determined with cAMP bound, and key residues that interact with different parts of the cyclic nucleotide have thus been identified (Yau & Chen 1995). CNGCs are activated by cyclic nucleotides with a Hill coefficient that is typically about 3; this is consistent with the prevailing view that the channels form as tetramers. Tetrameric channel formation has been shown for the homologous voltage-dependent K^+ channels (MacKinnon 1991, Shen et al 1994) and is strongly implied for CNGCs (Root & MacKinnon 1994). The regions of the pore of the CNGC that control ionic selectivity have been determined in considerable detail; the side chains of paired glutamate residues interact with permeating monovalent ions and also provide a binding site for external divalent cations which block the pore (Sather et al 1994, Root & MacKinnon 1993, 1994, Eisman et al 1994).

Opening of CFTR Cl^- channels requires their phosphorylation by A kinase, and the binding of ATP followed by its hydrolysis. In addition, the channels can be affected by non-hydrolysable analogues of ATP, suggesting that ATP can also bind in a regulatory role (Baukrowitz 1994). The nucleotide binding domains of these channels are well conserved, as they are in other members of the ABC (ATP-binding cassette) family of membrane proteins that includes P glycoprotein, adenylate cyclase and many other membrane transporters.

ATP binding sites in proteins

At the level of cloned genes, it might be possible by studying nucleotide binding sites in other proteins to deduce how they bind to and activate P2X receptors. The approach is somewhat validated by work on other classes of ligand-gated channels, either for individual amino acids or protein domains. For example, critical tyrosine residues, rather than negatively charged side chains, interact with the cationic head of acetylcholine whether it binds to nicotinic receptors (Tomaselli et al 1991) or to acetylcholinesterase (Sussman et al 1991). An amino acid binding protein within the periplasmic space of bacteria is homologous to extracellular regions of the AMPA glutamate receptor (and, incidentally, to regions within the N-terminal domain of the metabotropic glutamate receptors). This allows the sequence of the AMPA receptor domains to be superimposed upon the framework of the bacterial periplasmic protein, for which a crystal structure has been determined (Stern-Bach et al 1994).

The nucleotide binding site of many proteins has been determined at atomic resolution, first for NAD-dependent dehydrogenases (Branden & Tooze 1991). Recent analyses of such structures has led to the definition of several consensus sequences for peptide segments involved in nucleotide binding (Koonin 1993, Traut 1994, Yoshida & Amano 1995). In the majority of these proteins, ATP binds at a catalytic site (kinases, ATP synthases and adenyltransferases); by far the most common motif here is the one described by Walker et al (1982) (original Walker A: G(X4)GK(T/S)(X6)(I/V) and Walker B: R/K(X3)G(X3)L(ϕ4)D, where ϕ is any hydrophobic residue). In these cases, there is a β-strand/loop/α-helical structure, with the conserved lysine marking the beginning of the α-helix; this lysine commonly, but not always, interacts with the terminal phosphate of the ATP (Traut 1994). ATP may also act at regulatory sites on enzymes, from which the molecule unbinds intact, which is presumably more similar to the situation at the P2X receptor. The best characterized protein at which ATP has this kind of action is the regulatory subunit of aspartate carbamoyl-transferase, for which several residues involved in nucleotide binding have been identified (Zhang & Kantrowitz 1992, Wente & Schachman 1991) including a lysine (K94) which interacts with the terminal phosphate moiety. A simpler ATP binding domain—G(X3)G—is reported to mediate the regulatory action of ATP on receptor guanylate cyclases (Duda et al 1993) but no structural information is available.

There are no ATP binding domains of the Walker type within the predicted sequence of the P2X receptors (Fig. 3). It might be that these are not to be expected: in view of the extracellular location of this part of the protein and the likelihood that tertiary structure is dominated by disulfide bridges, the binding site might be formed by structural elements unlike those found among intracellular ATP binding proteins. On the other hand, some of the key features of other nucleotide binding proteins—glycine-rich loops and lysine residues—are present within the external loop and their involvement can be tested by mutagenesis. Several of these residues are strikingly conserved between P2X$_1$ and P2X$_2$ receptors (Fig. 3), as are some features of the overall secondary structure (Fig. 4A). For example, a glycine-rich region is conserved around residue 250, which may form a loop between an α-helix and a β-sheet (Fig. 4B). This part of the molecule has relatedness to a part of aldolase A that is involved in binding the C-6 phosphate group of fructose-1,6-bisphosphate (Freemont et al 1988, Sygusch et al 1987). Other ATP binding proteins on the outer surface of cells seem not to be related in sequence to P2X receptors; these include rat liver ectoATPase (CAM-105; Lin & Guidotti 1989), rat jejunum nucleoside transporter (Huang et al 1994) or *Saccharomyces cerevisiae* purine/cytosine permease (Weber et al 1990).

Pores in channels

The pore-lining residues of the nicotinic channel have been identified by the use of mutagenesis combined with functional studies of blocking ions. The most

```
vas   M A R R L Q D E L S A F F F E Y D T P R M V L V R N K K V G V I F R L I Q L V V   40
PC12  M V R R L A R G C W S A F W D Y E T P K V I V V R N R R L G F V H R M V Q L L I   40

vas   L V Y V I G W V F V Y E K G Y Q T S S D L - I S S V S V K L K G L A V T Q L Q G   79
PC12  L L Y F V W Y V F I V Q K S Y Q D S E T G P E S S I I T K V K G I T M S E - - -   77

vas   L G P Q V W D V A D Y V F P A H G D S S F V V M T N F I V T P Q Q T Q G H C A E   119
PC12  - - D K V W D V E E Y V K P P E G G S V V S I I T R I E V T P S Q T L G T C P E   115

vas   N P E G G I - - C Q D D S G C T P G K A E R K A Q G I R T G N C V P F N - G T V   156
PC12  S M R V H S S T C H S D D D C I A G Q L D M Q G N G I R T G H C V P Y Y H G D S   155

vas   K T C E I F G W C P V E V D D K I P S P A L L R E A E N F T L F I K N S I S F P   196
PC12  K T C E V S A W C P V E D G T S D N H F - L G K M A P N F T I L I K N S I H Y P   194

vas   R F K V N R R N L V E E V N G T Y M K K C L Y H K I Q H P L C P V F N L G Y V V   236
PC12  K F K F S K G N I A S Q K S   D Y L K H C T F D Q D S D P Y C P I F R L G F I V   233

vas   R E S G Q D F R S L A E K G G V V G I T I D W K C D L D W H V R H C K P I Y Q F   276
PC12  E K A G E N F T E L A H K G G V I G V I I I W N C D L D L S E S E C N P K Y S F   273

vas   H G L Y G E K N L - S P G F N F R F A R H F V Q N G T N R - R H L F K V F G I H   314
PC12  R R L D P K Y D P A S S G Y N F R F A K Y Y K I N G T T T T R T L I K A Y G I R   313

vas   F D I L V D G K A G K F D I I P T M T T I G S G I G I F G V A T V L C D L L L L   354
PC12  I D V I V H G Q A G K F S L I P T I I N L A T A L T S I G V G S F L C D W I L L   353

vas   H I L P K R H Y Y K Q K K F K Y A E D M G P G E G E H D P V A T S S T L G L Q E   394
PC12  T F M N K N K L Y S H K K F D K V R T P K H P S S - R W P V T L A L V L G Q I P   392

vas   N M R T S   399
PC12  P P P S H Y S Q D Q P P S P P S G E G P T L G E G A E L P L A V Q S P R P C S I   432

PC12  S A L T E Q V V D T L G Q H M G Q R P P V P E P S Q Q D S T S T D P K G L A Q L   47:
```

FIG. 3 Sequence comparison of P2X$_1$ (rat vas deferens) and P2X$_2$ (rat phaeochromocytoma cells) receptors. Boxes indicate similar and identical residues; similar is defined conservatively (E,D; R,K; A,S,T; Y,F,W; I,L,V,M). Filled circles indicate conserved cysteines and overlines indicate probable membrane-spanning domains.

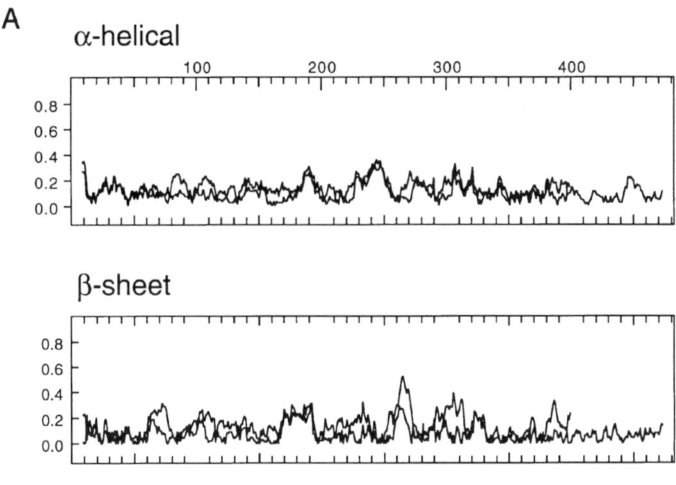

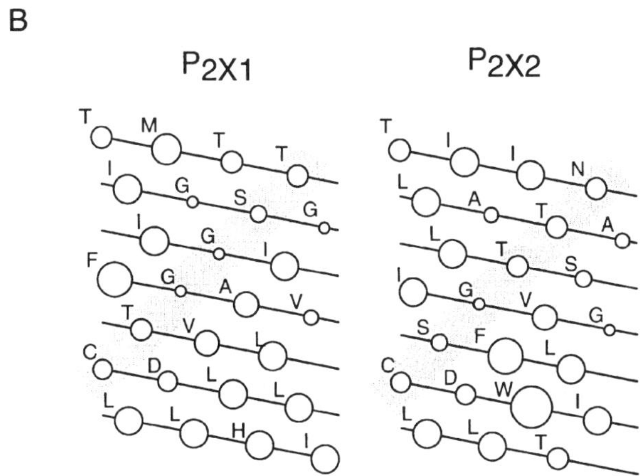

FIG. 4 Secondary structure comparison of $P2X_1$ (rat vas deferens) and $P2X_2$ (rat phaeochromocytoma cells) receptors. (A) Superimposed plots of predicted α-helical and β-sheeted parts of the molecules. (B) Helical net plots of the second transmembrane domain (beginning at T321 of $P2X_1$ and T320 of $P2X_2$, see Fig. 3). Circles are proportional to the sizes of the amino acids. The oblique shaded line indicates the band of small and/or polar residues that line one face of the helix.

complete work used cysteine substitution and blockade with methanesulfonate derivatives, which will work only where the cysteine side chain is accessible to the aqueous environment of the pore. This shows that the M2 helix lines the open pore, and that a segment near its centre is probably in an extended rather than helical conformation (Akabas et al 1994). This view agrees well with the most recent images of the receptor using cryo-electron microscopy, which show that during activation the five M2 helices of the pentamer bend at this position during channel activation (Unwin 1995).

The pore-forming region of the glutamate receptor involves the region originally called M2, because residues here can critically affect Ca^{2+} permeability and rectification (Keinanen 1990). This segment had been thought by some to form the second of four membrane-spanning domains but assignments of the sideness of different segments using antibodies and mutagenesis of glycosylation sites now indicate that it is intracellular (Hollmann 1994). Homology with K^+ channel pore regions has led to the suggestion that it may provide a channel-forming loop from within the cell (Wood et al 1995).

Many studies have focused on the pore of voltage-dependent K^+ channels, and their structural counterparts, the CNGCs. Each of the four subunits contributes one pore loop to the channel from the segment joining TM5 and TM6; four analogous segments play the same role in voltage-gated Na^+ and K^+ channels (Durell & Guy 1992, Sather et al 1994, Heginbotham et al 1994). The GYG of the polypeptide chain is thought to adopt an extended conformation, with a bend between two conserved threonine residues; according to this view, the K^+ ion is coordinated through the deep part of the pore by oxygen atoms provided by backbone carbonyls. This overall picture of pore formation is consistent with cystallographic data for maltoporin, a channel from the outer membrane of *E. coli* that allows maltose to enter the periplasmic space. Maltoporin has four loops nested within an 18-stranded antiparallel β-barrel, and these loops expose aromatic and ionizable residues to the narrowest constriction (about 0.5 nm) of the pore (Schirmer et al 1995). Overall analogies can be made between the loops lining the pore of channels with the loops that occur at the active sites of most enzymes (MacKinnon 1995).

The pore of the P2X receptor, in all cases so far studied (see Surprenant et al 1995), including the cloned $P2X_1$ and $P2X_2$ receptors (Evans 1995), is significantly permeable to both monovalent and divalent cations. The relative permeability of Ca^{2+} to Na^+ is about 4:1, but the fraction of the total current carried by Ca^{2+} in physiological conditions is quite small (probably 10–15%; Rogers & Dani 1995), because the extracellular Ca^{2+} concentration is so much lower than that of Na^+. The pore is significantly permeable to large organic cations such as Tris and *N*-methyl-D-glucosamine: the narrowest diameter of the pore has been estimated at 0.9–1.0 nm from measurements of the relative permeability of cations with a range of molecular sizes (Evans 1995). This is slightly larger than the pore of the nicotinic receptor channel. The relative permeability to Ca^{2+} ions is less than that seen for the NMDA receptor or the $\alpha 7$ nicotinic pentamer.

There are not obvious homologies between pore-lining sequences in known channels and regions of the P2X receptors. In the $P2X_1$ receptor, the region preceding TM2 shows a motif [F(X8)G(X5)P] that is conserved in many voltage-dependent K^+ channels, and it has been suggested that this part of the molecule might contribute to the pore (Brake et al 1994). It is among the most highly conserved regions between the two receptors. On the other hand, TM2 itself is strikingly amphipathic (Brake et al 1994). Figure 4B shows that if TM2 forms an α-helix, there is a clear line of small or polar residues along one face. A similar situation applies to the TM2 of the nicotinic receptor, where the polar residues are known to be exposed to the aqueous environment of the pore (Leonard et al 1988). Approaches similar to those which have been used for other channels might help to determine whether the 'pre-TM2 loop' or the 'TM2 helix' or neither contribute to the pore. These would include mutagenesis combined with a test of block by Ca^{2+}, other divalents, local anaesthetics and methanesulfonate derivates in the case of cysteine substitutions.

Conclusions

Nucleotide-binding proteins are among the best studied biological macromolecules; many have high-resolution structures available. The past few years has seen great advances in our knowledge of pore formation—both in mammalian cells and prokaryotes. These two lines of enquiry should converge to provide an understanding of how P2X receptor proteins function. Equally important will be the assessment of functional roles of the channel. The isolation and expression of many cDNAs, and the mapping of their RNAs and protein products, will quickly expand our horizons. The horizons will embrace tissues where ligand-gated channels were not previously thought to play important roles—these might include endocrine, immune and genito-urinary systems. The receptor family may encode proteins that are anion- rather than cation-selective, as for glutamate-activated channels. There may be family members that are activated by quite distinct transmitters, not even nucleotides. This is the case for channels in the glycine/GABA or 5-HT_3/nicotinic families. Gene cloning and functional studies on simpler organisms will enable advances in both the molecular and the physiological areas, given the fundamental importance of nucleotides in all life forms.

References

Akabas MH, Kaufmann, C, Archdeacon P, Karlin A 1994 Identification of acetylcholine receptor channel-lining residues in the entire M2 segment of the alpha subunit. Neuron 13:919–927

Baukrowitz T, Hwang TC, Gadbsy DC, Nairn AC 1994 Coupling of CFTR Cl channel gating to an ATP hydrolysis cycle. Neuron 12:473–482

Bean BP 1992 Pharmacology and electrophysiology of ATP-activated ion channels. Trends Pharmacol Sci 13:87–90

Berry MS, Cottrell GA 1979 Ionic basis of different synaptic potentials mediated by an identified dopamine-containing neuron in *Planorbis*. Proc R Soc Lond Ser B Biol Sci 203:427–444

Bertrand D, Galzi JL, Devillers-Thiery A, Bertrand S, Changeux JP 1993 Mutations at two distinct sites within the channel domain M2 alter calcium permeability of neuronal α7 nicotinic receptor. Proc Natl Acad Sci USA 90:6971–6975

Brake AJ, Wagenbach MJ, Julius D 1994 New structural motif for ligand-gated ion channels defined by an ionotropic ATP receptor. Nature 371:519–523

Buell G, Lewis C, Collo G, North RA, Surprenant A 1996 An antagonist-insensitive P_{2X} receptor expressed in epithelia and brain. EMBO J 15:55–62

Burnashev N, Monyer H, Seeburg PH, Sakmann B 1992a Divalent ion permeability of AMPA receptor channels is dominated by the edited form of a single subunit. Neuron 8:189–198

Burnashev N, Schoepfer R, Monyer H et al 1992b Control by asparagine residues of calcium permeability and magnesium blockade in the NMDA receptor. Science 257: 1415–1419

Canessa CM, Horisberger J-D, Rossier BC 1993 Epithelial sodium channel related to proteins involved in neurodegeneration. Nature 361:467–470

Canessa CM, Schild L, Buell G et al 1994 Amiloride-sensitive epithelial Na^+ channel is made of three homologous subunits. Nature 367:463–467

Cottrell GA, Green KA, Davies NW 1990 The neuropeptide Phe-Met-Arg-Phe-NH_2 (FMRFamide) can activate a ligand-gated ion channel in Helix neurons. Pflügers Arch 416:612–614

Cully DF, Vassilatis DK, Liu KK et al 1994 Cloning of an avermectin-sensitive glutamate-gated chloride channel from *Caenorhabditis elegans*. Nature 371:707–711

Duda T, Goraczniak RM, Sharma R 1993 Core sequence of ATP regulatory module in receptor guanylate cyclases. FEBS Lett 315:143–148

Dunn PM, Blakeley AGH 1988 Suramin: a reversible P2-purinoceptor antagonist in the mouse vas deferens. Br J Pharmacol 93:243–245

Durell SR, Guy HR 1992 Atomic scale structure and functional models of voltage-gated potassium channels. Biophys J 62:238–247

Edwards FA, Gibb AJ, Colquhoun D 1992 ATP receptor-mediated synaptic currents in the central nervous system. Nature 359:144–147

Eiselé J-L, Bertrand S, Galzi J-L, Devillers-Thiery A, Changeux JP, Bertrand D 1993 Chimaeric nicotinic–serotonergic receptor combines distinct ligand binding and channel specificities. Nature 366:479–483

Eismann E, Muller F, Heinemann SH, Kaupp UB 1994 A single negative charge within the pore region of a cGMP-gated channel controls rectification, Ca^{2+} blockage, and ionic selectivity. Proc Natl Acad Sci USA 91:1109–1113

Evans RJ 1995 Relative permeability to monovalent organic cations of heterologously expressed rat vas deferens and PC12 P_{2X} receptors. J Physiol 487:193

Evans RJ, Derkack V, Surprenant A 1992 ATP mediates fast synaptic transmission in mammalian neurons. Nature 357:503–505

Evans RJ, Surprenant A 1992 Vasoconstriction of guinea-pig submucosal arterioles following sympathetic nerve stimulation is mediated by the release of ATP. Br J Pharmacol 106:242–249

Evans RJ, Lewis C, Buell G, Valera S, North RA, Surprenant A 1995 Pharmacological characterization of heterologously expressed ATP-gated cation channels (P_{2X} purinoceptors). Mol Pharmacol 48:178–183

Firestein S, Zufall F 1994 The cyclic nucleotide gated channel of olfactory receptor neurons. Semin Cell Biol 5:39–46

Freemont PS, Dunbar B, Fothergill-Gilmore LA 1988 The complete amino acid sequence of human skeletal muscle fructose-bisphosphate aldolase. Biochem J 249:779–788

Galligan JJ, Bertrand PP 1995 ATP mediates fast synaptic potentials in enteric neurons. J Neurosci 14:7563–7571

Galzi JL, Deviller-Thiery A, Hussy N, Bertrand S, Changeuz JP, Bertrand D 1992 Mutations in the channel domain of a neuronal nicotinic receptor convert ion selectivity from cationic to anionic. Nature 359:500–505

Hardie RC 1989 A histamine-activated chloride channel involved in neurotransmission at a photoreceptor synapse. Nature 339:704–706

Heginbotham L, Abramson T, MacKinnon R 1992 A functional connection between the pores of distantly related ion channels as revealed by mutant K^+ channels. Science 258:1152–1155

Heginbotham L, Lu Z, Abramson T, MacKinnon R 1994 Mutations in the potassium channel signature sequence. Biophys J 66:1061–1067

Hollmann M, Caron C, Heinemann S 1994 N-glycosylation site tagging suggests a three transmembrane domain topology for the glutamate receptor GluR1. Neuron 13:1331–1343

Hong K, Driscoll M 1994 A transmembrane domain of the putative channel subunit MEC-4 influences mechanotransduction and neurodegeneration in *C. elegans*. Nature 367:470–472

Huang M, Chalfie M 1994 Gene interactions affecting mechanosensory transduction in *Caenorhabditis elegans*. Nature 367:467–470

Huang Q-Q, Yao SYM, Ritzel MWL, Paterson ARP, Cass CE, Young JD 1994 Cloning and functional expression of a complementary cDNA encoding a mammalian nucleoside transport protein. J Biol Chem 269:17757–17760

Hucho F, Oberthur W, Lotspeich F 1986 The ion channel of the nicotinic acetylcholine receptor is formed by the homologous helices MII of the receptor subunits. FEBS Lett 205:137–142

Jahr CE, Jessell TM 1983 ATP excites a subpopulation of rat dorsal horn neurons. Nature 304:730–733

Jan LY, Jan YN 1994 Potassium channels and their evolving gates. Nature 371:119–122

Keinanen K, Wisden W, Sommer B et al 1990 A family of AMPA-sensitive glutamate receptors. Science 249:556–560

Koonin EV 1993 A common set of conserved motifs in a vast variety of putative nucleic acid-dependent ATPases including MCM proteins involved in the initiation of eukaryotic DNA replication. Nucleic Acids Res 21:2541–2547

Krishtal OA, Marchenko SM, Pidoplichko VI 1983 Receptor for ATP in the membrane of mammalian sensory neurones. Neurosci Lett 35:41–45

Leonard RJ, Labarca C, Charnet P, Davidson N, Lester HA 1988 Evidence that the M2 membrane-spanning region lines the ion channel pore of the nicotinic receptor. Science 242:1578–1581

Lin S-H, Guidotti G 1989 Cloning and expression of a cDNA coding for a rat liver plasma membrane ecto-ATPase. The primary structure of the ecto-ATPase is similar to that of the human biliary glycoprotein I. J Biol Chem 264:14408–14414

MacKinnon R 1991 Determination of the subunit stoichiometry of a voltage-gated potassium channel. Nature 350:232–235

MacKinnon R 1995 Pore loops: an emerging theme in ion channel structure. Neuron 14:889–892

Misumi Y, Ogata S, Hirose S, Ikehara Y 1990 Primary structure of rat liver 5′-nucleotidase deduced from the cDNA. J Biol Chem 265:2178–2183

Ortells MO, Lunt GG 1995 Evolutionary history of the ligand-gated ion-channel superfamily of receptors. Trends Neurosci 18:121–127

Rogers M, Dani JA 1995 Comparison of quantitative calcium influx through NMDA, ATP and ACh receptor channels. Biophys J 68:501–506

Root MJ, MacKinnon R 1993 Identification of an external divalent cation-binding site in the pore of a cGMP-activated channel. Neuron 11:459–466

Root MJ, MacKinnon R 1994 Two identical non-interacting sites in an ion channel revealed by proton transfer. Science 265:1852–1856

Sather WA, Yang J, Tsien RW 1994 Structural basis of ion channel permeation and selectivity. Curr Opin Neurobiol 4:313–323

Schirmer T, Keller TA, Wang Y-F, Rosenbuch JP 1995 Structural basis for sugar translocation through maltoporin channels at 3.1 Å resolution. Science 267:512–514

Shen K-Z, Lagruitta A, Davies NW, Standen NB, Adelman JP, North RA 1994 Tetraethylammonium block of Slowpoke calcium-activated potassium channels: evidence for tetrameric channel formation. Pflügers Arch 426:440–445

Sneddon P, Westfall DP 1984 Pharmacological evidence that adenosine triphosphate and noradrenaline are co-transmitters in the guinea pig vas deferens. J Physiol 347: 561–580

Sprengel R, Seeburg PH 1995 Ionotropic glutamate receptors: In: North RA (ed) Ligand- and voltage-gated ion channels. CRC press, Boca Raton, FL, p 213–263

Stern-Bach Y, Bettler B, Hartley M, Sheppard PO, O'Hara PJ, Heinemann SF 1994 Agonist selectivity of glutamate receptors is specified by two domains structurally related to bacterial amino acid binding proteins. Neuron 13:1345–1357

Sukharev SI, Blount P, Martinac B, Blattner FR, Kung C 1994 A large-conductance mechanosensitive channel in *E. coli* encoded by *msc*L alone. Nature 368:265–268

Surprenant A 1996 Functional properties of native and cloned P2X receptors. In P2 purinoceptors: localization, function and transduction mechanisms. Wiley, Chichester (Ciba Found Symp 198) p 208–222

Surprenant A, Buell G, North RA 1995 P_{2X} receptors bring new structure to ligand-gated ion channels. Trends Neurosci 18:224–229

Sussman L, Harel M, Frolow F et al 1991 Atomic structure of acetylcholinesterase from *Torpedo californica*: a prototype acetylcholine binding protein. Science 253:872–879

Sygusch J, Beaudry D, Allaire M 1987 Molecular architecture of rabbit skeletal muscle alsolase at 2.7 Å resolution. Proc Natl Acad Sci USA 84:7846–7850

Tomaselli G, McLaughlin J, Jurman M, Hawrot E, Yellen G 1991 Mutations affecting agonist sensitivity of the nicotinic acetylcholine receptor. Biophys J 60:721–727

Traut TW 1994 The functions and consensus motifs of nine types of peptide segments that form different types of nucleotide-binding sites. Eur J Biochem 222:9–19

Unwin N 1995 Acetylcholine receptor channel imaged in the open state. Nature 373: 37–43

Valera S, Hussy N, Evans RJ et al 1994 A new class of ligand-gated ion channel defined by P_{2X} receptor for extracellular ATP. Nature 371:516–519

Walker JE, Saraste M, Runswick MJ, Gay NJ 1982 Distantly related sequences in the α and β subunits of ATP synthase, myosin, kinases and other ATP-requiring enzymes and a common nucleotide binding fold. EMBO J 8:949–951

Weber E, Rodriguez C, Chevallier MR, Jund R 1990 The purine–cytosine permease gene of *Saccharomyces cerevisiae*: primary structure and deduced protein sequence of the FCY2 gene product. Mol Microbiol 4:585–596

Wente SR, Schachman HK 1991 Different amino acid substitutions at the same position in the nucleotide-binding site of aspartate transcarbamoylase have diverse effects of the allosteric properties of the enzyme. J Biol Chem 266:833–839

Wood MW, VanDongen HMA, VanDongen AMJ 1995 Structural conservation of ion conduction pathways in K channels and glutamate receptors. Proc Natl Acad Sci USA 92:4882–4886

Yau K-W, Chen T-Y 1995 Cyclic nucleotide-gated channels. In: North RA (ed) Ligand- and voltage-gated ion channels. CRC press, Boca Raton, FL, p 307–335

Yoshida M, Amano T 1995 A common topology of proteins catalyzing ATP-triggered reactions. FEBS Lett 359:1–5

Zhang Y, Kantrowich ER 1992 Probing the regulatory site of *Escherichia coli* aspartate transcarbamoylase by site-specific mutagenesis. Biochemistry 31:792–798

Ziganshin AU, Hoyle CH, Bo X, Lambrecht G, Mutschler E, Baumert HG, Burnstock G 1994 PPADS selectively antagonizes P2X-receptor-mediated responses in the rabbit urinary bladder. Br J Pharmacol 110:1491–1495

Zimmermann H 1994 Signalling via ATP in the nervous system. Trends Neurosci 17:420–426

DISCUSSION

Leff: When you did the manipulations on the different receptor forms to introduce or remove desensitization, did you check the resulting agonist profiles?

North: We didn't check the agonist profiles. We have not seen marked differences in the ability of α,β-meATP to cause desensitization as compared with ATP, in the vas deferens form. We've not systematically looked at the ability of different agonists to cause desensitization.

Leff: I just wondered whether the agonist potency orders of the two chimeras became more similar as a result of doing these manipulations.

North: We have studied chimeras of the vas deferens and the PC12 forms, because they differ in desensitization and also in whether they are sensitive to α,β-meATP. The problem in interpreting this experiment with respect to the agonist is that α,β-meATP is *binding* to the PC12 form even though it's not an agonist. It is a weak antagonist; if you give enough α,β-meATP it will block the actions of ATP. That makes the experiment a little bit difficult to interpret. However, having said that, if you give the loop of the $P2X_1$ to the $P2X_2$ receptor, then α,β-meATP will become an agonist. You can certainly give it the sensitivity by giving it the outside loop.

Kennedy: Are you assuming that arginine, lysine and histidine are involved in the ATP binding site?

North: We're assuming that lysine is involved because in other nucleotide binding sites lysine is critical. It usually binds the γ phosphate, and more often than not arginine will not do that job. There are occasional examples where you find an arginine in that place, but in the majority it is a lysine.

Kennedy: Have you sequentially mutated out the lysines?

North: We're just studying this. We hope to find that the K to R substitutions don't change the sensitivity to ADP but reduce the sensitivity to ATP.

Barnard: Concerning the question of which region forms the channel, have you tried either mutating or swapping the region just before the region postulated to form the channel?

North: We have not swapped just the pre-M2 region alone, but we've swapped pre-M2 and M2 and we've swapped M2 alone. The only parameter we've really looked at as a result of these manipulations is the desensitization. In order to switch the desensitization completely, you need both pre-M2 and M2. M2 alone is not enough; you also need the highly conserved region ahead of M2. We've haven't swapped this region on its own yet; the problem with this concerns what would you look for, because the channels have the same Ca^{2+} permeability and the same general permeability. We can only go with available phenotypes.

We have made a few mutations in the region that is similar to the pore of K^+ channels, but the results were inconclusive (Valera et al 1994). One or two didn't express and some didn't have any effect.

Barnard: In the glutamate receptors, the usual view recently has become that the equivalent M2 pore region is forming a re-entrant β-loop (Hollman et al 1994). If there was such a structure in your receptors, that could vary quite a lot in sequence. For example the NMDA receptors and the kainate receptors don't have the same sequence there, yet it seems very hard to believe that both types of glutamate receptor are fundamentally different structures in the channel. Perhaps quite a few types of amino acid can be accepted in the β-structure.

North: According to the secondary structure predictions, the piece that immediately precedes M2 is strongly β-sheeted. It is also very highly conserved.

Barnard: But if you have mutated it without causing problems, that is obviously disappointing.

North: It is not, actually. The regions that we mutated originally are right on the border of pre-M2 and M2.

Barnard: It does seem to me that it is very plausible that pre-M2 is a β-sheet. It was already obvious by comparing the sequence by Brake et al (1994) published with yours (Valera et al 1994), that the sequences that both were citing to be a β-loop were entirely different. Perhaps the determinants of channel function are not just in that sequence but a part is in the M2 as well. It is quite likely, in analogy with the glutamate receptor, that it could be a re-entrant β-loop.

North: But if you make the analogy with the nicotinic receptor, you could argue that the pore is formed by the polar face of M2.

Barnard: But the nicotinic receptor has such a very different deduced transmembrane structure from the P2X receptors and 'M2' has quite a different meaning in the two.

Jacobson: I was wondering about overall similarities with other multimeric receptors and channels. Do you have any speculation about the multimeric nature of the P2X receptors? Are there going to be heteromultimers?

North: It's clear that the two that I've spoken about can form channels as homomultimers or as single subunits. Annmarie Surprenant has found that some but not all of these subunits can heteropolymerize. Because heteropolymerization can occur, we imagine that homopolymerization is the rule in the case of $P2X_1$ and $P2X_2$.

Jacobson: Have you thought about making constructs that have dimers covalently concatenated?

North: We plan to concatenate the cDNAs. It should work because they're joined at the intracellular ends; it works well for other channels.

Di Virgilio: Several times in this meeting the possible involvement of ATP receptors in cell death has been raised. I would like to speculate on the possible parallel between ATP as a neurotransmitter and glutamate. We know that glutamate is involved in excitation of neurons but also in excitotoxicity. Could you see any parallel between the activity of glutamate and ATP in the CNS?

North: There are obvious parallels. The excitotoxicity that glutamate produces results largely from Ca^{2+} entering neurons. Ca^{2+} will clearly enter through all the P2X channels that we've looked at so far. The relative permeability of Ca^{2+} to Na^+ is somewhat less than for the NMDA receptor—it is about 4 : 1. However, the relationship between ATP and P2X receptors and apoptosis on the one hand and glutamate cytotoxicity on the other hand, calls into question what if any is the relationship between excitotoxicty by glutamate and programmed cell death. There my knowledge fails me.

Di Virgilio: But, of course, if a cell wants to die it doesn't necessarily have to die by apoptosis.

North: If you're simply talking about cell death, I'm sure ATP will fill cells up with Ca^{2+}.

Abbracchio: P. Nicotera has shown that glutamate can induce apoptosis in cerebellar granule cells. Glutamate induced either early necrosis or delayed apoptosis in cultures of cerebellar granule cells. During and shortly after exposure to glutamate, a subpopulation of neurons died by necrosis. In these cells mitochondrial membrane potential collapsed. Neurons surviving the early necrotic phase recovered mitochondrial potential and energy levels and later underwent apoptosis. This suggests that excitotoxicity of glutamate involves either necrosis or apoptosis, and that necrosis, associated with extreme energy depletion, may simply reflect the failure of cells to carry out the 'default' apoptotic death program.

Illes: You implied that significant quantities of Ca^{2+} may enter the cells via the P2X receptor channel. A similar Ca^{2+} entry is known to occur via NMDA receptor channels after their stimulation by excitatory amino acids. After excessive stimulation of both receptors by their respective agonists, intracellular Ca^{2+} may rise and activate Ca^{2+}-dependent proteases, leading to irreparable cell damage. Ionotropic excitatory amino acid receptors are more or less ubiquitous in the CNS. In contrast, from electrophysiological mapping studies the distribution of P2X receptors in the brain seems to be much more limited in extent and they are restricted to certain areas. So I would not expect ATP release in connection with cell death to lead to a similar excitotoxicity as glutamate release induced by the same stimulus.

North: I think you should be careful about your estimates of P2X receptor distribution on the basis of electrophysiological results. These tend to show you that they are where you look for them. Certainly the mRNA distribution of some of the later P2X clones suggests that some of them are quite widespread throughout the CNS.

Brändle: Do you have any results for your clones with amiloride?

North: We tested $P2X_1$ and $P2X_2$ and they were insensitive up to 30 μM.

Surprenant: We saw nothing with $P2X_1$ and $P2X_2$. All the other antagonists more or less fell into two categories, so we gave up on amiloride.

Westfall: I find it somewhat ironic that the human urinary bladder is such a rich source of P2X receptor yet many claim that the human bladder receives only sparse purinergic innervation. Does anyone have any idea why there is so much receptor present in human bladder but we don't often see a clear contraction effect or an electrophysiological effect?

Burnstock: This is a fundamental question about the expression of receptors in effector cells. Are they controlled by genes independently from innervation? After denervation, receptors are upgraded (supersensitivity) and after agonist exposure, receptors are downgraded. In old age however, some receptors are upgraded, others are downgraded, others are unchanged. I believe in biological economy and feel intuitively that there aren't many receptors present in the body that aren't used for some purpose. I think that if receptors are present, they are likely to be important at some time during development or in some pathophysiological situations—perhaps after trauma, during regeneration or in disease. It is worth noting, too, that the normal levels of P2X receptors in the human bladder are lower than those you find in laboratory animals but their distribution is regional. The region generally available from the human bladder for experimentation is the tip of the detrusor cone: this happens to be very low in P2 purinoceptors so the interpretation can be misleading.

Surprenant: The human $P2X_1$ cDNA was obtained from bladder tissue taken from patients with severe cystitis. It may be that the $P2X_1$ receptor was tremendously up-regulated in these smooth muscle cells.

Westfall: In studies with John Bells a number of years ago, we found that in bladder strips from children there was often a more prominent purinergic innervation than in bladder strips from adults.

References

Brake AJ, Wagenbach MJ, Julius D 1994 New structural motif for ligand-gated ion channels defined by an ionotropic ATP receptor. Nature 371:519–523

Hollmann M, Caron C, Heinemann S 1994 N-glycosylation site tagging suggests a three transmembrane domain topology for the glutamate receptor GluR1. Neuron 13:1331–1343

Valera S, Hussy N, Evans RJ et al 1994 A new class of ligand-gated ion channel defined by P_{2X} receptor for extracellular ATP. Nature 371:516–519

P2 purinoceptors and pyrimidinoceptors of catecholamine-producing cells and immunocytes

P. Illes*, K. Nieber†, R. Fröhlich and W. Nörenberg

Institut für Pharmakologie und Toxikologie der Universität Freiburg, Hermann-Herder Strasse 5, D-79104 Freiburg, Germany

Abstract. ATP is a neuronal (co)transmitter. In addition, both ATP and UTP may exit damaged cells and thereby function as extracellular signal molecules. The targets of signalling may be the P2 (for ATP and UTP) and P1 (for the degradation product adenosine) receptors of, for instance, neurons and immunocytes. UTP may also act at separate pyrimidinoceptors. Catecholamine-producing cells (adrenal chromaffin cells and peripheral and central noradrenergic neurons) possess P2X and P2Y purinoceptors. ATP appears to be a fast excitatory neuro-neuronal transmitter of the noradrenergic coeliac and locus coeruleus neurons. This effect is mediated by P2X purinoceptors. P2Y purinoceptor-mediated slow excitatory synaptic potentials have not yet been demonstrated either in the peripheral or central nervous system. On the other hand, after neuronal injury microglial cells (brain immunocytes) are engaged in a process called 'synaptic stripping', i.e. the displacement of synaptic boutons from the neuronal surface. During this process microglial cells are in direct contact with the (co)transmitter ATP. Activation of P2X, P2Z and P2Y purinoceptors results in an elevated intracellular Ca^{2+} concentration in microglia and macrophages. Various functions of these cells are regulated by intracellular Ca^{2+} (e.g. cytokine production, phagocytosis) and may therefore be modulated by nucleotides. Since neuronal damage leads to the transformation of microglial cells to macrophages and, at the same time, to the efflux of nucleotides from the damaged cells, the requirements for a modulatory interaction are fulfilled.

1996 P2 purinoceptors: localization, function and transduction mechanisms. Wiley, Chichester (Ciba Foundation Symposium 198) p 110–129

Adenosine 5′-triphosphate (ATP) has been shown to activate P2 purinoceptors which can be differentiated from the adenosine-sensitive P1 purinoceptors by means of specific antagonists (Burnstock & Buckley 1985). The potency orders

Present address: *Institut für Pharmakologie und Toxikologie der Universität Leipzig, Härtelstrasse 16-18, D-04107 Leipzig and †Institut für Pharmazie der Universität, Abteilung Pharmakologie für Naturwissenschaftler, Brüderstrasse 34, D-04103 Leipzig, Germany.

of structural analogues of ATP (α,β-methylene ATP [α,β-meATP], 2-methylthioATP [2-MeSATP]) have been used to subdivide the P2 purinoceptors of multicellular preparations into the P2X (α,β-meATP > 2-MeSATP) and P2Y class (2-MeSATP > α,β-meATP). This approach did not take into account the fact that some P2 purinoceptor agonists (e.g. 2-MeSATP) are susceptible to degradation by ectonucleotidases (Kennedy & Leff 1995). When ectonucleotidases of isolated preparations are inhibited (Kennedy & Leff 1995) or cells are kept in culture systems (Illes & Nörenberg 1993), 2-MeSATP becomes a stronger agonist than α,β-meATP at the P2X purinoceptor. In spite of the identical agonist potency orders, P2Y purinoceptors are much more sensitive to 2-MeSATP than P2X purinoceptors (Kennedy & Leff 1995). Hence, P2X and P2Y purinoceptors represent different structural and functional entities, the former being a ligand-activated cationic channel, the latter a G protein-coupled receptor (Abbracchio & Burnstock 1994).

Additional G protein-coupled P2 purinoceptors belong to the P2U (sensitive to both ATP and UTP, and widely distributed), P2T (sensitive to ADP and localized on platelets) and P2D classes (sensitive to diadenosine polyphosphates, and localized on chromaffin cells and brain synaptosomes) (Fredholm et al 1994). Finally, ATP^{4-} activates P2Z purinoceptors of mast cells and macrophages which open membrane pores of up to 900 Da (Fredholm et al 1994). It is noteworthy that a separate receptor recognizing pyrimidine nucleotides, for example UTP, has been described in some tissues (Abbracchio & Burnstock 1994).

In order to activate membrane receptors of neurons or microglial cells, ATP has to leave the intracellular space by one of three mechanisms. Firstly, it is a (co)transmitter in neurons and is exocytotically released from the nerve terminals (Burnstock 1986). A similar exocytotic co-secretion of ATP with catecholamines takes place from the adrenal medulla into the circulating blood. Secondly, ATP passes the intact cellular membrane via specific channels also responsible for the extrusion of a variety of compounds including chemotherapeutic drugs (P-glycoprotein; Abraham et al 1993). Finally, ATP leaks from ischaemic, injured and dying cells via holes in the membrane (White & Hoehn 1991).

Effects of ATP and UTP on noradrenaline-producing cells

Adrenal medulla

Chromaffin cells of the adrenal medulla and sympathetic neurons derive from the same part of the neuronal crest, but their phenotype becomes neuro-endocrine during development. Both ATP and diadenosine polyphosphates are co-stored in secretory granules of chromaffin cells with adrenaline and noradrenaline, and are exocytotically released by the action of secretagogues (Pintor et al 1991).

In PC12 (rat phaeochromocytoma) cells, ATP and its structural analogues activated an inward current which reversed near 0 mV and exhibited inward rectification (Nakazawa et al 1990a). The single-channel conductance in excised patches was about 13 pS (Nakazawa et al 1990b). The ATP-induced whole-cell currents were blocked by the P2 purinoceptor antagonists suramin and Reactive blue 2 (Nakazawa et al 1991). Hence, a typical P2X purinoceptor appeared to be involved in the cationic conductance increase (Inoue & Nakazawa 1992).

ATP leads to catecholamine release in bovine chromaffin cells and rat PC12 cells. This is mostly due to the entry of extracellular Ca^{2+} via P2X purinoceptor channels and probably also via voltage-dependent Ca^{2+} channels (Rhoads et al 1993). However, ATP may also stimulate P2Y purinoceptors and mobilize intracellular Ca^{2+} (Sasakawa et al 1989). These receptors are coupled via a G protein to the enzyme phospholipase C (PLC), which generates the second messenger inositol 1,4,5-trisphosphate ($InsP_3$). Activation of the P2D purinoceptor by diadenosine tetraphosphate also increases cytosolic Ca^{2+} in chromaffin cells (Castro et al 1992).

Sympathetic ganglia

Although ATP caused inward currents in single neurons of both bullfrog and rat sympathetic ganglia, the ionic mechanisms of these effects were different. In paravertebral sympathetic neurons of bullfrogs, ATP indirectly inhibited the M current (a voltage-sensitive, non-inactivating K^+ conductance), probably via a G protein (Akasu et al 1983). It is likely that a P2Y-type receptor is involved in this effect. By contrast, in rat superior cervical neurons, ATP opened non-selective cationic channels, indicating the involvement of a P2X purinoceptor (Cloues et al 1993, Nakazawa 1994, Khakh et al 1995). This ATP-induced current showed characteristics similar to those of the current described in PC12 cells, with respect to both inward rectification (Nakazawa 1994) and single channel conductance (Nakazawa & Inoue 1993). Cibacron blue (one of the stereoisomers of Reactive blue 2) and suramin antagonized the effect of ATP in superior cervical neurons (Cloues et al 1993, Khakh et al 1995).

Neurons from rat superior cervical ganglia and guinea-pig coeliac ganglia both possess P2X purinoceptor channels, but their pharmacological sensitivities are different (Evans et al 1992, Khakh et al 1995). While 2-MeSATP is invariably a full agonist, it is about 10-times more potent in coeliac neurons than in superior cervical neurons (Khakh et al 1995). Moreover, α,β-meATP is a full agonist in coeliac neurons, but only a partial agonist in superior cervical neurons. Coeliac neurons form a fibre network in cell culture systems (Evans et al 1992). Focal electrical stimulation of these fibres evokes fast excitatory synaptic currents due to the activation of P2X purinoceptors. Hence, ATP is a neuro-neuronal transmitter of guinea-pig coeliac neurons.

When extracellular DC potentials were measured in rat superior cervical ganglia, ATP caused hyperpolarization, while α,β-meATP, 2-MeSATP and UTP caused depolarization (Connolly et al 1993). The rank order of agonist potencies was UTP$>$α,β-meATP$\gg$2-MeSATP. The ATP-induced hyperpolarization turned into depolarization after the blockade of P1 purinoceptors by 8-(p-sulfophenyl)-theophylline (Connolly et al 1993). This fact, in conjunction with the reported depolarizing effect of ATP in single superior cervical neurons (Cloues et al 1993), indicates that ATP is metabolized to adenosine, which causes hyperpolarization and thereby counteracts the depolarizing effect of ATP. It has been suggested that there are separate receptors for purine and pyrimidine nucleotides in this tissue, since responses to ATP but not to UTP were blocked by pyridoxalphosphate-6-azophenyl-2′,4′-disulfonic acid (PPADS; Connolly 1995) and Reactive blue 2 (Connolly & Harrison 1994). This proposal was strengthened by the fact that there was desensitization to both α,β-meATP and UTP, but no cross-desensitization between the two agonists (Connolly 1994).

Nucleus locus coeruleus

As measured by extracellular micro-electrodes, central noradrenergic neurons of the rat locus coeruleus (LC) fire in brain slice preparations at a constant rate (Tschöpl et al 1992). ATP caused no consistent effect when given alone, but increased the firing rate when given in the presence of the P1 purinoceptor antagonist 8-cyclopentyl-1,3-dipropylxanthine (DPCPX), indicating a balance between a direct excitatory P2 effect and an indirect (mediated by the degradation product, adenosine) inhibitory P1 effect.

α,β-meATP increased the firing of LC neurons in a concentration-dependent manner (Fig. 1). Whereas 2-MeSATP had a potency similar to that of α,β-meATP, UTP was inactive. Hence, central noradrenergic neurons, in contrast to their peripheral counterparts, do not possess either P2U purinoceptors or pyrimidinoceptors. Diadenosine polyphosphates facilitated the discharge of action potentials with agonist potencies of $Ap_5A > Ap_4A > Ap_3A$. In accordance with this finding, binding sites for diadenosine polyphosphates have been described on rat brain synaptosomes (Pintor et al 1993).

Subsequently, we searched for P2 purinoceptors on LC neurons by studying the interaction between various P2 purinoceptor agonists and antagonists. Suramin does not discriminate between P2X and P2Y receptors; low concentrations of PPADS prefer P2X (Windscheif et al 1994) whereas low concentrations of Reactive blue 2 prefer P2Y purinoceptors (Kennedy 1990). Under our experimental conditions suramin (30, 100 μM) and PPADS (30 μM) blocked the effect of α,β-meATP (Fig. 1B,C); Reactive blue 2 (30 μM) was also

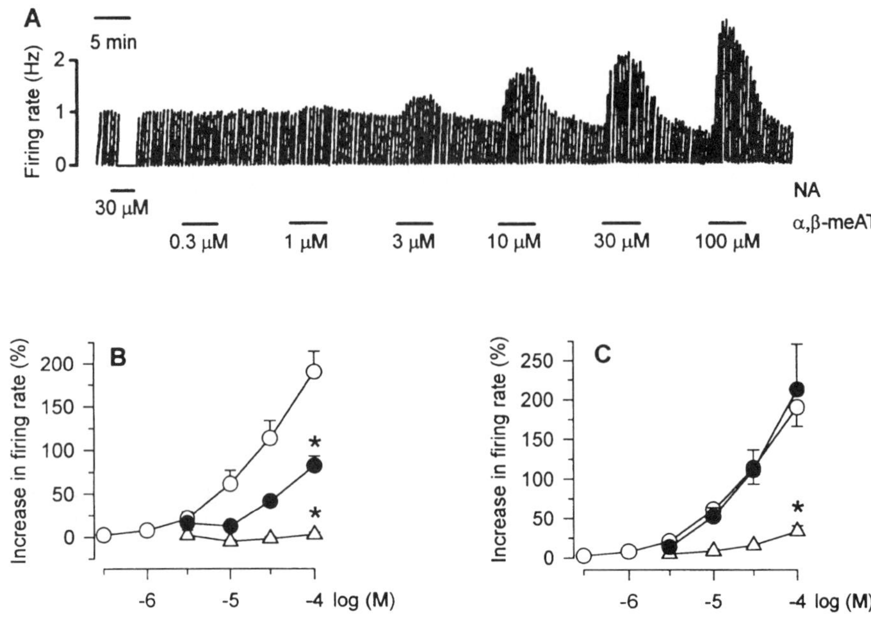

FIG. 1. Effect of α,β-meATP on the firing rate of locus coeruleus (LC) neurons, and interaction with suramin and PPADS. Pontine slices of the rat brain were prepared and the frequency of spontaneous action potentials was recorded extracellularly as consecutive 30 s samples. (A) Original tracing. Noradrenaline (NA; 30 μM) and α,β-meATP (0.3–100 μM) were present in the superfusion medium for 5 min every 10 min as indicated by the horizontal bars. (B) Concentration–response curves of α,β-meATP in the absence (○; $n=7$) and presence of 30 μM suramin (●; $n=6$) or 100 μM suramin (Δ; $n=5$). (C) Concentration–response curves of α,β-meATP in the absence (○; $n=7$) and presence of 10 μM PPADS (●; $n=6$) or 100 μM PPADS (Δ; $n=5$). Values are means ± SEM. All antagonists were present in the superfusion medium for 20 min before application of the lowest concentration of α,β-meATP. *$P<0.05$; significant differences from the effect of α,β-meATP (100 μM).

active. At the same time, of the three antagonists, only suramin blocked the effects of 2-MeSATP and Ap_4A. Thus, rather high concentrations of PPADS and Reactive blue 2, which almost certainly do not have any selectivity to either subtype of P2 purinoceptor, interacted with α,β-meATP, but not with 2-MeSATP or Ap_4A. Finally, suramin (100 μM), but not PPADS (30 μM) depressed the facilitatory effect of *N*-methyl-D-aspartate (NMDA) on the firing rate and neither suramin (100 μM) nor PPADS (30 μM) altered the facilitatory effect of substance P. Hence, PPADS is certainly a more specific antagonist of P2 purinoceptors than suramin.

Intracellular recordings in LC neurons suggested that α,β-meATP inhibits a resting K^+ conductance (probably via G protein activation) and, at the same time, opens non-selective cationic channels (Harms et al 1992). Direct measurements of ion currents confirmed these findings (Shen & North 1993) and led to the suggestion that LC neurons are endowed both with P2X and P2Y purinoceptors (Illes et al 1995).

After the blockade of excitatory amino acid receptors, the electrical stimulation of afferent fibres evoked a suramin-sensitive synaptic current in medial habenular neurons of the rat (Edwards et al 1992). This study is so far the only demonstration of the fast transmitter function of ATP in the CNS. In order to search for ATP as a neurotransmitter in other areas of the brain, we stimulated electrically the slice surface within the region of the LC. This procedure is known to evoke a synaptic depolarization (PSP) followed by hyperpolarization (IPSP) (Egan et al 1983). The PSP is due to the release of an excitatory amino acid transmitter from afferent fibres predominantly onto non-NMDA-receptors, and of γ-aminobutyric acid (GABA) onto $GABA_A$-receptors, of LC neurons (Cherubini et al 1988). A smaller part of the PSP is glycine-mediated (Williams et al 1991). The IPSP was found to be mostly due to the release of noradrenaline either from recurrent axon collaterals of the LC neurons themselves or from afferent fibres originating in the nucleus paragigantocellularis (Egan et al 1983, Williams et al 1991).

In the following experiments we aimed to demonstrate that a considerable part of the PSP is ATP-mediated. Kynurenic acid (500 μM), which blocks both NMDA and non-NMDA receptors, markedly reduced the PSP amplitudes, as did the $GABA_A$ receptor antagonist picrotoxin (100 μM) (Figs 2A, 3A). Suramin (30, 100 μM) produced a concentration-dependent and reversible further inhibition. The higher concentration (100 μM) of suramin depressed the PSP amplitude by about 45% (Fig. 2A). This effect of suramin (100 μM) persisted when both NMDA and non-NMDA receptors were blocked by phosphonopentanoic acid (AP-5; 50 μM) and 6-cyano-7-nitroquinoxaline-2,3-dione (CNQX; 10 μM), respectively, instead of kynurenic acid alone (500 μM). PPADS (30 μM) also inhibited the PSPs; its effect did not increase with the further addition of suramin (100 μM; Fig. 3A) and did not reverse on washout. Since suramin (100 μM; Fig. 2B) and PPADS (30 μM;Fig. 3B) almost abolished the effect of pressure-applied α,β-meATP, it was concluded that these P2 purinoceptor antagonists exclude a fraction of the PSP which is due to the release of ATP. ATP was suggested to be co-released together with noradrenaline from the recurrent axon collaterals of LC neurons. In support of this notion, guanethidine (10 μM), which is known to interrupt the propagation of action potentials to the varicosities, depressed both the PSP and IPSP amplitudes.

In accordance with the extracellular part of this study, suramin (100 μM) inhibited the depolarizing effect of pressure-applied NMDA, but potentiated the depolarizing effect of pressure-applied ($\pm$)-α-amino-3-hydroxy-5-methylisoxazole-4-propionic acid (AMPA). Kynurenic acid (500 μM) or CNQX (10 μM) markedly depressed but did not abolish the depolarizing response to pressure-applied AMPA. Since NMDA-antagonists only negligibly depress the PSP amplitude (Cherubini et al 1988), suramin should potentiate rather than depress an AMPA receptor-mediated residual fraction of the PSP. Hence, we conclude that all the available evidence supports the hypothesis that both suramin and PPADS decrease the PSP by selectively interacting with P2 purinoceptors. Therefore these antagonists are adequate tools to prove a co-transmitter function of ATP in the LC.

Effects of ATP and UTP on immunocytes

Macrophages and lymphocytes

Tetra-anionic ATP (ATP^{4-}), rather than $MgATP^{2-}$, permeabilizes the plasma membrane of mouse macrophages and the mouse macrophage-like cell line

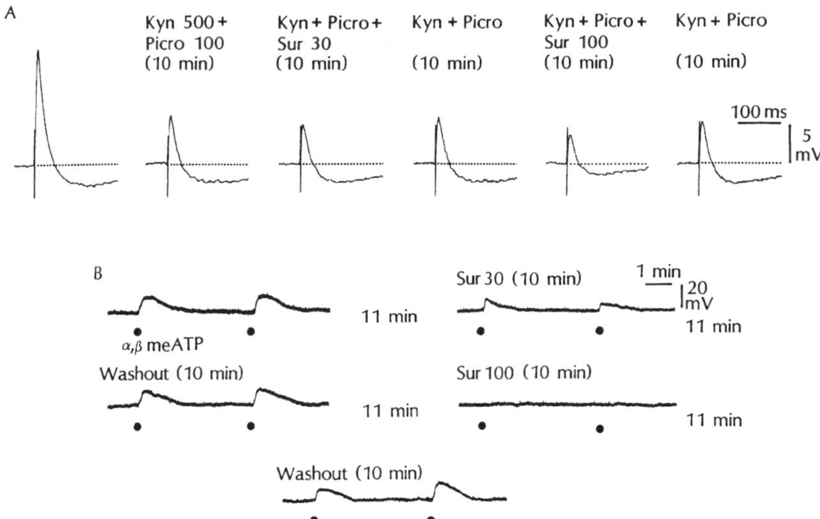

FIG. 2. Effects of suramin on locus coeruleus (LC) neurons. Pontine slices of the rat LC were prepared and the membrane potential was recorded with intracellular glass microelectrodes. (A) Synaptic potentials were evoked by electrical stimulation (0.1 Hz, 0.15 ms, 90 V) with bipolar tungsten electrodes inserted into the slice at a distance of about 100 μm from the site of recording. Dotted lines indicate the membrane potential. (B) Depolarization was evoked by the pressure application (33 kPa, 640 ms) of α,β-meATP (10 mM) from a micropipette (tip diameter, 10–20 μm) every 5 min. α,β-meATP was applied both before, during and after superfusion with suramin (30, 100 μM). Representative tracings in two different cells are shown in both A and B. The membrane potential was hyperpolarized by about 20 mV in order to abolish spontaneous firing. All concentrations are indicated in μM. Superfusion times with antagonists are shown in brackets. The intervals between the traces are indicated. Kyn, kynurenic acid; Picro, picrotoxin; Sur, suramin.

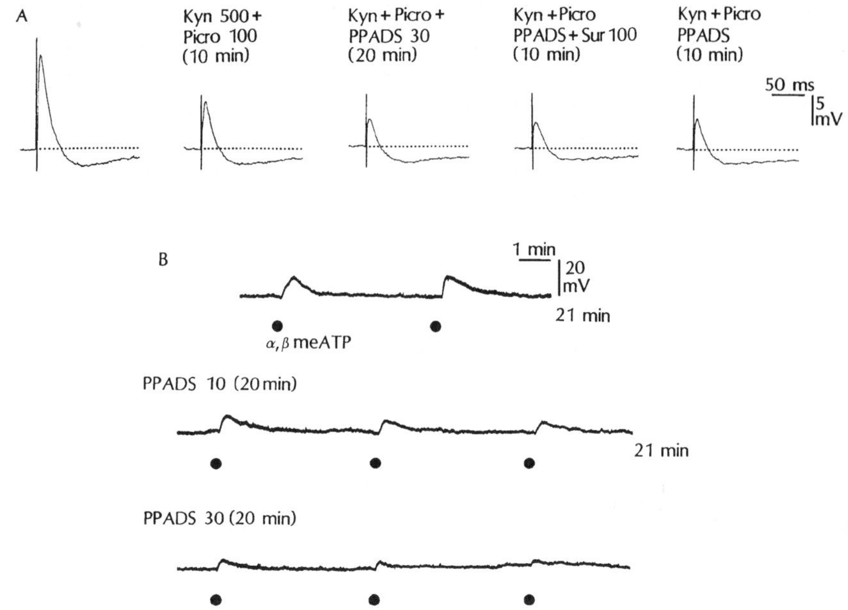

FIG. 3. Effects of PPADS on locus coeruleus (LC) neurons. Pontine slices of the rat brain were prepared and the membrane potential was recorded with intracellular glass microelectrodes. (A) Synaptic potentials were evoked by electrical stimulation (0.1 Hz, 0.5 ms, 115 V) with bipolar tungsten electrodes inserted into the slice at a distance of about 100 μm from the site of recording. Dotted lines indicate the membrane potential. (B) Depolarization was evoked by the pressure application (33 kPa, 400 ms) of α,β-meATP (10 mM) from a micropipette (tip diameter, 10–20 μm) every 5 min. α,β-meATP was applied both before and during superfusion with PPADS (30 μM). Representative tracings in two different cells are shown in both A and B. The membrane potential was hyperpolarized by about 20 mV in order to abolish spontaneous firing. All concentrations are indicated in μM. Superfusion times with antagonists are shown in brackets. The intervals between the traces are indicated. Kyn, kynurenic acid; Picro, picrotoxin; Sur, suramin.

J774.2 (Steinberg et al 1987). The resulting 800 Da pores allow not only the free passage of inorganic cations and anions, but also the entry of fluorescent dyes (Steinberg et al 1987). The opening of the large non-selective pores inevitably leads to an enhanced entry of Ca^{2+} into macrophages (Sung et al 1985), although the opening of smaller pores, sufficient for the passage of di- and monovalent organic cations only, may also subserve Ca^{2+} fluxes (Naumov et al 1992). An additional release of Ca^{2+} from intracellular stores, due to the stimulation by ATP of a phospholipase C-coupled P2Y (Murgia et al 1993) or P2U receptor (Alonso-Torre & Trautmann 1993), also occurs. Subsequently, the rise in intracellular Ca^{2+} concentration may activate Ca^{2+}-dependent K^+ channels (Hara et al 1990). It is interesting to note that the presence of P2Z purinoceptors was recently confirmed in human lymphocytes also (Wiley et al 1994).

Microglia

Microglial cells are the resident immunocytes of the brain and originate from blood monocytes/macrophages during early embryonic development (Thomas 1992, Jordan & Thomas 1988). Microglia are in a resting (ramified) state and can be driven into a macrophage-like (amoeboid) state by stimuli such as bacterial lipopolysaccharide (LPS). Microglia kept in a tissue culture system proliferate and exhibit inwardly rectifying K^+ channels (Kettenmann et al 1990). After LPS-treatment, the cells cease to proliferate; they secrete cytokines (e.g. interleukins 1 and 6), produce superoxide anions and express previously lacking outwardly rectifying K^+ channels (Nörenberg et al 1994a).

We have investigated the effects of ATP and UTP on non-proliferating, cultured rat microglia. Intermediate concentrations of ATP or its structural analogues (such as 100 μM ATPγS) caused, at a holding potential of -70 mV, a rapidly desensitizing inward current, which disappeared on repetitive application and simultaneously unmasked a slower outward current (Fig. 4A). Whereas lower concentrations of ATPγS evoked a biphasic current with a predominant outward component, higher concentrations of this agonist evoked the inward current component only (Fig. 4B). The early, desensitizing current was due to the stimulation of a P2X purinoceptor and the subsequent opening of non-selective cationic channels (Nörenberg et al 1994b, Langosch et al 1994). The late, non-desensitizing current was due to the stimulation of a P2Y purinoceptor and the subsequent opening of K^+ channels via G protein activation. The respective currents were studied in cells held near the reversal potential of either K^+ currents (-70 mV) or non-selective cationic currents (0 mV). Thereby, inward and outward currents could be studied in separation. It is interesting to note that the rank order of agonist potencies was the same at both receptor types (2-MeSATP $>$ ATP $\geqslant$ ATPγS), although the potency of ATP was almost two orders of magnitude higher at the P2Y than at the P2X purinoceptor (Fig. 5).

P2Z rather than P2X purinoceptors were expected to occur on microglia by analogy with peripheral immunocytes. However, this was apparently not the case, since the effect of 2-MeSATP (300 μM) was identical in both the presence and absence of extracellular Mg^{2+} (Nörenberg et al 1994b). The membrane potential of a microglial cell population showed two prominent peaks at -35 and -70 mV (Nörenberg et al 1994a). Single cells switched their membrane potentials between these preferred values (Fig. 6, middle panel). In the voltage-clamp mode of recording, 2-MeSATP (100 μM) caused, at a holding potential of -70 mV, mainly an inward current (Fig. 6Aa), but at a holding potential of -36 mV an outward current (Fig. 6Ab). In the current-clamp mode of

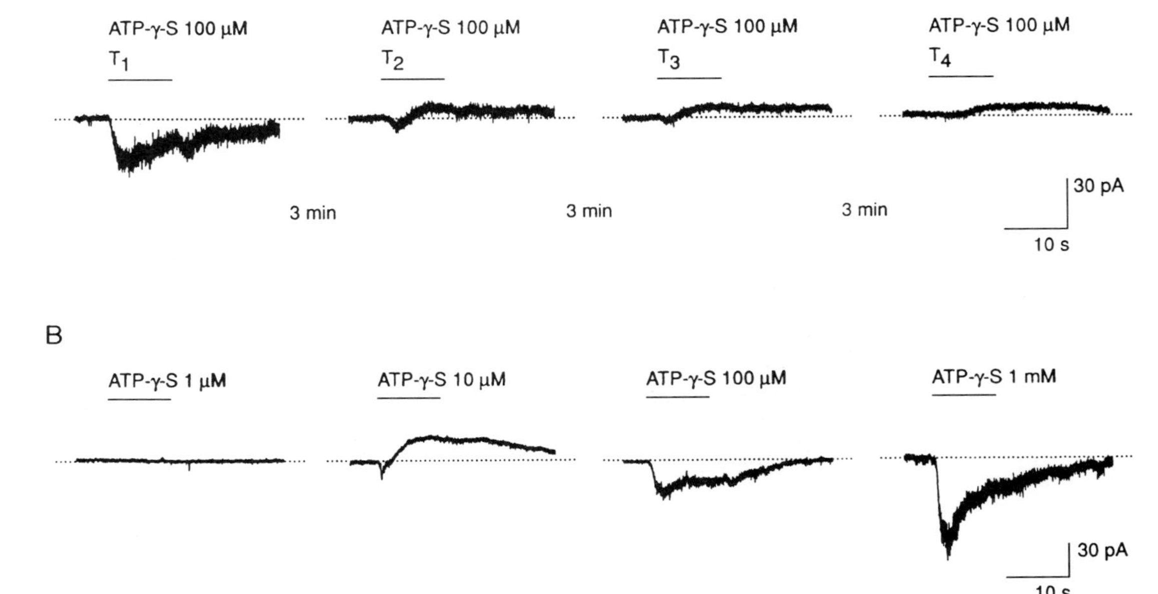

FIG. 4. Inward and outward currents activated by ATPγS in non-proliferating rat microglia. The holding potential was -70 mV. (A) Currents evoked by four consecutive applications (T_1–T_4) of ATPγS (100 μM). ATPγS was pressure-applied onto the same cell for 10 s and at 3 min intervals. (B) Currents evoked by increasing concentrations of ATPγS (1–1000 μM). Each concentration of ATPγS was pressure-applied to a different cell for 10 s. The dotted lines indicate the zero current levels.

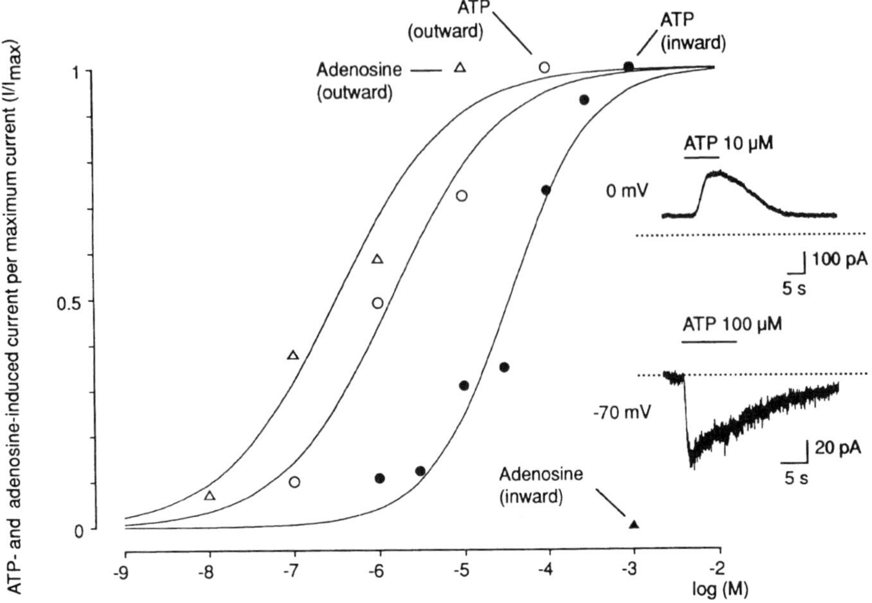

FIG. 5. Normalized concentration–response curves of ATP and adenosine with respect to the inward and outward current in non-proliferating rat microglia. The holding potential was −70 mV for the inward, and 0 mV for the outward current. Each concentration of each agonist was pressure-applied to a different cell for 10 s. The maximum inward current response to ATP was −25.0 ± 4.6 pA (at 1000 μM; $n = 5$). The maximum outward current responses to ATP and adenosine were 143.5 ± 25.4 pA (at 100 μM; $n = 6$) and 114.9 ± 10.2 pA (at 10 μM; $n = 8$), respectively. ○, outward current response to ATP ($n = 5$–9); ●, inward current response to ATP ($n = 5$–8); Δ, outward current response to adenosine ($n = 5$–13); ▲, inward current response to adenosine ($n = 5$). Values represent means ± SEM. Insets show original tracings of outward and inward currents evoked by ATP at 10 and 100 μM, respectively. The dotted lines indicate the zero current levels.

recording, a cell with a membrane potential of −78 mV responded to 2-MeSATP (300 μM) with a rapidly desensitizing depolarization (Fig. 6Ba). Another cell with a membrane potential of −35 mV responded with a long-lasting hyperpolarization to 2-MeSATP (300 μM; Fig. 6Bb). Apparently, P2X purinoceptor-mediated effects prevail at higher membrane potentials, while P2Y purinoceptor-mediated effects prevail at lower membrane potentials.

Although at a holding potential of −70 mV neither adenosine (100–1000 μM) nor UTP (100–1000 μM) evoked an inward current, both agonists (adenosine, 0.01–100 μM; UTP, 1–10 000 μM) evoked concentration-dependent outward currents at 0 mV. It was found that a P1 purinoceptor and a

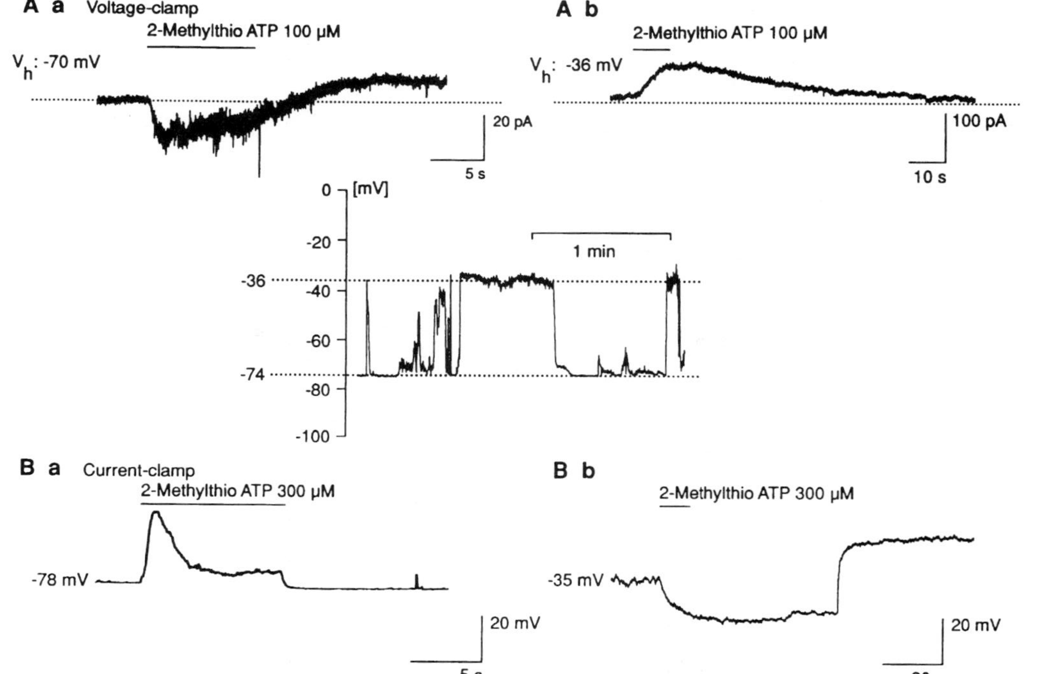

FIG. 6. Effects of 2-MeSATP on non-proliferating rat microglial cells. (*Middle panel*) Random fluctuations between two membrane potential states with the preferred values of -36 and -74 mV. (A) Current responses to 2-MeSATP (100 μM). (a) Inward current followed by outward current at a holding potential of -70 mV. (b) Outward current at a holding potential of 0 mV. (B) Membrane potential changes evoked by 2-MeSATP (300 μM). (a) Depolarization in a cell with a membrane potential of -78 mV. (b) Hyperpolarization in a cell with a membrane potential of -35 mV. Dotted lines indicate either the zero current potential or the two preferred values of membrane potential.

pyrimidinoceptor are involved in the respective effects, since the P1 purinoceptor antagonist 8-(p-sulfophenyl)-theophylline (100 μM) blocked exclusively the effect of adenosine, and the P2 purinoceptor antagonists suramin (300 μM) or Reactive blue 2 (50 μM) blocked the effect of 2-MeSATP only. In contrast, UTP effects were not depressed either by suramin (300 μM) or by Reactive blue 2 (50 μM).

Two independent lines of evidence indicate that the outward current evoked by UTP is carried by K^+ ions. Firstly, the replacement of K^+ by equimolar Cs^+ in the pipette solution abolished the UTP effect. Secondly, the current–voltage curve of UTP, as determined by fast voltage ramps, crossed the zero current level at about -80 mV—near the calculated reversal potential of K^+. The UTP-sensitive pyrimidinoceptor appeared to be coupled to a G protein, since UTP effects disappeared when the enzymically stable analogue guanosine 5′-*O*-(2-thiodiphosphate) (GDPβS; 200 μM) was included in the pipette solution instead of GTP (200 μM) which was normally present.

The effects of ATP and UTP differed from each other; only the UTP-induced current disappeared when the bath medium did not contain any added Ca^{2+}. This dependence of the UTP current on external Ca^{2+} led us to investigate whether microglial cells possess the so-called capacitative Ca^{2+} current (Putney & Bird 1993). Such a current (I_{CRAC}; for Ca^{2+} release-activated Ca^{2+}) has been described in the plasma membrane of many non-excitable cells, for example after the intracellular application of $InsP_3$ which releases Ca^{2+} from intracellular Ca^{2+}-storing organelles; the subsequent depletion of Ca^{2+} stores triggers Ca^{2+} entry across the plasma membrane (Hoth & Penner 1993). We have filled patch pipettes with a solution in which K^+ was replaced by equimolar Cs^+ in order to block outward K^+ currents; $InsP_3$ (10 μM) was also present in the solution. Under these conditions, soon after the solution gained access to the interior of microglial cells, an inward current developed, which disappeared in a nominally Ca^{2+}-free bath solution. This inward current may be I_{CRAC}. Hence, we hypothesized that the UTP-sensitive pyrimidinoceptor may activate, via the involvement of a G protein, the $InsP_3$-generating enzyme phospholipase C. The $InsP_3$-induced immediate rise in intracellular Ca^{2+} is, then, followed by a more sustained elevation of Ca^{2+}, due to the activation of I_{CRAC}. This capacitative Ca^{2+} current leads eventually to the opening of Ca^{2+}-activated K^+ channels. The presence of such channels was demonstrated in inside-out patches of rat microglial cells by the application of Ca^{2+} (0.5 μM) in a low (0.01 μM) Ca^{2+}-containing bath solution.

Acknowledgements

Experiments performed in the laboratory of P. I. were supported by grants of the Deutsche Forschungsgemeinschaft, Bonn (SFB 325, A2; Il 20/6-1; Il 20/7-1).

References

Abbracchio MP, Burnstock G 1994 Purinoceptors: are there families of P_{2X} and P_{2Y} purinoceptors? Pharmacol Ther 64:445–475

Abraham EH, Prat AG, Gerweck L et al 1993 The multidrug resistance (*mdr1*) gene product functions as an ATP channel. Proc Natl Acad Sci USA 90:312–316

Akasu T, Hirai K, Koketsu K 1983 Modulatory actions of ATP on membrane potentials of bullfrog sympathetic ganglion cells. Brain Res 258:313–317

Alonso-Torre SR, Trautmann A 1993 Calcium response elicited by nucleotides in macrophages. Interaction between two receptor subtypes. J Biol Chem 268: 18640–18647

Burnstock G 1986 The changing face of autonomic neurotransmission. Acta Physiol Scand 126:67–91

Burnstock G, Buckley NJ 1985 The classification of receptors for adenosine and adenine nucleotides. In: Paton DM (ed) Methods in pharmacology, vol 6. Plenum, London, p 193–212

Castro E, Pintor J, Miras-Portugal MT 1992 Ca^{2+} stores mobilization by diadenosine tetraphosphate, Ap_4A, through a putative P_{2Y} purinoceptor in adrenal chromaffin cells. Br J Pharmacol 106:833–837

Cherubini E, North RA, Williams JT 1988 Synaptic potentials in rat locus coeruleus neurones. J Physiol 406:431–442

Cloues R, Jones S, Brown DA 1993 Zn^{2+} potentiates ATP-activated currents in rat sympathetic neurons. Pflügers Arch 424:152–158

Connolly GP 1994 Evidence from desensitization studies for distinct receptors for ATP and UTP on the rat superior cervical ganglion. Br J Pharmacol 112:357–359

Connolly GP 1995 Differentiation by pyridoxal 5-phosphate, PPADS and IsoPPADS between responses mediated by UTP and those evoked by α,β-methylene-ATP on rat sympathetic ganglia. Br J Pharmacol 114:727–731

Connolly GP, Harrison PJ 1994 Reactive blue 2 discriminates between responses mediated by UTP and those evoked by ATP or α,β-methylene-ATP on rat sympathetic ganglia. Eur J Pharmacol 259:95–99

Connolly GP, Harrison PJ, Stone TW 1993 Action of purine and pyrimidine nucleotides on the rat superior cervical ganglion. Br J Pharmacol 110:1297–1304

Edwards FA, Gibb AJ, Colquhoun D 1992 ATP receptor-mediated synaptic currents in the central nervous system. Nature 359:144–147

Egan TM, Henderson G, North RA, Williams JT 1983 Noradrenaline-mediated synaptic inhibition in rat locus coeruleus neurones. J Physiol 345:477–488

Evans RJ, Derkach V, Surprenant A 1992 ATP mediates fast synaptic transmission in mammalian neurons. Nature 357:503–505

Fredholm BB, Abbracchio MP, Burnstock G et al 1994 Nomenclature and classification of purinoceptors. Pharmacol Rev 46:143–156

Hara N, Ichinose M, Sawada M, Imai K, Maeno T 1990 Activation of single Ca^{2+}-dependent K^+ channel by external ATP in mouse macrophages. FEBS Lett 267: 281–284

Harms L, Finta EP, Tschöpl M, Illes P 1992 Depolarization of rat locus coeruleus neurons by adenosine 5′-triphosphate. Neuroscience 48:941–952

Hoth M, Penner R 1993 Calcium release-activated calcium current in rat mast cells. J Physiol 465:359–386

Illes P, Nörenberg W 1993 Neuronal ATP receptors and their mechanism of action. Trends Pharmacol Sci 14:50–54

Illes P, Nieber K, Nörenberg W 1995 Neuronal ATP receptors. In: Belardinelli L, Pelleg A (eds) Adenine nucleotides: from molecular biology to integrative physiology. Kluwer Acad, Norwell, MA, p 77–84

Inoue K, Nakazawa K 1992 ATP receptor-operated Ca^{2+} influx and catecholamine release from neuronal cells. News Physiol Sci 7:56–59

Jordan FL, Thomas WE 1988 Brain macrophages: questions of origin and inter-relationship. Brain Res Rev 13:165–178

Kennedy C 1990 P_1 purinoceptor and P_2 purinoceptor subtypes—an update. Arch Int Pharmacodyn Ther 303:30–50

Kennedy C, Leff P 1995 How should P_{2X} purinoceptors be classified pharmacologically? Trends Pharmacol Sci 16:168–174

Kettenmann H, Hoppe D, Gottmann K, Banati R, Kreutzberg G 1990 Cultured microglial cells have a distinct pattern of membrane channels different from peritoneal macrophages. J Neurosci Res 26:278–287

Khakh BS, Humphrey PPA, Surprenant A 1995 Electrophysiological properties of P_{2X} purinoceptors in rat superior cervical, nodose and guinea-pig coeliac neurones. J Physiol 484:385–395

Langosch JM, Gebicke-Haerter PJ, Nörenberg W, Illes P 1994 Characterization and transduction mechanisms of purinoceptors in activated rat microglia. Br J Pharmacol 113:29–34

Murgia M, Hanau S, Pizzo P, Rippa M, Di Virgilio F 1993 Oxidized ATP. An irreversible inhibitor of the macrophage purinergic P_{2Z} receptor. J Biol Chem 268: 8199–8203

Nakazawa K 1994 ATP-activated current and its interaction with acetylcholine-activated current in rat sympathetic neurons. J Neurosci 14:740–750

Nakazawa K, Inoue K 1993 ATP- and acetylcholine-activated channels co-existing in cell-free membrane patches from rat sympathetic neuron. Neurosci Lett 163:97–100

Nakazawa K, Fujimori K, Takanaka A, Inoue K 1990a An ATP-activated conductance in pheochromocytoma cells and its suppression by extracellular calcium. J Physiol 428:257–272

Nakazawa K, Inoue K, Fujimori K, Takanaka A 1990b ATP-activated single-channel currents recorded from cell-free patches of pheochromocytoma PC12 cells. Neurosci Lett 119:5–8

Nakazawa K, Inoue K, Fujimori K, Takanaka A 1991 Effects of ATP antagonists on purinoceptor-operated inward currents in rat pheochromocytoma cells. Pflügers Arch 418:214–219

Naumov AP, Kuryshev YA, Kaznacheyeva EV, Mozhayeva GN 1992 ATP-activated Ca^{2+}-permeable channels in rat peritoneal macrophages. FEBS Lett 313:285–287

Nörenberg W, Gebicke-Haerter PJ, Illes P 1994a Voltage-dependent potassium channels in activated rat microglia. J Physiol 475:15–32

Nörenberg W, Langosch JM, Gebicke-Haerter PJ, Illes P 1994b Characterization and possible function of adenosine 5′-triphosphate receptors in activated rat microglia. Br J Pharmacol 111:942–950

Pintor J, Torres M, Miras-Portugal MT 1991 Carbachol induced release of diadenosine polyphosphates $-Ap_4A$ and Ap_5A- from perfused bovine adrenal medulla and isolated chromaffin cells. Life Sci 48:2317–2324

Pintor J, Diaz-Rey MA, Miras-Portugal MT 1993 Ap_4A and ADP-β-S binding to P_2 purinoceptors present on rat brain synaptic terminals. Br J Pharmacol 108:1094–1099

Putney JW, Bird GSJ 1993 The signal for capacitative calcium entry. Cell 75:199–201

Rhoads AL, Parui R, Vu N-D, Cadogan R, Wagner PD 1993 ATP-induced secretion in PC12 cells and photoaffinity labeling of receptors. J Neurochem 61:1657–1666

Sasakawa N, Nakaki T, Yamamoto S, Kato R 1989 Stimulation by ATP of inositol trisphosphate accumulation and calcium mobilization in cultured adrenal chromaffin cells. J Neurochem 52:441–447
Shen K-Z, North RA 1993 Excitation of rat locus coeruleus neurons by adenosine 5′-triphosphate: ionic mechanism and receptor characterization. J Neurosci 13:894–899
Steinberg TH, Newman AS, Swanson JA, Silverstein SC 1987 ATP^{4-} permeabilizes the plasma membrane of mouse macrophages to fluorescent dyes. J Biol Chem 262: 8884–8888
Streit WJ, Graeber MB, Kreutzberg GW 1988 Functional plasticity of microglia: a review. Glia 1:301–307
Sung SSJ, Young JDE, Origlio AM, Heiple JM, Kaback HR, Silverstein SC 1985 Extracellular ATP perturbs transmembrane ion fluxes, elevates cytosolic $[Ca^{2+}]$, and inhibits phagocytosis in mouse macrophages. J Biol Chem 260:3442–3449
Thomas WE 1992 Brain macrophages: evaluation of microglia and their functions. Brain Res Rev 17:61–74
Tschöpl M, Harms L, Nörenberg W, Illes P 1992 Excitatory effects of adenosine 5′-triphosphate on rat locus coeruleus neurones. Eur J Pharmacol 213:71–77
White TD, Hoehn K 1991 Release of adenosine and ATP from nervous tissue. In: Stone TW (ed) Adenosine in the nervous system. Academic Press, London, p 173–195
Wiley JS, Chen JR, Snook MB, Jamieson GP 1994 The P_{2Z} purinoceptor of human lymphocytes: actions of nucleotide agonists and irreversible inhibition by oxidized ATP. Br J Pharmacol 112:946–950
Williams JT, Bobker DH, Harris GC 1991 Synaptic potentials in locus coeruleus neurons in brain slices. Prog Brain Res 88:167–172
Windscheif U, Ralevic V, Bäumert HG, Mutschler E, Lambrecht G, Burnstock G 1994 Vasoconstrictor and vasodilator responses to various agonists in the rat perfused mesenteric arterial bed: selective inhibition by PPADS of contractions mediated via P_{2X}-purinoceptors. Br J Pharmacol 113:1015–1021

DISCUSSION

Burnstock: In the periphery, co-transmitters tend to act in synergy; they rarely have opposite effects. The co-transmission you have postulated appears to be breaking the rules. There's one exception in a blood vessel, where depending on the tone, one of the co-transmitters causes constriction and the other dilatation. I would suggest that in your schematic you should also to take into account the evidence that ATP is a co-transmitter with glutamate and with GABA and is perhaps doing things consistent with synergism. Also ATP is going to pour out of endothelial cells during hypoxia which would complicate your scheme and possibly contribute to the contentious issue as to whether co-transmitters do opposite things.

Illes: They can't be doing opposite things, because it's quite possible that ATP is degraded to adenosine which inhibits the excitability of LC neurons just as noradrenaline does.

Burnstock: That is a rather tortuous argument.

Illes: The other thing, about which I agree with you, is that there is also the possibility that glutamate and other excitatory transmitters are co-stored with

ATP. But under these special conditions in the LC we think that ATP is stored together with noradrenaline and is released from recurrent axon collaterals of the LC neurons themselves. Another possibility is that ATP is a co-transmitter of adrenaline released from the terminals of neurons originating in the nucleus paragigantocellularis and ending on the LC. In the first series of experiments we used the neurotoxin 6-OH-dopamine in order to selectively damage the terminals of catecholamine neurons. Unfortunately, 6-OH-dopamine led to a rather fast decrease of the PSP amplitudes, possibly by injuring the cell bodies of LC neurons from which the recording of synaptic potentials took place. Therefore, 6-OH-dopamine does not appear to be an adequate tool for our purposes. Next we took guanethidine, which decreased both the PSP and IPSP with a slow onset. The guanethidine-induced inhibition of the PSP amplitudes was smaller than the effect of suramin. If suramin was added on top of guanethidine, the same inhibition was obtained as by suramin alone. I think that this is pretty good evidence for a co-release of ATP with noradrenaline or adrenaline, although it goes against results from the periphery.

Starke: I agree that co-transmitters usually exert qualitatively similar effects. However, in the gut, neurons causing electrolyte secretion often contain transmitters with opposite effects on secretion. So there seems to be a situation where opposite effects of co-transmitters occur.

Harden: Have you tested UDP as an agonist? The reason I ask is that there is a glial tumour cell line that has a pyrimidine-specific receptor; it's not activated by adenine nucleotides. At this receptor, UDP is actually more potent than UTP by 30–50-fold.

Illes: No, we haven't done that.

Harden: Does pertussis toxin block the UTP response?

Illes: We haven't tested pertussis toxin, only GTPβS. We tested pertussis toxin with the ATP response, but not with UTP.

Burnstock: At some stage we ought to clear up this business of whether or not you can correctly call a receptor at which UTP is equally active with ATP (as it is with the P2U purinoceptor) a pyrimidine receptor. I don't believe you can. There are now a few examples where it looks as though only a pyrimidine is active and not ATP and this could be called a pyrimidine receptor, but I don't think one should call the P2U receptor a pyrimidine receptor.

Harden: I agree. I'm very excited about these data because we've been looking for another system where this pyrimidine-nucleotide-selective receptor exists.

Illes: The results I have shown are similar to the data obtained in superior cervical ganglia of rats, with the extracellular measurement of DC potentials. UTP and certain structural analogues of ATP depolarized this preparation. It was concluded that distinct receptors are involved in these effects, since α,β-meATP, but not UTP, was antagonized by suramin and Reactive blue 2 (Connolly et al 1993, Connolly & Harrison 1994).

Surprenant: In the early part of your talk with the LC you showed nice dose–response curves to α,β-meATP, 2-MeSATP and Ap_5A. Suramin more or less blocked everything, but PPADS didn't block 2-MeSATP, so what's it doing?

Illes: It is blocking α,β-meATP but not 2-MeSATP.

Harden: We published a paper comparing the P2Y receptor on turkey erythrocytes with the one on C6 glioma cells (Boyer et al 1994). The turkey erythrocyte receptor activates PLC, the receptor on C6 cells inhibits adenylate cyclase through G_i. PPADS competitively blocks the PLC-linked receptor with a K_i in the micromolar range. It has absolutely no effect on the G_i-coupled P2Y receptor, even though suramin competitively blocks this receptor with a K_i of 1 μM.

Surprenant: But Peter Illes was basically saying that all of the excitatory effects in the LC were mediated by the P2X receptor.

Illes: No. I was saying that in the LC there are both P2X and P2Y receptors on the same cell. Both of them depolarize LC neurons, although by different mechanisms. P2X receptors are ligand-activated cationic channels, whereas P2Y receptors operate via a G protein which closes K^+ channels (Illes et al 1995).

Barnard: Ken Harden, isn't there a difference between your pyrimidine nucleotide receptor and this one? This one requires external Ca^{2+}, and with yours you get a Ca^{2+} transient internally. Is yours totally insensitive to suramin?

Harden: It seems to be partially blocked by suramin. It seems to be a PLC-activating intracellular Ca^{2+} mobilizing receptor.

Barnard: Is it absolutely dependent on extracellular Ca^{2+}, as is this receptor described by Professor Illes?

Harden: We have not looked at that. My prediction is that it would be independent of extracellular Ca^{2+}.

Barnard: So it would be a different receptor then.

North: I would like to look a little more critically at the evidence that ATP actually contributes at all to the PSP. Why didn't you use more CNQX? I don't think that 10 μM is enough to completely block the glutamate-mediated PSP.

Illes: I think that is the concentration everybody uses.

North: If you go to 100 μM, does that further reduce the PSP?

Illes: 500 μM kynurenic acid blocks about 70% and 10 μM CNQX blocks about 80% of the glutamate-induced depolarization. So there is still a small part of the PSP which is, in the presence of these antagonists, mediated by excitatory amino acid receptors.

North: So there is a residual effect that is inhibited by suramin: I agree with that. What I'm saying is that it may be glutamate.

Illes: However, we have shown that 100 μM suramin potentiates AMPA effects although it blocks NMDA effects. In LC neurons the PSP is mediated, in the presence of external Mg^{2+}, almost exclusively by non-NMDA receptors

(Cherubini et al 1988). Hence, suramin should increase rather than decrease a residual glutamate component of the PSP.

North: You cannot compare application of exogenous transmitter with nerve stimulation release. The concentrations at the synapse are very high, and one often needs very high concentrations of antagonist to block them.

Illes: This may be right in quantitative terms. However, it is highly unlikely that an antagonist potentiates an exogenously applied agonist but inhibits the same agonist if it is released on nerve stimulation.

North: Let's ask about another component; the PPADS component.

Illes: We tested also 30 μM PPADS which was more selective to P2 receptors than suramin. 100 μM suramin depressed the facilitatory effects of α,β-meATP, NMDA and AMPA in experiments aimed at measuring the frequency of action potentials by extracellular microelectrodes, whereas 30 μM PPADS interfered with the effect of α,β-meATP only.

North: But 10 μM didn't do anything here.

Illes: No, we tested PPADS at 30 μM in the present intracellular experiments. 30 μM PPADS and 100 μM suramin produced a similar inhibition of the PSP. Moreover, the inhibitory effect of 100 μM suramin could not be increased by a further addition of 30 μM PPADS.

North: Was the PPADS effect quickly reversible?

Illes: No. It was not reversible at all after 10 min, but the suramin effect is certainly reversible within 10 min.

Boeynaems: Is anything known about the release of UTP from neurons or glial cells?

Illes: UTP is generally released from damaged cells. I don't know whether there is anything known about specific release mechanisms.

Burnstock: Endothelial cells certainly take up uridine and release UTP, just as they release ATP (see Saiag et al 1995).

Illes: Platelets do the same.

Miras-Portugal: We have been working on the vesicular nucleotide transporter. ATP, ADP, UTP, GTP and other nucleotides have a similar affinity for this transporter in chromaffin granules. They enter the vesicle as a function of their concentration in the cytosol. So during exocytosis the proportion of UTP released with respect to ATP is the same as in the cytosol.

Abbracchio: Concerning the question of whether UTP is physiologically released, I'm not aware of definite evidence supporting this, at least in the brain, under any conditions. But huge amounts of UTP are likely to be released in the brain following trauma and ischaemia, as a breakdown product of nucleic acid degradation from dying cells. So UTP might not be released physiologically, but may come out during ischaemic hypoxia and may therefore participate in the response of the brain to this condition.

Illes: That's what I would expect.

Edwards: There are complications you might come across in voltage recording of this kind. Even in the presence of all the blockers there is still a considerable residual voltage change. This means you are going to get voltage changes in the dendrites and the axons, which may well lead to voltage-activated effects. I really think you need to use voltage clamp and, as far as possible, isolate the different currents pharmacologically. Otherwise it is going to be difficult to be very convincing.

Illes: First, in these experiments we will be using all types of antagonist you can think of in the bath solution. Second, we shall repeat these experiments in patch-clamp. We have found it very easy to do patch-clamp in striatum, but so far we haven't succeeded in LC. With the LC you have the added problem that for patch-clamp experiments you need brain slices of very young animals. There is electrical coupling between individual cells in these preparations which interferes with the successful clamping of single cells. However, we will definitely be able to switch-clamp LC neurons by using single microelectrodes.

Starke: One comment on pyrimidine receptors. When we postulated specific pyrimidine receptors eight years ago, we were naturally very interested to find a source of pre-formed UTP. The only known rich source in the body known at that time was the storage granules in platelets. So UTP might be released from platelets and then act on the smooth muscle cells of blood vessels where we found pyrimidine receptors to cause vasoconstriction. What is a pyrimidine receptor? It may be defined as a receptor where only UTP acts and ATP does little or nothing. If ATP acts, then there should be no cross-desensitization between ATP and UTP. Peter mentioned the work by Connolly et al (1993) who showed that the pyrimidine receptors are suramin resistant, another difference from most purinoceptors.

References

Boyer JL, John IE, Jacobson KA, Harden TK 1994 Differential effects of P_2 purinoceptor agonists on phospholipase C-coupled and adenylyl cyclase-coupled P_{2Y} purinoceptors. Br J Pharmacol 113:614–620

Cherubini E, North RA, Williams JT 1988 Synaptic potentials in rat locus coeruleus neurones. J Physiol 406:431–442

Connolly GP, Harrison PJ 1994 Reactive blue 2 discriminates between responses mediated by UTP and those evoked by ATP or α,β-methylene-ATP on rat sympathetic ganglia. Eur J Pharmacol 259:95–99

Connolly GP, Harrison PJ, Stone TW 1993 Action of purine and pyrimidine nucleotides on the rat superior cervical ganglion. Br J Pharmacol 110:1297–1304

Illes P, Nieber K, Nörenberg W 1995 Neuronal ATP receptors. In: Bellardinelli L, Pelleg A (eds) Adenine nucleotides: from molecular biology to integrative physiology. Kluwer Acad, Norwell, MA, p 77–84

Saiag B, Bodin P, Shacoori V, Catheline M, Rault B, Burnstock G 1995 Uptake and flow-induced release of uridine nucleotides from isolated vascular endothelial cells. Endothelium 2:279–285

Trophic actions of extracellular ATP on astrocytes, synergistic interactions with fibroblast growth factors and underlying signal transduction mechanisms

Joseph T. Neary

Laboratory of Neuropathology, Research Service 151, VA Medical Center, 1201 NW 16 St and Departments of Pathology, and Biochemistry and Molecular Biology, University of Miami School of Medicine, Miami, FL 33125, USA

A major aim of current neuroscience research is to determine the trophic factors required for, and the mechanisms underlying, neural regeneration. Following neural injury, there are increases in extracellular levels of a number of growth factors. One class of compounds that have been studied in great detail are the polypeptide growth factors. For example, it is well established that nerve growth factor and brain-derived neurotrophic factor promote survival of neurons and neurite extension, while other polypeptide growth factors such as the fibroblast growth factors (FGFs), epidermal growth factor and platelet-derived growth factor stimulate astrocyte proliferation. The trophic actions of the polypeptide growth factors are initiated by binding to receptor tyrosine kinases. Less is known about another class of trophic agents, generally smaller molecules that stimulate receptors coupled to heterotrimeric G proteins. In this category are extracellular nucleotides and nucleosides, and recent studies have suggested that these compounds, either alone or in combination with polypeptide growth factors, may exert trophic effects in the brain during development and/or following injury (Neary et al 1996). In the latter case, the release of nucleotides upon tissue injury, hypoxia or cell death may contribute to gliosis, the hypertrophic and hyperplastic response of astrocytes observed in many common neurological conditions such as trauma, stroke, seizure, and degenerative and demyelinating disorders.

Reactive gliosis is characterized by the generation and elongation of astrocytic processes, an increase in the expression of glial fibrillary acidic protein (GFAP) and, in some types of injury, by cellular proliferation (Norenberg 1994). Several studies from different groups using different glial preparations have now documented that gliotic-like responses can be evoked by extracellular ATP. For example, extracellular ATP increases DNA synthesis and astrocyte proliferation in chick astroblasts (Rathbone et al 1992), primary cultures of rat cerebral cortical astrocytes (Neary & Norenberg 1992, Neary et al 1994a) and neuronal/glial primary cultures of rat striatum (Abbracchio et al 1994). In addition, short-term (1 h) treatment of primary cultures of rat cerebral cortical astrocytes with extracellular ATP induces stellation and an increase in the content of GFAP (Neary & Norenberg 1992, Neary et al 1994a). In mixed neuronal/glial primary cultures of rat striatum, the P2 purinoceptor agonist α,β-methylene ATP stimulated elongation of astrocytic processes (Abbracchio et al 1994, 1995). Results from agonist and antagonist studies (Rathbone et al 1992, Abbracchio et al 1994, 1995), together with the relatively slow breakdown of ATP and ADP by astrocyte ectoATPase and ectoADPase (Lai & Wong 1991, Neary et al 1994a), indicate that the mitogenic and morphogenetic effects of extracellular ATP are mediated by P2 purinoceptors.

Extracellular ATP can also interact synergistically with polypeptide growth factors. For example, ATP greatly enhances mitogenesis induced by polypeptide growth factors in astrocytes (Neary et al 1994b) and in other cells (Wang et al 1990). As shown in Fig. 1A, ATP increases DNA synthesis in astrocytes (as measured by [^{3}H]thymidine incorporation) approximately twofold, whereas FGF-2 (also known as basic fibroblast growth factor, bFGF) increased DNA synthesis 14-fold; when ATP was added concurrently with FGF-2, a 52-fold increase was observed (Neary et al 1994b). A synergistic interaction was also observed by measuring bromodeoxyuridine (BrdU) incorporation into the DNA of dividing cells (Fig. 1B). The synergistic interaction between FGF-2 and ATP was mediated by P2 rather than P1 purinoceptors; thus, breakdown of ATP to adenosine is not required for the synergistic effect of ATP on FGF-2-induced DNA synthesis. ATP also synergistically potentiated DNA synthesis induced by FGF-1 in astrocytes.

Which signal transduction mechanisms underlie the trophic actions of extracellular ATP on astrocytes? Mitogen-activated protein (MAP) kinases are key elements of signal transduction pathways involved in cell growth (for reviews see Ahn et al 1992, Davis 1993, Avruch et al 1994). These cytoplasmic enzymes transduce signals from both receptor tyrosine kinases and heterotrimeric G protein-coupled receptors to the nucleus, resulting in the activation and/or induction of transcription factors, thereby leading to gene expression and cell growth. MAP kinases are part of a protein kinase cascade; upon activation of receptor tyrosine kinases or heterotrimeric G protein-coupled receptors, a sequence of events occurs including stimulation of a MAP kinase kinase kinase (Raf or MEK kinase), which in turn phosphorylates a MAP kinase activator known as MAP kinase kinase (MEK) which then

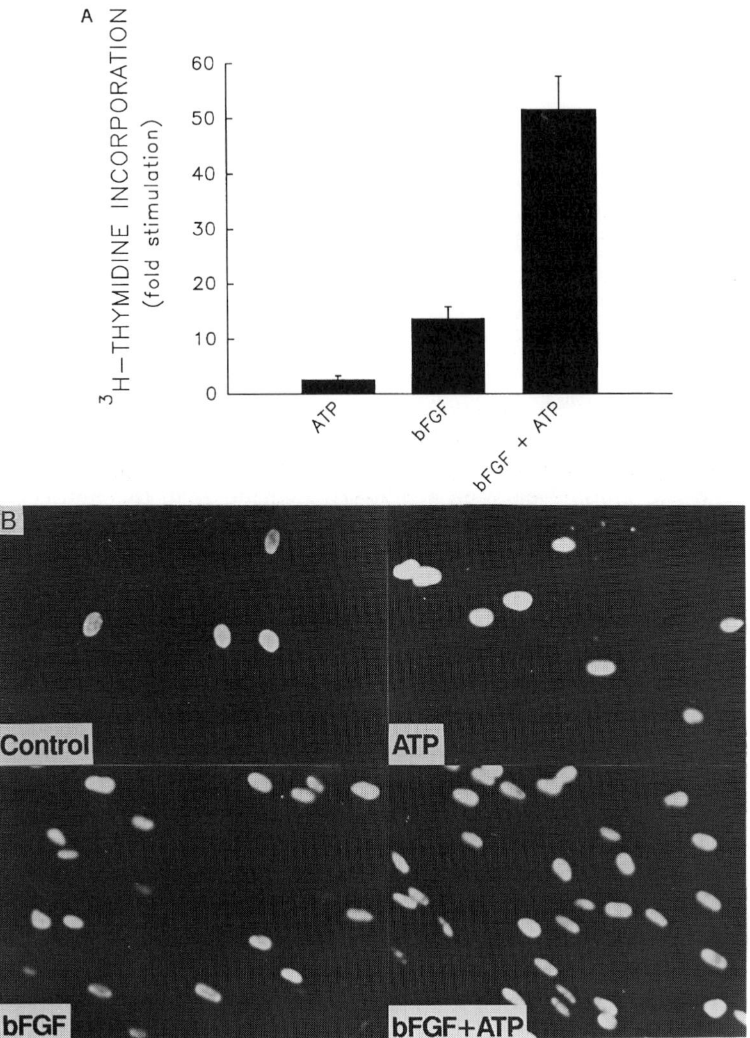

FIG. 1. Synergistic activation of DNA synthesis by extracellular ATP and FGF-2. (A) [^{3}H]thymidine incorporation. ATP (100 μM), FGF-2 (50 ng/ml), or ATP (100 μM) + FGF-2 (50 ng/ml) were added to quiescent cultures of rat cerebral cortical astrocytes. After 18 h, [^{3}H]thymidine (0.5 μCi/ml; 73 Ci/mmol) was added for an additional 4 h, and [^{3}H]thymidine incorporation was measured as previously described (Neary et al 1994b). Values given are the means $\pm$ SEM from a minimum of three independent experiments, each of which was conducted in triplicate or quadruplicate. (B) BrdU incorporation. Quiescent cultures were treated with the same concentrations of ATP and FGF-2 as in (A), and the number of cells incorporating BrdU was compared with the total number of cells in each field, as previously described (Neary et al 1994b). In the fields shown above, the total numbers of cells were approximately equal; the results shown are representative of seven independent experiments.

phosphorylates MAP kinases. Because of the important role of MAP kinases in cell growth and because of the mitogenic and morphogenic effects of ATP on astrocytes, we hypothesized that P2 purinoceptor stimulation may activate MAP kinases in astrocytes. We found that treatment of primary cultures of rat cerebral cortical astrocytes with 100 μM ATP for 15 min caused a three- to fourfold increase in MAP kinase activity (Neary & Zhu 1994). Time course and dose–response studies revealed that activation was rapid (1.5 min), peaked at 10–15 min and declined to near baseline by 1 h; activation was also noted at 1 μM ATP with maximum stimulation at 100 μM ATP. Stimulation of MAP kinase activity by ATP was mediated by P2 purinoceptors because the effect was inhibited by a P2 purinoceptor antagonist (suramin) but not by a P1 purinoceptor antagonist [8-(para-sulfonphenyl)-theophylline]; moreover, adenosine, 2-chloroadenosine or cyclohexyladenosine were ineffective in stimulating MAP kinase activity in astrocytes. Activation was observed with 2-methylthioATP and UTP, but not α,β-methylene ATP or ITP, thereby suggesting that $P2Y_1$ (P2Y) and $P2Y_2$ (P2U) purinoceptors are linked to MAP kinase in astrocytes (Neary et al 1995).

Which signalling elements couple P2 purinoceptors to MAP kinase in astrocytes? A role for protein kinase C (PKC) in coupling the P2 purinoceptor to MAP kinase signalling pathways was indicated because inhibition of PKC by Ro 31-8220 or down-regulation of PKC by chronic phorbol ester treatment markedly reduced the ability of extracellular ATP to activate MAP kinase (Neary et al 1995). As shown in Fig. 2, treatment of astrocytes with ATP followed by cell lysis and anion exchange chromatography on a Resource Q column (Pharmacia Biotech) resulted in two peaks of MAP kinase activity. The earliest-eluting peak consisted mainly of MAP kinase isoform p42 and the second peak contained the p44 isoform, as identified by immunoblotting with an anti-MAP kinase antibody (Santa Cruz Biotechnology). To down-regulate PKC, we treated cells for 24 h with 100 nM 12-*O*-tetradecanoylphorbol 13-acetate (TPA) (Neary et al 1988) prior to application of extracellular ATP; under these conditions, activation of both MAP kinase isoforms by extracellular ATP was greatly reduced (Fig. 2). These findings indicate that PKC is upstream of MAP kinase in the P2 purinoceptor signalling pathway.

FGF-2 also activates MAP kinase in astrocytes, but it does so by a mechanism distinct from that of extracellular ATP (Neary & Zhu 1994). For example, the activation of astrocytic MAP kinase by FGF-2 is mediated by a Ras/Raf pathway, because inhibition of Raf by protein kinase A blocked the activation of MAP kinase by FGF-2. However, activation of protein kinase A did not significantly reduce the stimulation of MAP kinase by extracellular ATP. Preliminary results from studies conducted in collaboration with Dr J. Avruch and colleagues at the Massachusetts General Hospital (Boston, MA) suggest that FGF-2 does indeed activate Raf-1 in astrocytes, whereas neither Raf-1, B-Raf, nor MEK kinase appear to be appreciably activated by

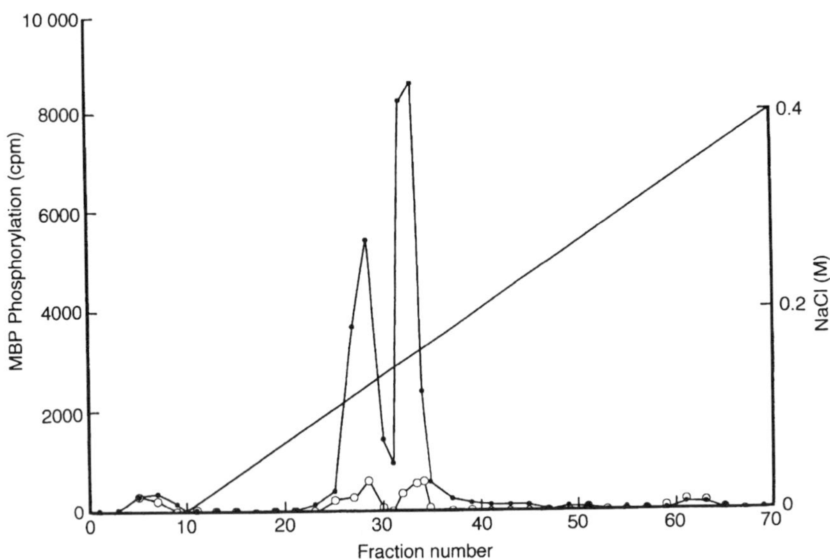

FIG. 2. Stimulation of mitogen-activated protein (MAP) kinases by extracellular ATP and reduction in protein kinase C-depleted astrocytes. Quiescent cultures of rat cerebral cortical astrocytes were treated with 100 μM ATP for 15 min (●—●) or with 100 nM TPA for 24 h prior to application of 100 μM ATP for 15 min (○—○). Homogenates were centrifuged, and supernatants containing equivalent amounts of protein were applied to 1 ml Resource Q (Pharmacia Biotech) anion exchange column. Fractions (1 ml) were collected at a flow rate of 0.5 ml/min, and proteins were eluted with a linear, 60 ml NaCl gradient (0–400 mM). Fractions were assayed for MAP kinase activity using myelin basic protein (MBP) as substrate as previously described (Neary & Zhu 1994). Similar results were obtained by chromatography on a Hi-Trap Q (Pharmacia Biotech) anion exchange column.

extracellular ATP. This suggests the involvement of a novel MEK activator in the P2 purinoceptor/MAP kinase signalling pathway in astrocytes.

The trophic effects of extracellular ATP are likely to be mediated by changes in gene expression. Are there downstream signalling elements that link the P2 purinoceptor/MAP kinase pathway to alterations in gene expression? MAP kinases have been reported to activate transcription factors (Pulverer et al 1991, Gille et al 1992), DNA binding proteins that are stimulated by numerous extracellular stimuli, including growth factors. These proteins convert the transient signals generated by the stimulation of cell surface receptors into long-term changes in gene expression. c-Fos and c-Jun are two well-studied transcription factors whose expression is increased following CNS injury (Morgan et al 1987, Dragunow et al 1990, de Felipe et al 1993). Fos and Jun families of transcription factors can form functional, heterodimeric transcription complexes known as AP (activator protein)-1 (Angel & Karin 1991), which are also increased following experimental brain injury (Dash et al 1995). GFAP, a marker for astrocyte differentiation and gliosis, contains in its promoter a binding site of AP-1 complexes (Masood et al 1993). To determine if extracellular ATP could activate AP-1 complexes in astrocytes, we conducted gel mobility shift assays with nuclear extracts from astrocytes treated with extracellular ATP (Neary et al 1994c, Zhu et al 1995). Figure 3 shows that exposure of cultures to 100 μM ATP results in a marked increase in AP-1 complexes. The DNA–protein interaction is sequence-specific because the shifted bands are markedly reduced when excess unlabelled AP-1 oligonucleotide

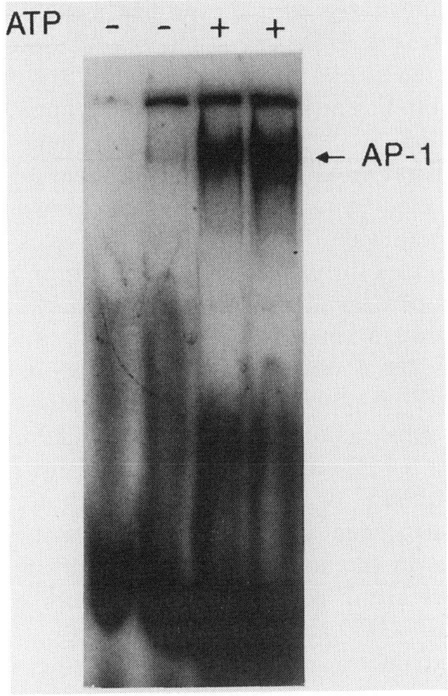

FIG. 3. Extracellular ATP increases AP-1 complex formation in astrocytes. Quiescent cultures of rat cerebral cortical astrocytes were treated with or without 100 μM ATP for 1 h. Cells were then lysed, nuclei were isolated, nuclear extracts were prepared, and AP-1 binding was measured by gel mobility shift assays using ^{32}P-labelled AP-1 consensus oligonucleotide (Promega Corp.) as previously described (Dash et al 1995). The samples shown in the autoradiogram are (from left to right) [^{32}P]AP-1 consensus oligonucleotide alone; nuclear extract from untreated cells plus [^{32}P]AP-1 consensus oligonucleotide; duplicate lanes of nuclear extracts from ATP-treated cells plus [^{32}P]AP-1 consensus oligonucleotide. The arrow indicates the position of the protein/AP-1 consensus oligonucleotide complex. Equivalent amounts of protein were applied in each condition. The results shown are representative of 10 independent experiments.

is included in the binding assay, whereas SP-1 and AP-2 oligonucleotides do not significantly affect the binding. Time course studies reveal that AP-1 binding is increased at 10 min, is maximal at 1 h, and declines thereafter. One of the proteins in the AP-1 heterodimer complex was identified as c-Fos by Shift–Western blotting. FGF-2 also stimulates AP-1 complexes in astrocytes, although the maximum effect is observed at 3 h rather than at 1 h. The formation of AP-1 complexes evoked by extracellular ATP and FGF-2 is consistent with the observations of Abbracchio et al (1995), who have shown that levels of c-Fos and c-Jun are increased by α,β-methylene ATP and by FGF-2.

To see whether the increases in AP-1 complexes stimulated by ATP and FGF-2 are mediated by induction of protein synthesis or by a post-translational mechanism, we used the protein synthesis inhibitor, cycloheximide (Zhu et al 1995). Cycloheximide reduces ATP-evoked AP-1 binding at 10 min and 1 h by 25% and 38%, respectively. By contrast, FGF-2 does not stimulate AP-1 binding at 10 min, but at 1 h cycloheximide reduces FGF-2-evoked AP-1 binding by 80%. These findings indicate that whereas the FGF-2-induced increase in AP-1 is mainly dependent on protein synthesis, the ATP-evoked AP-1 binding is mediated by both protein-synthesis-dependent and -independent pathways. This suggests that part of the ATP-evoked increase in AP-1 complex formation is due to activation by a post-translational mechanism such as protein phosphorylation, perhaps as catalysed by MAP kinases or related protein kinases.

On the basis of these observations, the signal transduction pathways outlined in Fig. 4 are proposed as a working hypothesis for the mechanisms that mediate the trophic actions of extracellular ATP and for the synergistic interactions between the P2 purinoceptor and FGF receptor pathways. In this model, stimulation of $P2Y_1$ or $P2Y_2$ purinoceptors leads to activation of MAP kinases by a PKC-dependent signalling pathway; in turn, the stimulation of these kinases results in the activation of functional transcription complexes with subsequent increase in gene expression and cell growth. As mentioned previously, ATP and FGF-2 activate the MAP kinase cascade in astrocytes by distinct signal transduction mechanisms. This, together with the utilization of different mechanisms for stimulating the formation of AP-1 transcriptional complexes, may underlie the synergistic enhancement of FGF-2-induced mitogenesis by extracellular ATP.

The trophic actions of extracellular ATP, alone or in combination with polypeptide growth factors such as FGF-2, may have important implications in CNS development and/or injury. The role of astrocytes in axonal guidance is well established, and the release of ATP, perhaps from cells programmed to die, may influence this process. Regarding brain injury, the role of gliosis in regeneration is unclear. Several reports have suggested that reactive astrocytes and/or the glial matrix form an impediment to neuronal sprouting, while others have suggested that it may enhance the regeneration process (for reviews see

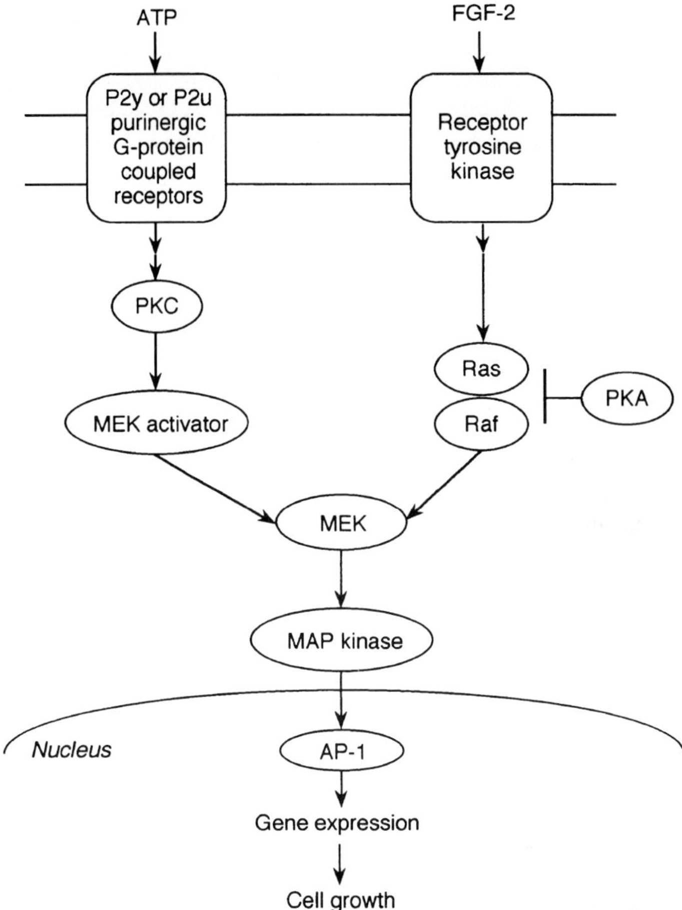

FIG. 4. A model for P2 purinoceptor and FGF-2 receptor signal transduction mechanisms underlying the trophic actions of extracellular ATP and synergistic interaction with FGF-2 in cerebral cortical astrocytes. P2 purinoceptor and FGF-2 receptor pathways converge at the MAP kinase cascade. It should be noted that Ras is recruited to the membrane upon receptor activation but is offset here for purposes of illustration. The vertical bar from PKA to Ras/Raf represents an inhibitory pathway. As indicated in the text, ATP stimulates AP-1 formation by both protein-synthesis-dependent and -independent pathways whereas FGF-2-evoked AP-1 complex formation occurs mainly by protein synthesis; this difference, perhaps together with distinct mechanisms upstream of MAP kinase, may contribute to the synergism between the P2 purinoceptor and FGF-2 receptor signalling pathways. MEK, mitogen-activated protein kinase kinase; MAP, mitogen-activated protein; AP, activator protein; PKC, protein kinase C; PKA, protein kinase A.

Eng et al 1987, Hatten et al 1991, Norenberg 1994). It appears that although astrocytes are needed for axonal growth and guidance, an excess of reactive astrocytes may inhibit neuronal regeneration. An understanding of the signalling mechanisms through which extracellular ATP acts, either alone or in concert with other trophic factors, may provide insights into the pathogenesis of neural injury and may suggest a novel approach for promoting neuronal regeneration.

Acknowledgements

The author is grateful to Drs M. D. Norenberg, S. W. Whittemore, P. Dash, J. Avruch, Q. Zhu, and to L. White, S. Jorgensen, C. Peterson and K. Akong for their contributions to the studies described here, to Dr G. Lawton, Roche Research Centre, Hertfordshire, UK, for the generous gift of Ro 31-8220, and to Judy Neary for assistance with computer-aided graphics. This work was supported by the Department of Veterans Affairs.

References

Abbracchio MP, Saffrey MJ, Höpker V, Burnstock G 1994 Modulation of astroglial cell proliferation by analogues of adenosine and ATP in primary cultures of rat striatum. Neuroscience 59:67–76

Abbracchio MP, Ceruti S, Saffrey MJ et al 1995 Activation of P2-purinoceptors induces differentiation of astroglial cells in rat brain primary cultures. Pharmacol Res 31:195S

Ahn NG, Seger R, Bratlien RL, Krebs EG 1992 Growth factor-stimulated phosphorylation cascades: activation of growth factor-stimulated MAP kinase. In: Interactions among cell signalling systems. Wiley, Chichester (Ciba Found Symp 164) p 113–131

Angel P, Karin M 1991 The role of Jun, Fos and the AP-1 complex in cell-proliferation and transformation. Biochim Biophys Acta 1072:129–157

Avruch J, Zhang X-F, Kyriakis JM 1994 Raf meets Ras: completing the framework of a signal transduction pathway. Trends Biochem Sci 19:279–283

Dash PK, Moore AN, Dixon CE 1995 Spatial memory deficits, increased phosphorylation of the transcription factor CREB, and induction of the AP-1 complex following experimental brain injury. J Neurosci 15:2030–2039

Davis RJ 1993 The mitogen-activated protein kinase signal transduction pathway. J Biol Chem 268:14553–14556

deFelipe C, Jenkins R, O'Shea R, Williams TSC, Hunt SP 1993 The role of immediate early genes in the regeneration of the central nervous system. Adv Neurol 59:263–271

Dragunow M, Goulding M, Faull RLM, Ralph R, Mee E, Frith R 1990 Induction of c-*fos* mRNA and protein in neurons and glia after traumatic brain injury: pharmacological characterization. Exp Neurol 107:236–248

Eng LF, Reier PJ, Houle JD 1987 Astrocyte activation and fibrous gliosis: glial fibrillary acidic protein immunostaining of astrocytes following intraspinal cord grafting of fetal CNS tissue. Prog Brain Res 71:439–455

Gille H, Sharrocks AD, Shaw PE 1992 Phosphorylation of transcription factor p62[TCF] by MAP kinase stimulates ternary complex formation at c-fos promoter. Nature 358:414–417

Hatten ME, Liem RKH, Shelanski ML, Mason CA 1991 Astroglia in CNS injury. Glia 4:233–243

Lai K-M, Wong PCL 1991 Metabolism of extracellular adenine nucleotides by cultured rat brain astrocytes. J Neurochem 57:1510–1515
Masood K, Bresnard F, Su Y, Brenner M 1993 Analysis of a segment of the human glial fibrillary acidic protein gene that directs astrocyte-specific transcription. J Neurochem 61:160–166
Morgan JI, Cohen DR, Hempstead JL, Curran T 1987 Mapping patterns of c-*fos* expression in the central nervous system after seizure. Science 237:192–197
Neary JT, Norenberg MD 1992 Signaling by extracellular ATP: physiological and pathological considerations in neuronal–astrocytic interactions. Prog Brain Res 94:145–151
Neary JT, Zhu Q 1994 Signaling by ATP receptors in astrocytes. NeuroReport 5: 1617–1620
Neary JT, Norenberg LOB, Norenberg MD 1988 Protein kinase C in primary astrocyte cultures: cytoplasmic localization and translocation by a phorbol ester. J Neurochem 50:1179–1184
Neary JT, Baker L, Jorgensen SL, Norenberg MD 1994a Extracellular ATP induces stellation and increases GFAP content and DNA synthesis in primary astrocyte cultures. Acta Neuropathol 87:8–13
Neary JT, Whittemore SR, Zhu Q, Norenberg MD 1994b Synergistic activation of DNA synthesis in astrocytes by fibroblast growth factor and extracellular ATP. J Neurochem 63:490–494
Neary JT, Zhu Q, Bruce JH, Moore AN, Dash PK 1994c Synergistic activation of AP-1 complexes by ATP and FGF-2 in astrocytes. Soc Neurosci Abstr 20:1501
Neary JT, Zhu Q, Akong K, Peterson C 1995 Signaling from P2-purinoceptors to MAP kinase in astrocytes involves protein kinase C. Soc Neurosci Abstr 21:581
Neary JT, Rathbone MP, Cattabeni F, Abbracchio MP, Burnstock G 1996 Trophic actions of extracellular nucleotides and nucleosides on glial and neuronal cells. Trends Neurosci 19:13–18
Norenberg MD 1994 Astrocyte responses to CNS injury. J Neuropath Exp Neurol 53:213–220
Pulverer BJ, Kyriakis JM, Avruch J, Nikolakaki E, Woodgett JR 1991 Phosphorylation of c-*Jun* mediated by MAP kinases. Nature 353:670–674
Rathbone MP, Middlemiss PJ, Kim J-L et al 1992 Adenosine and its nucleotides stimulate proliferation of chick astrocytes and human astrocytoma cells. Neurosci Res 13:1–17
Wang D, Huang N-N, Heppel LA 1990 Extracellular ATP shows synergistic enhancement of DNA synthesis when combined with agents that are active in wound healing or as neurotransmitters. Biochem Biophys Res Commun 166:251–258
Zhu Q, Dash PK, Neary JT 1995 Extracellular ATP-evoked activation of AP-1 complexes in astrocytes. Soc Neurosci Abstr 21:582

DISCUSSION

Di Virgilio: We have also seen co-mitogenic effects of ATP in human lymphocytes stimulated with anti-CD_3 antibodies. Under our conditions, however, UTP was completely inactive. For this reason—and for a number of related reasons I'll mention in my paper (Di Virgilio et al 1996, this volume)—we believe that it is not the G protein-coupled receptor that has this co-stimulatory effect, but that it may be an ionotropic receptor.

Neary: Would you predict that it is a P2X receptor?

Di Virgilio: In fibroblasts, these ionotropic receptors haven't yet been characterized, although it is known that mouse fibroblasts are depolarized by ATP. I don't know if ionotropic P2 receptors exist at all in astrocytes.

Surprenant: In electrophysiological experiments we've observed an endogenous P2X receptor in fibroblasts from some peripheral ganglia (unpublished results). We've not examined fibroblasts from other tissues.

Neary: As indicated, we have evidence that P2Y and P2U receptors are coupled to the MAP kinase pathway in astrocytes, but when we try to stimulate P2X receptors with α,β-methylene ATP we don't see any activation of MAP kinase.

Burnstock: But there are some P2X purinoceptors that are totally unresponsive to α,β-methylene ATP.

Kennedy: In collaboration with Drs Anne Graham and Robin Pleuin, we have found that ATP and UTP are equipotent at activating MAP kinase in the EAhy 926 endothelial cell line. The conclusion we came to was that it was a P2U receptor.

Edwards: In practice, do you think that this sort of thing can happen in the CNS? ATP doesn't hang around: ectonucleotidases break it down very rapidly, so much so that you can't get a response by putting ATP on a slice.

Neary: In astrocyte cultures, the breakdown of ATP and ADP to AMP and adenosine by ectonucleotidases is a relatively slow process (Lai & Wong 1991, Neary et al 1994), whereas MAP kinase is stimulated within a minute by extracellular ATP.

Edwards: It would have to be a subsecond stimulation to avoid this problem.

Westfall: How do you know ATP lasts that long?

Edwards: It may be that the ATP is not the transmitter that is going to activate these P2X receptors at all. But assuming that it is, putting ATP into the slice you get a slight increase in noise and no current. When we try to pump it on fairly locally, so it's a fast but not a very fast application, it is broken down fast enough so that we don't get any current at all.

Westfall: You may not get any current, but you don't know for sure that it is being broken down.

Edwards: But α,β-methylene ATP does give a current, so the receptors are present. Also, we have evidence that adenosine is being produced, which suggests that ATP is being produced and broken down.

Westfall: We often assume that these compounds are not there because we lose the response, but without actually doing specific measurements you have to be careful about the conclusions you draw.

Edwards: It's also possible that ectonucleotidases are not expressed everywhere or, for that matter, in all stages of development. It may be that you only really get them coexpressed with P2X receptors, for example, where the cell would die if you had lots of ATP around.

Burnstock: That's a very important point. I wouldn't be surprised to find, when people finally do the analysis, that the ectoATPase enzymes involved and/or their levels are different in fast synapses with P2X purinoceptors from slower events where there are P2Y purinoceptors. Nobody has looked at this possibility systematically yet.

References

Di Virgilio F, Ferrari D, Falzoni S et al 1996 P2 purinoceptors in the immune system. In: P2 purinoceptors: localization, function and transduction mechanisms. Wiley, Chichester (Ciba Found Symp 198) p 290–305

Lai K-M, Wong PCL 1991 Metabolism of extracellular adenine nucleotides by cultured rat brain astrocytes. J Neurochem 57:1510–1515

Neary JT, Baker L, Jorgensen SL, Norenberg MD 1994 Extracellular ATP induces stellation and increases glial fibrillary acidic protein content and DNA synthesis in primary astrocyte cultures. Acta Neuropathol 87:8–13

Trophic roles of P2 purinoceptors in central nervous system astroglial cells

Maria P. Abbracchio, Stefania Ceruti, Chiara Bolego, Lina Puglisi, Geoffrey Burnstock* and Flaminio Cattabeni

*Facoltà di Farmacia, Istituto di Scienze Farmacologiche, Università di Milano, Via Balzaretti 9, I-20133 Milan, Italy and *Department of Anatomy and Developmental Biology, University College London, Gower Street, London WC1E 6BT, UK*

In the nervous system, purines have well characterized functions as neurotransmitters and neuromodulators (Burnstock 1993) that are mediated by specific receptors for nucleosides (P1 purinoceptors) or nucleotides (P2 purinoceptors) (Abbracchio & Burnstock 1994, Fredholm et al 1994). However, growing evidence supports another fundamental role for these compounds as trophic factors and endogenous regulators of cell growth and differentiation in both development and adulthood. Both ATP and adenosine are believed to participate in brain responses to trauma and ischaemia, including regeneration of damaged neurites, activation of microglial cells and reactive astrogliosis (Neary & Norenberg 1992, Rathbone et al 1992, Abbracchio et al 1994, Neary et al 1996). Here we discuss the experimental evidence implicating P2 purinoceptors in the induction of reactive astrocytes.

Astrocytes, reactive astrogliosis and P2 purinoceptors

Astroglial cells play key roles in normal brain function and development, and in the pathology of the nervous system (for reviews, see Eddleston & Mucke 1993, Abbracchio et al 1995a). Astrocytes respond to various types of injury with vigorous and rapid astrogliosis, a reaction characterized by both increased astrocytic proliferation and cellular hypertrophy, as shown by increased cellular size and presence of longer and thicker astrocytic processes which stain intensively for the astroglial-specific marker, glial fibrillary acidic protein (GFAP). Although there is still debate as to the exact significance of this phenomenon, the functional importance of reactive astrogliosis in post-traumatic brain recovery has been recently underlined by the demonstration that activated astrocytes are needed for axonal regrowth and guidance *in vitro* (Hatten et al 1991) and that suppression of GFAP expression by antisense mRNA also obliterates the formation of stable astrocytic processes in response to neuronal signals (Weinstein et al 1991). In keeping with a possible role for

ATP in the onset and maintenance of reactive astrogliosis, we have investigated the functional effects of the relatively hydrolysis-resistant ATP analogue α,β-methylene ATP (α,β-meATP) on astroglial cells in mixed neuron–glia primary cultures of rat striatum. Exposure of cells to this compound results in concentration-dependent stimulation of DNA synthesis and consequent increase of astroglial cell number (Abbracchio et al 1994). These effects are blocked by the P2 purinoceptor antagonist suramin, suggesting a role for these receptors in astroglial cell proliferation. Exposure of cells to α,β-meATP also results in morphological changes to astrocytes, as shown by marked 'stellation' of GFAP-positive cells, in agreement with previous reports by other authors (Neary et al 1994a). Since emission and elongation of GFAP-positive processes is believed to represent an index of the astrocytic differentiation known to occur in every type of astrogliosis, even in the absence of astrocytic proliferation (e.g. fetal anoxia, Hatten et al 1991), we have further characterized this effect in comparison with known triggers for astrogliosis, such as the polypeptide growth factor fibroblast growth factor 2 (FGF-2; also known as basic fibroblast growth factor, bFGF).

Exposure of cultures to either FGF-2 or α,β-meATP in a serum-free medium results in concentration-dependent increase of the mean length of GFAP-positive processes (Abbracchio et al 1995b). Interestingly, the maximal effect obtained with the ATP analogue was quantitatively comparable with the maximal effect induced by the polypeptide growth factor. Exposure of cultures of another ATP analogue (2-methylthioATP) also results in significant elongation of astrocytic processes, although to a lesser extent than that induced by α,β-meATP. Suramin specifically antagonizes α,β-meATP-induced elongation of processes without affecting the action of FGF-2, suggesting that the effects induced by the purine analogues are indeed mediated by activation of P2 purinoceptors (Abbracchio et al 1995b). Brief challenge of cultures with either agent rapidly and transiently induces the nuclear accumulation of both Fos and Jun (Abbracchio et al 1995c), two well-studied transcription factors that have been shown to increase in response to brain injury (see Neary et al 1996 and references therein). Also in this case, suramin specifically antagonizes α,β-meATP induction of primary response genes without affecting FGF-2 responses (C. Bolego, S. Ceruti, L. Puglisi, F. Cattabeni & M. P. Abbracchio, unpublished results). Together, these results suggest that both ATP and polypeptide growth factors can promote the maturation of astrocytes towards a more differentiated phenotype characterized by the presence of longer astrocytic processes; induction of primary response genes participates in the cascade of cellular events activated by the two agents.

These data are also consistent with results obtained by other authors. Neary et al (1994b) have demonstrated that ATP and FGF-2 are both mitogenic for rat cortical astrocytes in culture. More recently, ATP was also shown to activate astrocytic mitogen-activated protein (MAP) kinases via P2 purinoceptors (Neary & Zhu 1994). The MAP kinase cascade is a key element of signal transduction pathways involved in cellular proliferation and differentiation by polypeptide growth factors (for reviews, see Davis 1993, Avruch et al 1994). MAP kinases have been reported to regulate Jun by post-translational phosphorylation (Pulverer et al 1991) and to also affect c-*fos* induction via phosphorylation of p62, a key transcription factor that, together with the serum response factor (SRF), participates to the activation of the serum-response element (SRE) of the c-*fos* promoter (Gille et al 1992). Both these mechanisms could contribute to the induction of primary response genes by these agents and to the formation of AP-1 complexes which in turn stimulate long-term changes in gene expression mediating the gliotic response to these agents. To support this hypothesis, ATP and FGF-2 have been shown to induce the formation in astrocytes of AP-1 complexes, functional DNA binding protein heterodimers consisting of Fos and Jun families of transcription factors (Neary et al 1994c).

Conclusions

On the basis of these results, a picture emerges for astrocytic activation where both 'classical' polypeptide growth factors (e.g. FGF-2) and 'novel' trophic agents (e.g. ATP), following interaction with their specific extracellular receptors, can activate distinct and independent transduction pathways that merge at the MAP kinase cascade (Fig. 1). Tyrosine kinase receptors activate the Ras/Raf kinase pathway; conversely, ATP signalling to MAP kinase has been suggested to involve a P2Y-like purinoceptor and to be mediated by MEK kinase (Neary & Zhu 1994), like other G protein-coupled receptors (Lange-Carter et al 1993). Subsequent activation of MAP kinase may lead to rapid and transient induction of primary response genes (e.g. c-*fos* and c-*jun*), which in turn may regulate late-response genes mediating long-term phenotypic changes of astroglial cells (Fig. 1). The GFAP promoter contains a binding site for AP-1 complexes; it may be speculated that the above mechanisms are the means by which both ATP and FGF-2 increase GFAP expression in astrocytes. In addition, ATP-induced gliosis may involve modulation of growth factor synthesis and/or release, on the basis of preliminary indications in astrocytoma cells showing that suppression of FGF-2 expression by antisense cDNA transfection also abolishes α,β-meATP-induced elongation of GFAP-positive processes (S. Ceruti, G. Redekop, C. C. G. Naus, F. Cattabeni & M. P. Abbracchio, unpublished results). A full elucidation of the molecular mechanisms responsible for ATP-induced formation of reactive astrocytes and of the interactions with polypeptide growth factors may disclose new exciting opportunities to regulate the gliotic response and thereby influence the outcome of neurological conditions in stroke patients.

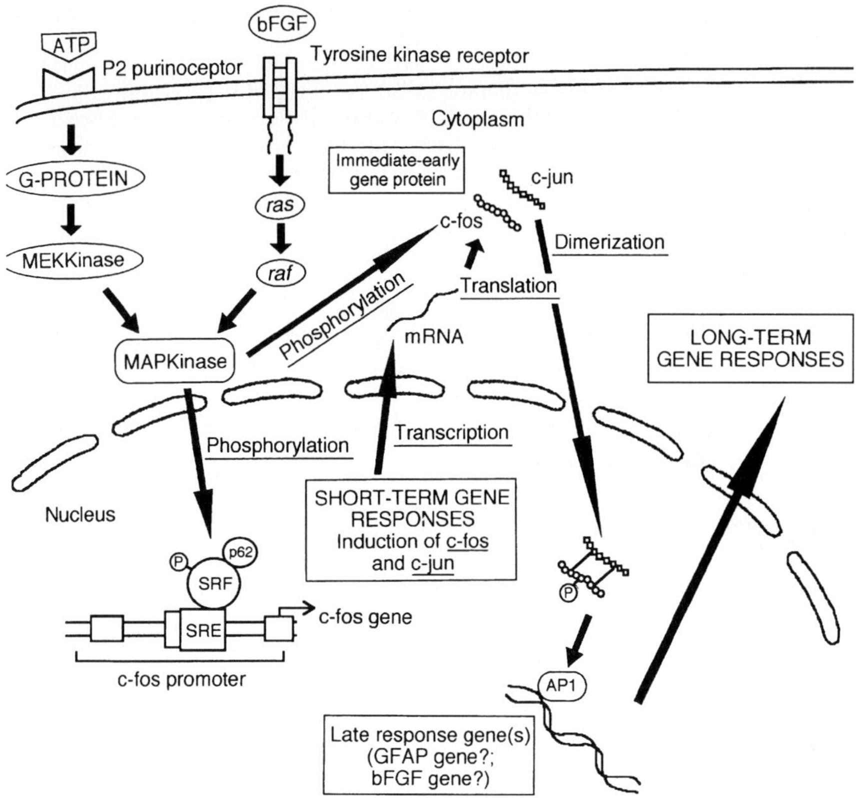

FIG. 1. Proposed pathways for astrocytic activation by polypeptide growth factors and ATP. P2 purinoceptor and tyrosine kinase receptor transduction pathways merge at the MAP kinase cascade, which regulates primary response genes both by promoting their transcription and by post-translational phosphorylation. Induction of c-*fos* and c-*jun* can lead to the formation of active protein heterodimers (the AP-1 complexes) which, by binding to specific DNA sequences, can in turn regulate the transcription of late-response genes mediating long-term phenotypic changes of astroglial cells (e.g. GFAP and FGF-2 [bFGF] genes). See text for further details.

References

Abbracchio MP, Burnstock G 1994 Purinoceptors: are there families of P_{2X} and P_{2Y} purinoceptors? Pharmacol Ther 64:445–475

Abbracchio MP, Saffrey MJ, Höpker V, Burnstock G 1994 Modulation of astroglial cell proliferation by analogues of adenosine and ATP in primary cultures of rat striatum. Neuroscience 59:67–76

Abbracchio MP, Ceruti S, Burnstock G, Cattabeni F 1995a Purinoceptors on glial cells of the central nervous system: functional and pathological implications. In: Belardinelli L, Pelleg A (eds) Adenosine and adenine nucleotides: from molecular biology to integrative physiology. Kluwer Acad, Norwell, MA, p 271–280

Abbracchio MP, Ceruti S, Langfelder R, Cattabeni F, Saffrey MJ, Burnstock G 1995b Effects of ATP analogues and basic fibroblast growth factor on astroglial cell differentiation in primary cultures of rat striatum. Int J Dev Neurosci 13: 685–693

Abbracchio MP, Ceruti S, Saffrey MJ et al 1995c Activation of P2-purinoceptors induces differentiation of astroglial cells in rat brain primary cultures. Pharmacol Res 31:195S

Avruch J, Zhang X-F, Kyriakis JM 1994 Raf meets Ras: completing the framework of a signal transduction pathway. Trends Biochem Sci 19:279–283

Burnstock G 1993 Physiological and pathological roles of purines: an update. Drug Dev Res 28:195–206

Davis RJ 1993 The mitogen-activated protein kinase signal transduction pathway. J Biol Chem 268:14553–14556

Eddleston M, Mucke L 1993 Molecular profile of reactive astrocytes—implications for their role in neurologic diseases. Neuroscience 54:15–36

Fredholm BB, Abbracchio MP, Burnstock G et al 1994 Nomenclature and classification of purinoceptors. Pharmacol Rev 46:143–156

Gille H, Sharrocks AD, Shaw PE 1992 Phosphorylation of transcription factor $P62^{TCF}$ by MAP kinase stimulates ternary complex formation at c-*fos* promoter. Nature 358:414–417

Hatten ME, Liem RKH, Shelanski ML, Mason CA 1991 Astroglia in CNS injury. Glia 4:233–243

Lange-Carter CA, Pleiman CM, Gardner AM, Blumer KJ, Johnson GL 1993 A divergence in the MAP kinase regulatory network defined by MEK kinase and Raf. Science 260:315–319

Neary JT, Norenberg MD 1992 Signaling by extracellular ATP: physiological and pathological considerations in neuronal–astrocytic interactions. Prog Brain Res 94:145–151

Neary JT, Zhu Q 1994 Signaling by ATP receptors in astrocytes. NeuroReport 5:1617–1620

Neary JT, Baker L, Jorgensen SL, Norenberg MD 1994a Extracellular ATP induces stellation and increases glial fibrillary acidic protein content and DNA synthesis in primary astrocytic cultures. Acta Neuropathol 87:8–13

Neary JT, Whittemore SR, Zhu Q, Norenberg MD 1994b Synergistic activation of DNA synthesis in astrocytes by fibroblast growth factors and extracellular ATP. J Neurochem 63:490–494

Neary JT, Zhu Q, Bruce JH, Moore AN, Dash PK 1994c Synergistic activation of AP-1 complexes by ATP and bFGF in astrocytes. Soc Neurosci Abstr 29:1501

Neary JT, Rathbone MP, Cattabeni F, Abbracchio MP, Burnstock G 1996 Trophic actions of extracellular nucleotides and nucleosides on glial and neuronal cells. Trends Neurosci 19:13–18

Pulverer BJ, Kyriakis JM, Avruch J, Nikolakaki E, Woodgett JR 1991 Phosphorylation of c-*Jun* mediated by MAP kinases. Nature 353:670–674

Rathbone MP, DeForge S, DeLuca B et al 1992 Purinergic stimulation of cell division and differentiation: mechanisms and pharmacological implications. Med Hypoth 37:213–219

Weinstein DE, Shelanski ML, Liem R 1991 Suppression by antisense mRNA demonstrates a requirement for the glial fibrillary acidic protein in the formation of stable astrocytic processes in response to neurons. J Cell Biol 112:1205–1213

DISCUSSION

Burnstock: We are trying to explain long-term trophic effects, which are very different from short-term physiological effects, in terms of the purinoceptors that we now know. It is just possible that there are totally different kinds of purinoceptors involved in trophic events that have longer-lasting attachments. I am not convinced that it is wise to try and interpret trophic events only in terms of the purinoceptors that we currently know are involved in fast physiological events.

North: Activation of seven transmembrane domain receptors need not be thought of only in terms of fast physiological events (time-scale of seconds). There are other examples where the agonists have similar long-term trophic or mitogenic effects.

Burnstock: They might be the same, but I think we are jumping to conclusions too quickly.

Surprenant: Along those lines, it seems that you are showing evidence of a P2X mediation. How do you see an ion channel turning on genes?

Abbracchio: Actually, even though it is responsive to α,β-meATP, I'm not convinced that this is necessarily a P2X receptor. In doing further pharmacological characterization of this receptor we were able to show responses to 2-methylthioATP as well. More importantly, we are now trying to understand if the purinoceptor involved is an ion channel or a G protein-linked receptor. This is a major point that requires clarification.

Surprenant: My understanding is that α,β-meATP is one of the few 'clean' drugs in the purinoceptor field. At least it doesn't seem to activate anything else.

Abbracchio: Another reason to rule out the involvement of P2X purinoceptors in such trophic effects is that there is no definite evidence for P2X receptors on astrocytes, whereas there is definite evidence for the expression of G protein-coupled P2Y and P2U receptors.

Kennedy: There are plenty of examples of G protein-coupled receptors that can interact with mitogenic pathways via the interaction of $\beta\gamma$ subunits of G proteins with Ras (see Clapham & Neer 1993).

Burnstock: I don't find it difficult to accept the idea that G protein-mediated receptors are involved in trophic events, but I find it more difficult for the P2X receptors because these are involved in transmission, i.e. opening channels for fast activities. You don't need that kind of mechanism for trophic responses, although I realize that my argument is rather teleological.

Neary: Generally speaking, the MAP kinases are not activated by Ca^{2+}. There may be one or two cases where their activity is enhanced by Ca^{2+}, but the MAP kinases are distinct from those kinases that are directly activated by second messengers such as Ca^{2+} or cAMP. Thus, you wouldn't need to invoke a Ca^{2+} channel pathway to stimulate MAP kinase.

Reference

Clapham DE, Neer EJ 1993 New roles for G protein $\beta\gamma$ dimers in transmembrane signalling. Nature 365:403–406

Transduction mechanisms of P2Z purinoceptors

J. S. Wiley, J. R. Chen, M. S. Snook, C. E. Gargett and G. P. Jamieson

Haematology Department, Austin Hospital, Studley Road, Heidelberg, Victoria 3084, Australia

Abstract. The ability of extracellular ATP to increase the cation permeability of a variety of fresh and cultured cells has been known for decades, but evidence of a separate class of P2 purinoceptor, termed P2Z, which mediates this effect has only recently been obtained. Several features of the P2Z purinoceptor clearly distinguish it from other P2 purinoceptors and show that it is a ligand-gated ion channel. P2Z purinoceptors are highly selective for the ATP^{4-} species and addition of Mg^{2+} in excess over ATP closes the channel. The most potent agonist is 3′-*O*-(4-benzoyl)benzoyl ATP which has a 10-fold lower EC_{50} than ATP. Ca^{2+} is the preferred permeant for the P2Z ion channel although it will pass ions up to the size of ethidium$^+$(314 Da) in lymphocytes or fura-2 (813 Da) in macrophages. The inhibitors of the P2Z purinoceptor or its associated ion channel include suramin, amiloride analogues, high extracellular Na^+ concentrations and 2′,3′-dialdehyde ATP (oxidized ATP), which blocks irreversibly. Occupancy of P2Z purinoceptors stimulates a phospholipase D activity, which may be involved in membrane remodelling. Moreover, extracellular ATP causes loss of the glycosylated adhesion molecule L-selectin from the surface of human lymphocytes by enzymic cleavage, suggesting a possible role for P2Z purinoceptors in intercellular interactions.

1996 P2 purinoceptors: localization, function and transduction mechanisms. Wiley, Chichester (Ciba Foundation Symposium 198) p 149–165

Extracellular ATP has been shown to induce a wide range of biological responses in a variety of cell types, including cells of haemopoietic origin. Initial studies of the effects of extracellular ATP showed it was able to permeabilize cells as diverse as Ehrlich ascites tumour cells (Hempling et al 1969), dog erythrocytes (Parker & Snow 1972) and renal tubular epithelial cells (Rorive & Kleinzeller 1972). Indeed, much of the early interest in extracellular ATP concerned its use to change the intracellular ion composition and nucleotide content of susceptible cells (Dubyak & El-Moatassim 1993, Heppel et al 1985, Weisman et al 1989a). However, after Burnstock proposed the concept of P2 purinoceptors (Burnstock 1978, Burnstock & Kennedy 1985), the permeabilizing effect of ATP was gradually recognized to be mediated by a specific P2 purinoceptor subclass known as P2Z or the ATP^{4-} receptor (Cockcroft & Gomperts 1980, Gordon 1986). This P2Z purinoceptor subtype is a receptor-operated ion channel with a wide distribution in cells of haemopoietic origin, such as macrophages (Sung et al 1985, Steinberg & Silverstein 1987, Greenberg et al 1988), mast cells (Tatham & Lindau 1990), fibroblasts (Weisman et al 1989b), lymphocytes from both mouse and human (Wiley & Dubyak 1989, El-Moatassim et al 1989, Pizzo et al 1991), dog erythrocytes (Parker & Snow 1972), erythroleukaemia cells (Chahwala & Cantley 1984), as well as parotid and salivary gland acinar cells (Soltoff et al 1992, Sasaki & Gallacher 1990). However, in the majority of these cell types, other P2 purinoceptors (either P2Y or P2U) are co-expressed with the P2Z purinoceptor.

There are few studies which have characterized the P2Z purinoceptor, since few cell types express the P2Z subtype in isolation. An exception is the lymphocyte, which shows an ATP-induced increase in cytosolic Ca^{2+} that depends on the presence of extracellular Ca^{2+}, indicating that ATP opens a ligand-gated ion channel (Fig. 1). Moreover, lymphocytes show no ATP-induced release of internal Ca^{2+} (a P2Y effect) and no response to UTP, ADP or α,β-methylene ATP—agonists which are specific for the P2U, P2T and P2X purinoceptors, respectively. These results confirm that lymphocytes possess only P2Z purinoceptors and are an ideal cell type to study the characteristics of this receptor and its associated ion channel. A second advantage of lymphocytes is that large numbers can be obtained from patients with chronic lymphocytic leukaemia (CLL), in whom lymphocyte counts attain values up to 100×10^9/L in peripheral blood (i.e. 25- to 100-fold normal) due to clonal proliferation and accumulation of mature cells of B phenotype (i.e. surface immunoglobulin-positive, $CD19^+$, $CD20^+$ and $CD5^+$). Lymphocytes from peripheral blood of normal subjects also express P2Z purinoceptors but at a much lower level than in lymphocytes from most patients with CLL (Wiley & Dubyak 1989). Increasing maturity of cells in the monocytic lineage also leads to greater expression of P2Z purinoceptors. Thus peripheral blood monocytes possess few P2Z purinoceptors, but *in vitro* culture of these cells for 7–14 days to mimic their maturation into macrophages up-regulates P2Z purinoceptor expression (Hickman et al 1994).

Agonists for the P2Z purinoceptor

It is well established that the P2Z purinoceptor expressed on mast cells, macrophages and lymphocytes requires the ATP^{4-} species as an agonist (Cockroft & Gomperts 1980, Steinberg & Silverstein 1987, Wiley et al 1990). Thus acidification of extracellular pH reduces the ATP-dependent cation influx by converting ATP^{4-} to $HATP^{3-}$ species. Also the channel formed by ATP^{4-}

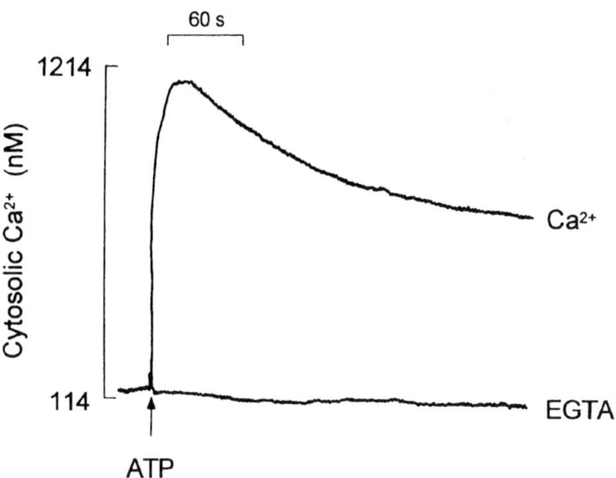

FIG. 1. ATP-induced increase in cytosolic Ca^{2+} of lymphocytes. Fura-2 loaded cells were suspended at 2×10^6/ml in isotonic KCl medium containing either 1.0 mM Ca^{2+} or 0.1 mM EGTA. Addition of 1.0 mM ATP is as indicated (arrow).

closes rapidly after the addition of Mg^{2+}, which forms a Mg–ATP^{2-} complex, thereby decreasing the concentration of ATP^{4-}. The P2Z purinoceptor shows a rank order of agonist potency that clearly distinguishes it from other P2 purinoceptors. In a murine macrophage cell line (expressing P2Y, P2U and P2Z), 3′-*O*-(4-benzoyl)benzoyl ATP (BzATP) powerfully stimulated influx of Ca^{2+} from the medium without triggering release of internal Ca^{2+} and this effect was attributed to a specific occupancy of P2Z purinoceptors (El-Moatassim & Dubyak 1992). BzATP is also a far more potent agonist than ATP for P2Z purinoceptors in human lymphocytes, and produced twofold greater maximal Ba^{2+} influx compared with ATP (Fig. 2). Indeed, BzATP stimulated Ba^{2+} influx with an EC_{50} of 8 μM, compared with 89 μM for ATP, in lymphocytes suspended in K^+ media (Wiley et al 1994). BzATP has also been reported to be greater than 10-fold more potent than ATP in stimulating $^{45}Ca^{2+}$ influx via the P2Z purinoceptor of rat parotid acinar cells (Soltoff et al 1992). In nearly all studies of the P2Z purinoceptor, the increment in permeant fluxes produced by extracellular ATP showed a sigmoid dependence on ATP concentration and Hill analysis gave '*n*' values between 2 and 2.5 (Fig. 2). This cooperative effect of ATP in stimulating permeant influx suggests that ATP must bind to multiple sites on the receptor in order to open the channel.

Selectivity of the P2Z purinoceptor ion channel

In lymphocytes, macrophages and mast cells, the P2Z ion channel conducts a wide range of inorganic and organic cations. In lymphocytes, ATP increased

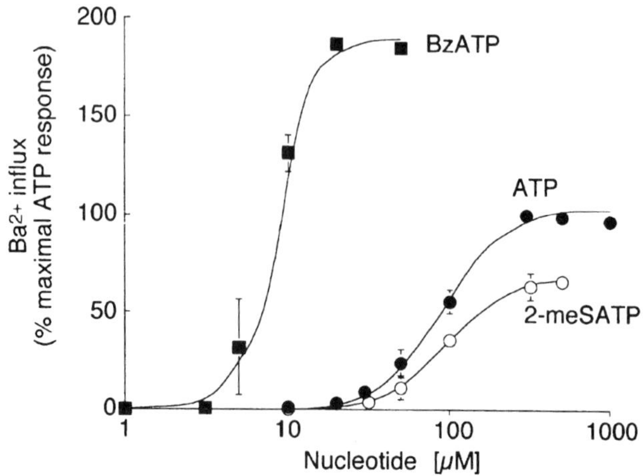

FIG. 2. Dose–response curves of Ba^{2+} influx in human lymphocytes for various nucleotides and analogues. The increment of Ba^{2+} influx in isotonic KCl medium produced by the nucleotides (●) ATP, (■) BzATP and (○) 2-methylthioATP was expressed as a percentage of the maximal response to 1.0 mM ATP in KCl media which was defined as 100% response. Maximal rates of Ba^{2+} influx and EC_{50}s for agonists were calculated by non-linear regression analysis. Means ± SEM from experiments on three donors are shown.

the fluxes of K^+, Rb^+, Na^+, Li^+, Ca^{2+}, Sr^{2+}, Ba^{2+} and Mn^{2+} as well as the weak bases $NH_4{}^+$, monomethylamine$^+$ and dimethylamine$^+$ (Wiley & Dubyak 1989, Wiley et al 1990, 1992, 1993, Pizzo et al 1991, El-Moatassim et al 1990, Chen et al 1994). In the macrophage cell line J774, ATP stimulated the efflux of K^+ and Rb^+, influx of Na^+ and Ca^{2+} and uptake of the fluorescent molecules lucifer yellow (anionic, 457 Da) and fura-2 (anionic, 636 Da) (Steinberg & Silverstein 1987, Steinberg et al 1987). Mast cells and fibroblasts also develop large pores on exposure to high concentrations of ATP with loss of both nucleotides and phosphorylated intermediates as well as entry of solutes of up to 900 Da (Tatham & Lindau 1990, Saribas et al 1993). In comparison, human lymphocytes show an upper limit of 314–414 Da for cations which traverse the ATP-induced channel (Wiley et al 1993). Figure 3 shows that lymphocytes in a KCl medium were impermeable to ethidium$^+$ (314 Da) but after addition of ATP the ethidium$^+$ was taken up at a rate which was linear over 5 min. In contrast, lymphocytes were impermeable to a larger fluorescent dye, propidium^{2+} (414 Da), in either the absence or the presence of ATP at concentrations as high as 5.0 mM. It is uncertain whether these differences between lymphocyte and mast cell or macrophage reflect different isoforms of the P2Z ion channel or whether there are different conductance states of the channel in different cells. The latter possibility was proposed for P2Z in mast cells (Tatham & Lindau 1990) and is strongly supported by recent voltage-clamp experiments by Dubyak and colleagues on *Xenopus* oocytes injected with macrophage mRNA (Nuttle & Dubyak 1994). In these cells BzATP, a selective agonist for the expressed P2Z purinoceptors, rapidly activated an inward current which could be carried by small inorganic cations such as Na^+, Li^+ or K^+. Within one minute, however, a further steady increase in conductance was observed and this second phase of increased conductance could be carried by larger organic cations (*N*-methyl glucamine$^+$ or Tris$^+$) as well as Na^+, Li^+ or K^+. This second phase of BzATP-induced current increase was inhibited by temperature reduction to 20 °C or by addition of amiloride analogues (see below). Whether this transition from a smaller to a greater conductance state ('channel to pore transition') involves physical association of monomeric subunits to form an oligomeric pore or whether some enzymic activity is involved is presently unknown.

Whatever the size pass of permeants for P2Z channel, it is clear that it shows considerable selectivity for Ca^{2+} over monovalent cations. Thus ATP-induced whole-cell currents in human B-lymphocytes studied by the voltage-clamp technique showed a permeability sequence $P_{Ca} : P_K : P_{Na} = 35:2:1$ (Bretschneider et al 1995). We have used an entirely different technique, time-resolved flow cytometry, to show that Ca^{2+} has a higher apparent affinity than other cations for the P2Z ion channel. The ATP-induced influx of ethidium$^+$ was measured by flow cytometry into CLL lymphocytes suspended in KCl media containing other permeant cations, either Ca^{2+}, Sr^{2+}, Ba^{2+} (each 0.1–2.0 mM) or Na^+ (1–20 mM). Since these divalent ions complex with ATP^{4-}, the total ATP added was adjusted to maintain free ATP^{4-} at 230 μM (Brooks & Storey

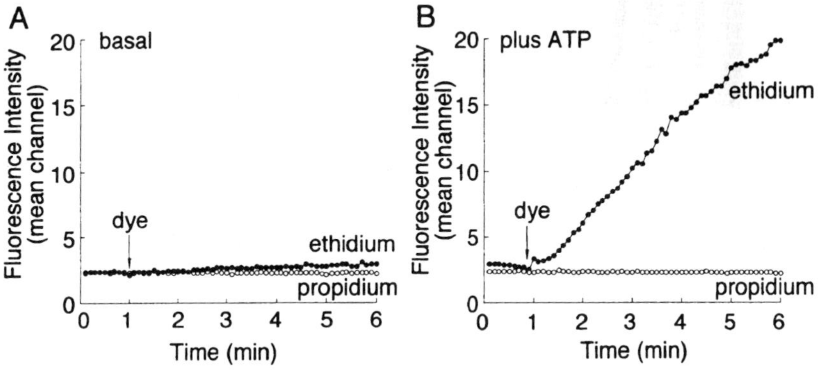

FIG. 3. Extracellular ATP-induced uptake of ethidium$^+$ in lymphocytes measured by flow cytometry. Cells were suspended at 10^6/ml and mean cell-associated fluorescence intensity was measured at 6 s intervals. Cells were incubated for 5 min at 37 °C prior to flow cytometric analysis: (a) without any addition to KCl medium; and (b) with 0.5 mM ATP added to KCl medium. Ethidium bromide (25 μM) or propidium iodide (25 μM) were added as indicated by arrow.

1992), a concentration sufficient to give near maximal response. Figure 4 shows the marked inhibition of ATP-induced ethidium$^+$ uptake observed on increasing the extracellular Ca^{2+}, with half-maximal inhibition occurring at 0.5 mM $CaCl_2$ (300 μM free ionized Ca^{2+}). Control experiments showed that fluorescence output by ethidium from digitonin-permeabilized cells was not influenced by the presence of 2 mM $CaCl_2$ in the medium. Ba^{2+}, Sr^{2+} and Na^+ also inhibited ATP-induced ethidium$^+$ influx (Fig. 4) with half-maximal inhibitory concentrations of 1.5 mM Ba^{2+}, 1.5 mM Sr^{2+} and 15 mM Na^+. This approach yields apparent inhibitory constants for these cations for the ATP-induced ethidium influx and it is striking that the ratio of inhibitory constants for $Ca^{2+} : Na^+ = 30 : 1$, which is consistent with the channel being selective for Ca^{2+} over monovalent cations.

Inhibition by extracellular Na^+

The entry of various permeants through the P2Z purinoceptor ion channel is attenuated or absent in NaCl media compared with Na^+-free media such as KCl, choline chloride or *N*-methyl glucamine chloride. This inhibitory effect of extracellular Na^+ has been shown for both human leukaemic lymphocytes (Wiley et al 1992, 1993, 1994, Chen et al 1994) and mouse thymocytes (Pizzo et al 1991), and with the various permeants—Ca^{2+}, Ba^{2+}, Rb^+, NH_4^+ and ethidium$^+$. Some of this inhibitory effect of Na^+ may result from competition between Na^+ ions and other ions for passage through the channel. However, Na^+ may also interact with modifying sites on the P2Z purinoceptor to inhibit channel opening or occupancy of the receptor.

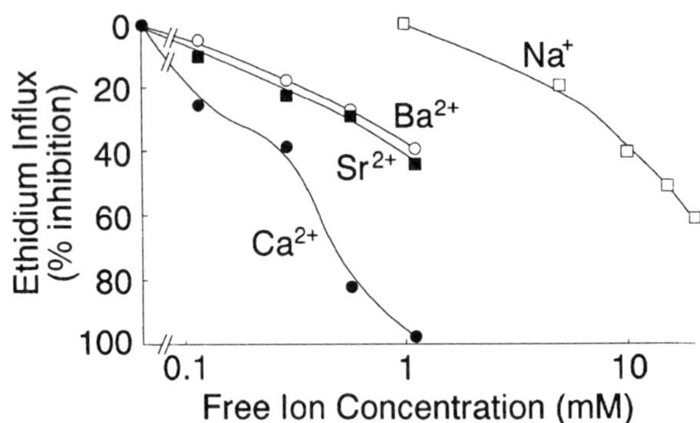

FIG. 4. Inhibition of ATP induced ethidium$^+$ uptake by extracellular Ca^{2+}, Sr^{2+}, Ba^{2+} (0.1–2.0 mM) or Na^+ (1–20 mM). CLL lymphocytes were incubated with ATP (maintaining ATP^{4-} at 230 μM) for 5 min at 37 °C prior to flow cytometric analysis. Ethidium bromide (25 μM) was added and its initial rate of influx was measured by flow cytometry. Percentage inhibition of influx was calculated relative to ATP-induced ethidium influx in isotonic KCl medium without other competing cations.

Inhibitors of the P2Z purinoceptor and its channel

The most widely used inhibitors of P2 purinoceptors are suramin and reactive blue 2, which are sulfonic acid derivatives of naphthalene and anthrone, respectively. Numerous studies have documented their inhibitory effects on P2 purinoceptors of all subclasses, but it is notable that their inhibitory potency towards lymphocyte P2Z purinoreceptors is weak, with IC_{50} values of 60 μM for suramin and 70 μM for reactive blue 2, respectively (Wiley et al 1993). Although suramin—which has a history as a trypanosomal agent dating back to 1916—has been used in many studies, it is notoriously non-specific and blocks the interaction of many growth factors with their receptors (e.g. fibroblast growth factor 2, platelet-derived growth factor, interleukin 3) as well as inhibiting a wide variety of enzymes, including ectoATPase (Stein 1993, Crack et al 1994). Recent clinical trials have demonstrated some promise for suramin against prostate cancer, but its lack of specificity does not allow P2 purinoceptors to be implicated in the responses.

Among the most potent inhibitors of the P2Z purinoceptor and/or its ionic channel are the amiloride analogues such as 5-(*N*, *N*-hexamethylene) amiloride (HMA) or 5-(*N*-ethyl-*N*-isopropyl) amiloride, which inhibit permeant fluxes in CLL lymphocytes as well as in *Xenopus* oocytes injected with macrophage mRNA (Wiley et al 1990, 1992, Nuttle & Dubyak 1994). In general, amiloride analogues even at high concentrations (40–100 μM) block only 70–85% of the flux of monovalent or divalent cations through the P2Z channel, despite having inhibitor constants (IC_{50}s) of around 2–10 μM (Wiley et al 1990, 1993). This incomplete inhibition of permeant fluxes has been attributed to the P2Z channel existing in two conductance states; an initial small conductance channel insensitive to amiloride analogues and a pore of larger conductance whose formation is time-dependent and can be blocked by these analogues (Nuttle & Dubyak 1994). Another useful reagent is the 2′,3′-dialdehyde derivative of ATP, also called oxidized ATP, which reacts with lysine residues at the ATP binding site to form a covalent Schiff base. This reaction has been demonstrated to give total and irreversible inhibition of P2Z purinoceptors on mouse macrophages while 2′,3′-dialdehyde ATP was without effect on either P2Y or P2U purinoceptors present on these cells (Murgia et al 1993). In contrast 5′-fluorosulfonylbenzoyladenosine, another irreversible inhibitor of many ATP-requiring enzymes, inhibited only 60–90% of the ion fluxes mediated by P2Z purinoceptor agonists (Wiley et al 1994).

Activation of phospholipase D by P2Z purinoceptor agonists

A new role for P2Z purinoceptors has been suggested by the work of El-Moatassim & Dubyak (1992, 1993) who described the activation of phospholipase D by agonists for the P2Z purinoceptor. Phospholipase D hydrolyses plasma membrane phosphatidylcholine to produce choline and phosphatidic acid which may act as a second messenger as well as having a role in membrane remodelling. Although phospholipase D can be activated in many cell types by a range of receptor agonists it was previously considered that this activation was indirect and resulted from the agonist activating the phosphoinositide signalling pathway, and producing 2,3-diacylglycerol and inositol-1,4,5-trisphosphate. Indeed, the ability of phorbol esters and Ca^{2+} ionophores to activate phospholipase D in diverse cell types supported the concept that phospholipase D activation involved protein kinase C and/or an elevation of cytosolic Ca^{2+} levels. However, Dubyak and colleagues showed that BzATP, a specific agonist for P2Z purinoceptors, stimulated a rapid and large increase in phosphatidylcholine-selective phospholipase D in a murine macrophage cell line, BAC1.2F5, and this stimulation was independent of the presence of extracellular Ca^{2+}. Various agonists were used to exclude the involvement of other P2 purinoceptor subtypes and inhibitors of P2Z purinoceptors (e.g. 2′,3′-dialdehyde ATP) were shown to abolish the effect of BzATP or ATP on phospholipase D activity. Under some conditions the activation of phospholipase D could be dissociated from the opening of the ionic channel by P2Z agonists. Thus when macrophages were suspended in isosmotic choline, BzATP failed to activate phospholipase D, although this agonist opened the P2Z ionic channel (El-Moatassim & Dubyak 1993). Presumably, under these conditions choline exerted end-product inhibition on phospholipase D activity.

Using human leukaemic lymphocytes (which only have the P2Z purinoceptor subclass), our laboratory has confirmed the activation of phospholipase D by ATP or BzATP. Inclusion of proven P2Z purinoceptor antagonists such as suramin (300 μM) or a high-Na^+ medium inhibited ATP-activated phospholipase D by 80% or more, while pretreatment of lymphocytes with 2′,3′-dialdehyde ATP irreversibly abolished ATP-induced phospholipase D activity (Gargett et al 1996). Thus, in both macrophages and lymphocytes, agonists for P2Z purinoceptors stimulate phospholipase D by a mechanism which is as yet unknown. However, at least in macrophages, simple occupancy of the P2Z purinoceptor by agonist appears sufficient to activate phospholipase D. Murine macrophages exhibit high levels of a phospholipase D directed towards the phosphatidylinositol moiety of a complex glycolipid, glycosylphosphatidyl inositol (GPI), which anchors some proteins into the plasma membrane (Xie & Low 1994). However, it is not known whether it is this GPI-specific form of phospholipase D which can be activated by agonists for the P2Z purinoceptor and which may function as an ecto-enzyme.

P2Z agonists down-regulate L-selectin on lymphocytes

Leukocytes possess an array of adhesion molecules which mediate their reversible interactions with other cell types such as endothelial cells. The initial step in these interactions involves leukocyte selectins, termed L-selectins (CD62L), which form low-affinity bonds to carbohydrate epitopes on endothelial cells as a first step in the migration of leukocytes through the vascular wall (Ager 1994). L-selectins are constitutively expressed on most classes of lymphocytes of both B- and T-lineage (Springer 1995) as well as on B-lymphocytes from patients with CLL. Exposure of lymphocytes to the phorbol ester 12-*O*-tetradecanoylphorbol 13-acetate down-regulates L-selectin within minutes (Prystas et al 1993, Spertini et al 1991). However, the physiological stimulus for L-selectin modulation is not known.

Our laboratory has shown that addition of ATP (500 μM) to normal peripheral blood lymphocytes or CLL lymphocytes caused a rapid loss of L-selectin from these cells that is largely complete within 5 min at 37 °C (Fig. 5). In contrast, ADP, adenosine or UTP gave no loss of L-selectin (Fig. 5). The characteristics of this ATP-induced selectin loss indicated that this effect was mediated by P2Z purinoceptors. First, addition of molar excess of Mg^{2+} over ATP abruptly halted further loss of L-selectin. Moreover, L-selectin loss required a concentration of 50–100 μM ATP, which is equivalent to the ATP concentrations that activate P2Z purinoceptors. ATP-induced loss of L-selectin was inhibited in media with high Na^+ concentrations compared with Na^+-free media. Finally, BzATP was a more potent agonist than ATP in

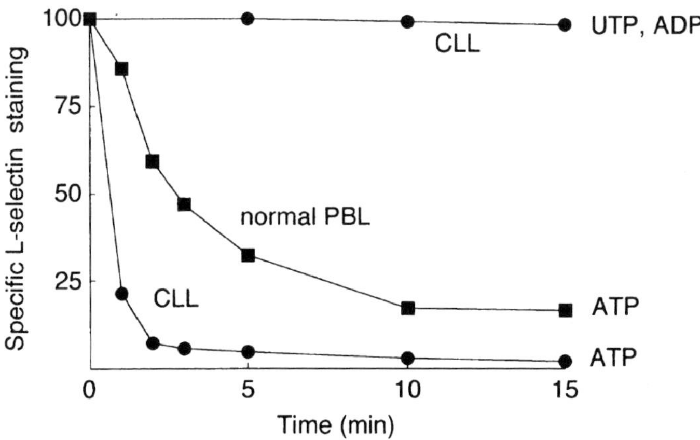

FIG. 5. Effect of ATP on loss of L-selectin from normal peripheral blood lymphocytes (PBL) and chronic lymphocytic leukaemia (CLL) lymphocytes. Cells were incubated with 1 mM ATP, UTP or ADP for up to 15 min at 37 °C in isotonic KCl media plus 1 mM Ca^{2+}. Cells were taken at time intervals between 0 and 15 min and immediately mixed with four volumes of medium containing 2 mM $MgCl_2$. Cells were then incubated with monoclonal antibody to L-selectin followed by fluorescein-conjugated sheep anti-mouse antibody. The mean channel of cell-associated fluorescence was then measured by flow cytometry.

causing L-selectin loss, while UTP, ADP and adenosine were ineffective (G. P. Jamieson, P. J. Thurlow & J. S. Wiley, unpublished results 1995).

How occupancy of P2Z purinoceptors leads to loss of L-selectin is unknown. One possibility is that L-selectin is anchored to the membrane by a GPI linkage and that ATP-stimulated phospholipase D converts the membrane-anchored form into soluble L-selectin. It is therefore of interest that many cells of haemopoietic origin express a GPI-specific phospholipase D which may function as an ecto-enzyme to cleave cell-associated GPI-anchored proteins (Xie & Low 1994). However, there is only limited evidence that leukocyte L-selectin is GPI-anchored to the plasma membrane. Alternatively, P2Z purinoceptor agonists may activate a protease by its interaction with P2Z purinoceptor. Whatever mechanism is involved, P2Z purinoceptors have a role of modulating the expression of certain adhesion molecules which are important in intercellular interactions.

Acknowledgement

Studies from this laboratory were supported by a grant from the Australian National Health and Medical Research Foundation.

References

Ager A 1994 Lymphocyte recirculation and homing: roles of adhesion molecules and chemoattractants. Trends Cell Biol 4:326–333

Bretschneider F, Klapperstück M, Löhn M, Markwardt F 1995 Nonselective cationic currents elicited by extracellular ATP in human and B-lymphocytes. Pflügers Arch 429:691–698

Brooks SPJ, Storey KB 1992 Bound and determined: a computer program for making buffers of defined ion concentrations. Anal Biochem 201:119–126

Burnstock G 1978 A basis for distinguishing two types of purinergic receptor. In: Straub RW, Bolis L (eds) Cell membrane receptors for drugs and hormones: a multidisciplinary approach. Raven Press, New York, p 107–118

Burnstock G, Kennedy C 1985 Is there a basis for distinguishing two types of P_2 purinoceptor? Gen Pharmacol 16:433

Chen JR, Jamieson GP, Wiley JS 1994 Extracellular ATP increases NH_4^+ permeability in human lymphocytes by opening a P_{2Z} purinoceptor operated ion channel. Biochem Biophys Res Comm 202:1511–1516

Chahwala SB, Cantley LC 1984 Extracellular ATP induces ion fluxes and inhibits growth of friend erythroleukemia cells. J Biol Chem 259:13717–13722

Cockcroft S, Gomperts BD 1980 The ATP^{4-} receptor of rat mast cells. Biochem J 188:789–798

Crack BE, Beukers MW, McKechnie KCW, IJzerman AP, Leff P 1994 Pharmacological analysis of ecto-ATPase inhibition: evidence for combined enzyme inhibition and receptor antagonism in P_{2X} purinoceptor ligands. Br J Pharmacol 113:1432–1438

Dubyak GR, El-Moatassim C 1993 Signal transduction via P_2-purinergic receptors for extracellular ATP and other nucleotides. Am J Physiol 265:577C–606C

El-Moatassim C, Maurice T, Mani JC, Dornand J 1989 The $[Ca^{2+}]_i$ increase induced in murine thymocytes by extracellular ATP does not involve ATP hydrolysis and is not related to phosphoinositide metabolism. FEBS Lett 242:391–396

El-Moatassim C, Mani JC, Dornand J 1990 Extracellular ATP^{4-} permeabilizes thymocytes not only to cations but also to low-molecular-weight solutes. Eur J Pharmacol 181:111–118

El-Moatassim C, Dubyak GR 1992 A novel pathway for the activation of phospholipase D by P_{2Z} purinergic receptors in BAC1.2F5 macrophages. J Biol Chem 267:23664–23673

El-Moatassim C, Dubyak GR 1993 Dissociation of the pore-forming and phospholipase D activities stimulated via P_{2Z} purinergic receptors in BAC1.2F5 macrophages. Product inhibition of phospholipase D enzyme activity. J Biol Chem 268:15571–15578

Gargett CE, Cornish EJ, Wiley JS 1996 Phospholipase D activation by P2Z purinoceptor agonists in human lymphocytes is dependent on divalent cation influx. Biochem J 313:529–535

Gordon JL 1986 Extracellular ATP: effects, sources and fate. Biochem J 233:309–319

Greenberg S, Di Virgilio F, Steinberg TH, Silverstein SC 1988 Extracellular nucleotides mediate Ca^{2+} fluxes in J774 macrophages by two distinct mechanisms. J Biol Chem 263:10337–10343

Hempling HG, Stewart CC, Gasic G 1969 Effect of exogenous ATP on the volume of TA3 ascites tumor cells. J Cell Physiol 73:133–140

Heppel LA, Weisman GA, Friedberg I 1985 Permeabilization of transformed cells in culture by external ATP. J Memb Biol 86:189–196

Hickman SE, Elkhoury J, Greenberg S, Schieren I, Silverstein SC 1994 P2Z adenosine triphosphate receptor activity in cultured human monocyte-derived macrophages. Blood 84:2452–2456

Murgia M, Hanau S, Pizzo P, Rippa M, Di Virgilio F 1993 Oxidized ATP—an irreversible inhibitor of the macrophage purinergic-P_{2Z} receptor. J Biol Chem 268:8199–8203

Nuttle LC, Dubyak GR 1994 Differential activation of cation channels and non-selective pores by macrophage P_{2Z} purinergic receptors expressed in *Xenopus* oocytes. J Biol Chem 269:13988–13996

Parker JC, Snow RL 1972 Influence of external ATP on permeability and metabolism of dog red blood cells. Am J Physiol 56:888–893

Pizzo P, Zanovello P, Bronte V, Di Virgilio F 1991 Extracellular ATP causes lysis of mouse thymocytes and activates a plasma-membrane ion channel. Biochem J 274:139–144

Prystas EM, Parker CJ, Holguin MH, Bohnsack JF 1993 Aberrant glycosylation of L-selectin on the lymphocytes of chronic lymphocytic leukemia. Leukemia 7:1355–1362

Rorive G, Kleinzeller A 1972 The effect of ATP and Ca^{2+} on the cell volume in isolated kidney tubules. Biochem Biophys Acta 274:226–239

Saribas AS, Lustig KD, Zhang X, Weisman GA 1993 Extracellular ATP reversibly increases the plasma membrane permeability of transformed mouse fibroblasts to large macromolecules. Anal Biochem 209:45–52

Sasaki T, Gallacher DV 1990 Extracellular ATP activates receptor-operated cation channels in mouse lacrimal acinar cells to promote calcium influx in the absence of phosphoinositide metabolism. FEBS Lett 264:130–134

Soltoff SP, McMillan MK, Talamo BR 1992 ATP activates a cation-permeable pathway in rat parotid acinar cells. Am J Physiol 262:934C–940C

Spertini O, Freedman AS, Belvin MP, Penta AC, Griffin JD, Tedder TF 1991 Regulation of leukocyte adhesion molecule-1 (TQ1, Leu-8) expression and shedding by normal and malignant cells. Leukemia 5:300–308

Springer TA 1995 Traffic signals on endothelium for lymphocyte recirculation and leukocyte emigration. Annu Rev Physiol 57:827–872

Stein CA 1993 Suramin: a novel antineoplastic agent with multiple potential mechanisms of action. Cancer Research 53:2239–2248

Steinberg TH, Silverstein SC 1987 Extracellular ATP^{4-} promotes cation fluxes in the J774 mouse macrophage cell line. J Biol Chem 262:3118–3122

Steinberg TH, Newman AS, Swanson JA, Silverstein SC 1987 ATP^{4-} permeabilizes the plasma membrane of mouse macrophages to fluorescent dyes. J Biol Chem 262: 8884–8888

Sung SJ, Young JD, Origlio AM, Heiple JM, Kaback HR, Silverstein SC 1985 Extracellular ATP perturbs transmembrane ion fluxes, elevates cytosolic [Ca^{2+}], and inhibits phagocytosis in mouse macrophages. J Biol Chem 260:13442–13449

Tatham PER, Lindau M 1990 ATP-induced pore formation in the plasma-membrane of rat peritoneal mast cells. J Gen Physiol 95:459–476

Weisman GA, Lustig KD, Friederg I, Heppel LA 1989a Permeabilizing mammalian cells to macromolecules. Methods Enzymol 171:857–861

Weisman GA, De BK, Pritchard RS 1989b Ionic dependence of the extracellular ATP-induced permeabilization of transformed mouse fibroblasts: role of plasma membrane activities that regulate cell volume. J Cell Physiol 138:375–383

Wiley JS, Dubyak GR 1989 Extracellular adenosine triphosphate increases cation permeability of chronic lymphocytic leukemic lymphocytes. Blood 73:1316–1323

Wiley JS, Jamieson GP, Mayger W, Cragoe EJ Jr, Jopson M 1990 Extracellular ATP stimulates an amiloride-sensitive sodium influx in human lymphocytes. Arch Biochem Biophys 280:263–268

Wiley JS, Chen JR, Wiley MJ, Jamieson GP 1992 The ATP^{4-} receptor-operated ion channel of human lymphocytes: inhibition of ion fluxes by amiloride analogs and by extracellular sodium ions. Arch Biochem Biophys 292:411–418

Wiley JS, Chen JR, Jamieson GP 1993 The ATP^{4-} receptor-operated channel (P_{2Z} class) of human lymphocytes allows Ba^{2+} and $ethidium^{+}$ uptake: inhibition of fluxes by suramin. Arch Biochem Biophys 305:54–60

Wiley JS, Chen JR, Snook MB. Jamieson GP 1994 P_{2Z} purinoceptor of human lymphocytes: actions of nucleotide agonists and irreversible inhibition by oxidised ATP. Br J Pharmacol 112:946–950

Xie MS, Low MG 1994 Expression and secretion of glycosylphosphatidylinositol-specific phospholipase D by myeloid cell lines. Biochem J 297:547–554

DISCUSSION

Burnstock: It's nice that you finished by talking about the physiological significance of P2Z purinoceptors. Could you expand a little for those of us who are not in the adhesion field? What exactly will it do in practical terms? Is there any therapeutic potential in this?

Wiley: There's at least one study that shows that after leukocytes have transmigrated through the vessel wall, much of their selectin has been lost (Ager & Wood 1994). In other words, it appears that L-selectin down-regulation or shedding is a prerequisite for the movement of white cells out of the circulation into the tissues. Therapeutically, if one were to cause shedding of selectin, you would abolish the leukocyte rolling and thus possibly have an inhibitory effect on the transmigration of leukocytes. This may have an anti-inflammatory effect, but it is premature to suggest a therapeutic application.

Leff: I wondered about the balance between L- and E-selectin function in that situation. My understanding is that E-selectin is involved in leukocyte rolling. Antibody work suggests that you can block rolling under flow with anti-E-selectin antibodies, but I wasn't sure about the L-selectin.

Wiley: The definitive experiment hasn't been done. What you really want to do is to induce L-selectin shedding and then assay it under flow conditions to show with appropriate chemotactic stimuli whether or not the leukocytes will transmigrate. We've done it in a static system, but the problem with using static systems for transmigration is that the dominant molecular interactions are those of the β integrins, such as $\alpha_4\beta_1$ with VCAM-1.

Boeynaems: Did you actually measure the release of soluble L-selectin induced by ATP?

Wiley: Yes, it's actually lost from the surface into the plasma. That correlates with the well known observation that there is a plasma level of L-selectin of about 5–10 μg/ml, which must be a result of continuous shedding of selectin by leukocytes.

Di Virgilio: Is this effect reversible? In other words, will the cultured leukocytes express their selectins again?

Wiley: That's an experiment we're currently doing. One might expect that after the leukocyte has transmigrated, L-selectin will be resynthesized and inserted into the membrane.

Abbracchio: Can you speculate about the mechanism by which Mg^{2+} closes the P2Z channel? Is it similar to what happens to the *N*-methyl-D-aspartate receptor for glutamate?

Wiley: The definitive experiment in lymphocytes hasn't been done. But in the *Xenopus* model, it does look as though Mg^{2+} actually blocks channel conductance of Ba^{2+}, because it partially inhibits the inward current. To some extent this is analogous to the NMDA channel, where Mg^{2+} can actually block ion permeation.

North: You interpret the actions of Mg^{2+} as being complexed in with the ATP. This is your evidence for ATP^{4-} being the preferred ligand. But you can't be sure that Mg^{2+} isn't simply acting by blocking a channel, in which case the interpretation that ATP^{4-} is in some way preferred is groundless.

Wiley: The only evidence is from the *Xenopus* model, and there the pore may be slightly aberrant.

North: It would be very easy to test in *Xenopus*, because you can look at the inward currents and the outward currents: if it's blocking the channel you will see a difference. If it's simply complexing ATP there will be no difference.

Wiley: That is a good point.

North: The reason I raised it is because some P2X receptors expressed heterologously are clearly blocked by Mg^{2+}.

Harden: By what mechanism do you think the activation of phospholipase D occurs? Can you see this activation in cell-free preparations?

Wiley: We've not used cell-free preparations. In terms of the actual mechanism, it's very likely that this is a G protein-coupled system, where the P2Z is coupled through a G protein to phospholipase D, but firm evidence for that is lacking. Another possibility is that the purinoceptor itself has intrinsic phospholipase D activity. I guess we'll know when the primary sequence comes out.

Surprenant: What is the evidence that this effect is mediated through P2Z? Why could this not be mediated through a P2Y-like receptor? It seems your only evidence against involvement of a P2Y type receptor is lack of action of UTP.

Wiley: BzATP activates phospholipase D with an EC_{50} which is many fold lower than for ATP. Other evidence is from inhibitor studies. For example, 2′,3′-dialdehyde ATP, which is without effect on the P2Y or P2U purinoceptors, knocked out 99% of the ATP-stimulated phospholipase D. These two lines of evidence clearly implicate P2Z directly activating the phospholipase D.

Burnstock: Activation of P2Z purinoceptors does keep channels open for a long time compared with P2X purinoceptors.

Surprenant: But how exactly is the phospholipase activated if it is not by increased intracellular Ca^{2+} due to Ca^{2+} entry through the P2Z channel? Presumably the P2Z receptor is some type of ion channel or large 'pore'. Are you suggesting that this ion channel sometimes couples to G proteins and sometimes does not?

Wiley: El-Moatassim & Dubyak (1992) say that simple agonist occupancy of the P2Z receptor stimulates phospholipase D. It's an area that we're working on at the moment, and I think it's a bit premature for me to say anything. But cytosolic Ca^{2+} elevations do not activate phospholipase D. We've incubated human lymphocytes with thapsigargin and raised the bulk cytosolic Ca^{2+} up over 1 μM: in these conditions there is no activation of phospholipase D whatsoever.

Harden: If the activation is anything like the activation of phospholipase D by G protein-coupled receptors, it's going to be very complicated. It's not a straightforward heterotrimeric G protein coupling. There are small molecular weight GTP binding proteins like Arf involved, and it is probably multistep and multicomponent.

Surprenant: But the real problem here is that I thought we were talking about a pore: if it couples to a G protein how can it also be a pore?

Boucher: I believe that many of these ion channels are going to turn out to be multifunctional. Certainly CFTR is a better regulator of Na^+ channels than it is a Cl^- channel in certain systems. It also regulates an anion channel that has many features very similar to the P2Z receptor, that is it regulates an outward rectifying Cl^- channel (ORCC) that is activated by ATP and BzATP. I don't know if the ORCC is inhibited by dimethyl chloride—I'd be curious to know if that binding site has been identified yet in Na^+ proton exchangers. But I think the idea that some of these channels are multifunctional is very likely to be correct.

Wiley: That is the big enigma at the moment: is the P2Z receptor a multifunctional single protein, or is it coupled through small G proteins to other molecules, such as phospholipase D?

Boucher: Coupling to the cytoskeleton is a feature of a number of membrane-associated proteins. It can be very complicated with these pore-forming proteins.

Di Virgilio: These effects are not probably not due to P2Y receptors from lymphocytes. In non-activated human and mouse lymphocytes, we were never able to see any Ca^{2+} released from stores with ATP, UTP or any other nucleotides. On the contrary, these lymphocytes do have internal Ca^{2+} stores that can be released by other agents. Therefore, our impression is that lymphocytes only have a nucleotide-activated channel, not a receptor linked to $InsP_3$ generation.

Leff: Early on in your paper you showed data on Ba^{2+} influx with 2-methylthioATP and ATP itself, showing them to be what looked like partial agonists with respect to BzATP. Did you do interaction experiments to back up the statement that the system only contains P2Z receptors? That's one way of telling whether there is another receptor there or not. Did you test the partial agonists as antagonists of BzATP?

Wiley: No, we've not done that.

Leff: The obvious supplementary question is: have you used ARL67156 to eliminate ectoATPase influence in this system? Do you think it is an influence in this system?

Wiley: The ectoATPase is going to require divalent cations for maximal activity. We've not investigated whether Ba^{2+} stimulates ectoATPase, but I would be very surprised if ectoATPase was active enough to make any difference to the Ba^{2+} influx curves, because the initial Ba^{2+} uptake is over in 30 s after you add the Ba^{2+}. If it was a 5 min incubation it might be a different story, but in 30 s I think the ATPase won't make a great deal of difference.

Barnard: Weren't you using defolliculated oocytes? They don't have ectoATPase to any significant extent (Ziganshin et al 1995).

Wiley: Yes, in our *Xenopus* oocyte work we used totally defolliculated oocytes.

Barnard: What is presumed to be the function of the opening of this large channel in macrophage activity? Is it related to phagocytosis?

Di Virgilio: I don't think there is any definitive evidence for a clear-cut physiological role for this receptor in lymphocytes, macrophages or any immune cell. We have some indications that this pore could be involved in macrophage–macrophage communication, for instance during granulomatosis inflammations. There are some recent data suggesting that it might also be involved in cytokine production (Perregaux & Gabel 1994).

Barnard: How would a large pore be required for cytokine production?

Di Virgilio: I don't think that the link is between the large pore itself and cytokine passage through the membrane. Instead, the link is most likely between the changes in ion homeostasis and activation of the enzyme responsible for maturation of some cytokines, for instance IL-1β.

Starke: Would ATP, acting through the P2Z receptor, release mediators from mast cells?

Di Virgilio: I think so; there are good data supporting this.

Boucher: This may seem like a bizarre question, but it is dominating a lot of the cystic fibrosis research: can P2Z receptors carry ATP?

Wiley: No, I don't believe so.

Di Virgilio: I think they can. There are some early data by Bastien Gomperts showing that when he activates the P2Z receptor of mast cells, ATP also leaves the cell from the cytoplasm (Cockcroft & Gomperts 1979).

Hickman: Tatham & Lindau (1990) also demonstrated that GTP effluxes from mast cells treated with ATP and Gary Weisman's group described efflux of intracellular nucleotides from murine 3T6 cells in response to ATP or BzATP (Gonzalez et al 1989). However, Sung et al (1985) were unable to detect efflux of nucleotides from J774 cells in response to ATP. In mast cells the GTP efflux was dependent on the extracellular ATP concentration used to elicit the response. Higher ATP concentrations were required to allow loss of nucleotides than were required for cation fluxes. There was also a delay of several minutes after addition of high concentrations of ATP before the intracellular nucleotide loss was detected. These observations led to the suggestion that the ATP-induced pores may vary in size within the same cell membrane.

Wiley: It does depend on which cell you are talking about. If it is lymphocytes, the answer is no. If you're talking about mast cells and macrophages, which have a higher size exclusion limit for the P2Z pore, the answer might be yes.

Burnstock: My guess is that there will be more than one P2Z purinoceptor.

Harden: Going back to my question from yesterday: if it takes millimolar concentrations to open the P2Z receptor, what difference does it make that you're not letting ATP out? And, therefore, what is the physiological significance of P2Z receptors *vis-à-vis* the extracellular concentration of ATP if that is the primary regulator?

Wiley: It does take 30 s or more before the pore opens to the extent that ATP can come out of macrophages and mast cells, so the initial flux will consist only of small inorganic cations. The pathway by which ATP exits from endothelial cells, which don't possess P2Z purinoceptors, is much more problematic. It could be through the CFTR or the P glycoprotein—those are two candidates that have been proposed in recent months. The jury is still out on how ATP gets out of these cell types, but certainly it does get out of endothelial cells and will probably be shown to have a physiological significance in terms of selectin shedding by leukocytes.

References

Ager A, Wood A 1994 Downregulation of L-selectin on lymphocytes following binding to high endothelium: a pre-requisite for transendothelial migration? Lymphology 27:187–192

Cockcroft S, Gomperts BD 1979 ATP-induced nucleotide permeability in rat mast cells. Nature 279:541–542

El-Moatassim C, Dubyak GR 1992 A novel pathway for the activation of phospholipase D by P_{2Z} purinergic receptors in BAC1.2F5 macrophages. J Biol Chem 267:23664–23673

Gonzalez FA, Ahmed AH, Lustig KD, Erb L, Weisman G 1989 Permeabilization of transformed mouse fibroblasts by 3′-*O*-(4-benzoyl)benzoyl adenosine 5′-triphosphate and desensitization of the process. J Cell Physiol 139:109–115

Perregaux D, Gabel CA 1994 Interleukin 1β maturation and release in response to ATP and nigericin. J Biol Chem 269:15195–15203

Sung SJ, Young JD, Origlio AM, Heiple JM, Kaback HR, Silverstein SC 1985 Extracellular ATP perturbs transmembrane ion fluxes, elevates cytosolic [Ca^{2+}], and inhibits phagocytosis in mouse macrophages. J Biol Chem 260:13442–13449

Tatham PER, Lindau M 1990 ATP-induced pore formation in the plasma-membrane of rat peritoneal mast cells. J Gen Physiol 95:459–476

Ziganshin AU, Ziganshin LE, King BF, Burnstock G 1995 Characteristics of ecto-ATPase of Xenopus oocytes and the inhibitions of actions of suramin on ATP breakdown. Eur J Physiol 429:412–418

The diverse series of recombinant P2Y purinoceptors

E. A. Barnard, T. E. Webb, J. Simon and S. P. Kunapuli*

*Molecular Neurobiology Unit, Division of Basic Medical Sciences, Royal Free Hospital School of Medicine, Rowland Hill Street, London NW3 2PF, UK and *Department of Physiology, Temple University Medical School, Philadelphia, PA 19140, USA*

Abstract. A cDNA encoding a P2Y purinoceptor was originally cloned from chick brain and the bovine and human homologues have recently been obtained. These are seven-transmembrane-domain polypetides, i.e. G protein-coupled receptors. When activated by agonists, this P2Y receptor mobilizes intracellular Ca^{2+} and has been shown to be coupled to inositol-1,4,5-trisphosphate formation. Its pharmacology has been established in several expression systems, using both ligand binding and functional responses: 2-methylthioATP has the highest potency of nucleotides and derivatives tested, while UTP and α,β-methylene ATP are inactive. This was hence assigned as a new subtype of the pharmacologically defined P2Y receptors, $P2Y_1$. $P2Y_1$ receptors are exceptionally abundant in the brain. A P2U receptor reported by others can be designated $P2Y_2$. Another P2 receptor subtype, $P2Y_3$, now cloned as a cDNA from the brain and expressed in oocytes and in transfected cells, shows a quite different ligand potency profile to the first two. A fourth subtype is expressed primarily in certain haemopoietic cells and in cardiac muscle. A putative fifth subtype is expressed only in T lymphocytes, upon activation. Yet other P2Y subtypes are indicated by recent cloning studies. The amino acid sequences of all of these P2 receptors, while displaying some homology, are strikingly diverse: they form a separate and unusual new family in the G protein-coupled receptor main superfamily.

1996 P2 purinoceptors, localization, function and transduction mechanisms. Wiley, Chichester (Ciba Foundation Symposium 198) p 166–188

The receptors activated by extracellular ATP are, for their most abundant type, metabotropic, in the G protein-coupled class. Receptors whose properties indicate membership in this class were inferred earlier from physiological studies on peripheral tissues, falling into several main categories on the basis of their rank orders of potency for congeners of ATP:

(1) The originally recognized P2Y purinoceptors have an apparent potency order of 2-MeSATP > ATP = ADP, with α,β-meATP, β,γ-meATP and

UTP virtually inactive (where 2-MeSATP is 2-methylthioATP, and α,β- or β,γ-meATP is α,β- or β,γ-methylene ATP). This receptor type was recognized (with some variations in the relative potencies) in a variety of tissues, including most smooth muscles (Burnstock & Kennedy 1985), and arterial, venous and capillary endothelial cells (Allsup & Boarder 1990), endocrine and exocrine glandular cells including hepatocytes (Keppens & De Wulf 1993), and some others. Similar P2Y purinoceptors have also been detected on astrocytes, where they are suggested to mediate stellation in response to extracellular ATP (Neary & Norenberg 1992). We would not now consider all of these cases to represent the same receptor subtype.

(2) The P2T purinoceptor, as found on platelets, other cells of the megakaryocyte lineage and a few other sites, has ADP as its preferred ligand, with ATP or 2-chloroATP (2-Cl-ATP) acting as an apparent antagonist. ADP appears to activate several platelet responses, including shape change, adherence, aggregation, secretion and inhibition of cAMP accumulation (Hourani & Hall 1996, this volume).

(3) The P2U nucleotide receptor possesses the potency order ATP = UTP > ADP >> α,β-meATP, 2-MeSATP (O'Connor et al 1991). This receptor type has been found on airway, hepatic and some other epithelia, vascular smooth muscle and endothelial cells, neutrophils, pituitary cells and elsewhere, and may play a role in wound healing, cell growth and control of some metabolic pathways or the maintenance of vascular tone (O'Connor et al 1991, Keppens & De Wulf 1993, Boarder et al 1995).

(4) There may be a pyrimidine-specific P2 nucleotide receptor, activated by UTP and UDP and not by adenosine-containing nucleotides. Evidence for this type was derived from responses to nucleotides found on a rat glioma cell line by Lazarowski & Harden (1994), who also cite some indications previously reported from other sources.

(5) Dinucleotide or P2D receptors: see Miras-Portugal et al (1996, this volume).

According to a recent consensus sanctioned by the International Union of Pharmacology Committee on Nomenclature (Fredholm et al 1994), the term 'P2Y receptors' now designates the series of metabotropic nucleotide receptors, so that all of these are to be assigned (eventually) as $P2Y_1$ to $P2Y_n$. In parallel, the classification 'P2X receptors' designates the series of ionotropic receptors for extracellular nucleotides, i.e. the cation channels gated by ATP (discussed fully by North 1996 and Surprenant 1996, this volume). The distinction between the P2Y and P2X types is no longer to be made, therefore, on different ligand selectivities as in their original definition, but on the more fundamental basis of the transduction feature employed—an intrinsic ion channel or a G protein-coupled system. This basis for the division brings the receptors for extracellular nucleotides into line with the same duality of transduction that is well established as the basis for primarily classifying the receptors for the 'dual-action' classical transmitters (acetylcholine, γ-aminobutyric acid, glutamate, serotonin and, at least in invertebrates, histamine and dopamine). This recognition for the ATP case and the consequent revised definition of P2X receptors on that basis had already been made in recent years (Barnard 1992a,b, Edwards & Gibb 1993, Abbracchio & Burnstock 1994). For the P2Y receptors, the original meaning of that term (see (1) above) is broadened, therefore, so that all of the subtypes, (including P2U, P2T and P2D) are termed P2Y receptors in the new nomenclature (Fredholm et al 1994). If type (4) above is confirmed as a molecular entity that has no recognition of adenosine nucleotides, then it would be more logical to term the entire P2Y series 'P2Y nucleotide receptors' instead of 'P2Y purinoceptors' with type (4) as another member of that same class.

A series of P2Y receptors is now being identified (from several laboratories) as a set of different gene products with distinct pharmacologies. This will be reviewed here. It is becoming clear from these results that the subtypes of the metabotropic P2 receptors (reviewed by Abbracchio & Burnstock 1994) which have been based on different agonist potency series or selectivity on intact tissues or cells will require re-evaluation. A lack of correspondence of these with the molecular subtypes can arise in many cases, from several sources:

(a) It can be very difficult to dissect out a single pharmacology when the tissue observed contains more than one P2 receptor type, as is now known to be the most usual case. The agonist potency series, or an agonist or antagonist selectivity, found on a given source may not match, therefore, any actual molecular subtype. This requires analyses at the individual sequence level. For example, we have found, with Dr Christian Frelin (Sophia Antipolis, France) (cf. Feolde et al 1995), by cloning the cDNAs that can be derived from cultured capillary endothelial cells (of rat brain), that these cells contain the rat P2U receptor ($P2Y_2$) and a rat $P2Y_1$ receptor inert to UTP (in both cases shown by sequence; Webb et al 1996b) so that on the endothelium the observed nucleotide potency order will be the sum of these two activities.

(b) Ectonucleotidase activity is generally present on tissues and can selectively destroy some agonists, greatly distorting some agonist rank orders (Kennedy & Leff 1995), unless countered. This interference has been avoided in our studies on expressed recombinant P2Y receptors, firstly because in oocyte expression the defolliculation and the absence of Mg^{2+} during assays (as we have used; Simon et al 1995a) together completely remove all of the considerable ectonucleotidase activity of the folliculated oocytes (Ziganshin et al 1995). In any case, the bulk concentration of nucleotides at the oocyte surface has been maintained unchanged, by superfusion and replacement of the recording chamber volume every 6 seconds (Simon et al 1995a), below the response time in oocyte voltage-clamp. Secondly, for the agonist-evoked internal Ca^{2+} release in P2Y receptor-transfected mammalian cells, for the brief transient phase after rapid introduction of the agonist it can be calculated that any enzymic breakdown would be negligible. Measurements reported on such cells of second messenger levels, which are usually accumulated in a long equilibration with the agonist, are, however, prone to this error. Thirdly, in ligand binding assays, which also employ a long exposure to nucleotides, breakdown was prevented (as we have verified) by omitting divalent cations and Na^+ from the assay medium: the Ca^{2+}/Mg^{2+}-independent fraction of ectonucleotidase has an extremely low affinity for the nucleotides ($K_m = 7$ mM for ATP was measured for the case studied by Ziganshin et al [1995]) and even that activity is Na^+-dependent.

(c) Much ATP is released from tissues *in vitro* and from dying cells in cultures and can desensitize all classes of P2Y receptors. A pre-treatment with the phosphatase apyrase to destroy this accummulation in the medium and allow the receptors to recover has been found necessary and effective in our measurements of agonist-evoked Ca^{2+} transients on transfected cells expressing $P2Y_1$ or $P2Y_3$ receptors (Henderson et al 1995, Webb et al 1996a) and likewise in recording inositol-1,4,5-trisphosphate ($InsP_3$) increases (Filtz et al 1994, Parr et al 1994). This treatment should generally be tested for all types of P2Y responses.

The $P2Y_1$ receptor

A cDNA encoding a 362-residue novel protein having the predicted seven-transmembrane-domain (7TM) pattern of all G protein-coupled receptors was originally cloned from embryonic chick brain and confirmed to be a P2Y-type purinoceptor by oocyte expression (Webb et al 1993). The relative responses to agonists were very broadly similar to those of the classically defined P2Y receptor as deduced from tissue responses (see (1) above) and the subtype was designated as $P2Y_1$. The $P2Y_1$ receptor is assigned to the general P2Y receptor series as defined above. In a detailed study of $P2Y_1$ receptor expression (Simon et al 1995a) the agonist potency order in the oocyte was found to be as shown in Table 1. The P2Y antagonists suramin and Reactive blue 2 (RB-2) were both strongly inhibitory. After transfection of the $P2Y_1$ cDNA into COS-7 cells, application of agonists produced transients of internal Ca^{2+} increase which are associated with $InsP_3$ formation (Simon et al 1995a). 2-MeSATP was again the most potent agonist (Fig. 1), with UTP and α,β-meATP inactive and the increase entirely suppressible by suramin. No change in cAMP levels is produced by the agonists.

3′-deoxyATPαS (dATPαS) was found to be a suitable ligand for the study of the receptor–agonist binding in membranes of these cells. It was shown to be a partial agonist (Fig. 1), as are several related ATP derivatives (Table 1). dATPαS can be used labelled with ^{35}S at very high specific activity, has a high affinity (~10 nM at 25 °C) for the $P2Y_1$ receptor, with binding fully displaceable by the agonists and antagonists at this receptor (Table 1), and is more stable than ATP. The relative potencies of ligands in binding, measured by this competition, are (within the experimental errors) approximately parallel to those measured functionally (Table 1). An agreement of the actual values derived from binding on the COS cells and from Cl^- channel coupling in the oocyte would not be expected, since the intrinsic affinity is not necessarily determined in the latter system, the values are from different cell types and, also, because most of the agonists are partial (Table 1).

By cross-hybridization and PCR amplification based upon the chick $P2Y_1$ receptor cDNA (Webb et al 1993), $P2Y_1$ sequences have been isolated from turkey brain (Filtz et al 1994), mouse and rat insulinoma cells (Tokuyama et al 1995), bovine aortic endothelial cells (Henderson et al 1995), rat brain and human erythroid leukaemic (HEL) cells (Ayyanathan et al 1996) and human placenta (Léon et al 1996). The agonist profile was obtained for the expressed bovine $P2Y_1$ receptor and is broadly similar to that of chicken $P2Y_1$, with 2-MeSATP still extremely potent (EC_{50} = 40 nM). However, ADP is about equally potent, in contrast with the 25-fold difference in the chick, thought to be due to a true species difference (14% of amino acids being changed between the two).

The $P2Y_1$ receptor mRNA in the brain (discussed below) shows a distinctive localization in the cortex, cerebellum (Purkinje and granule cell layers) and many other brain regions (Webb et al 1994). This correlates well with the autoradiographic distribution across brain sections of [^{35}S]dATPαS binding sites (displaceable by 2-MeSATP but not by α,β-meATP) in both chick and rat (J. Simon, unpublished results). In Northern blots (Webb et al 1993), apart from the high level of the chicken $P2Y_1$ mRNA in the brain, it is present also in the spinal cord, spleen, gastrointestinal tract and skeletal muscle of the adult chicken, but was not detectable in the heart, kidney or liver.

In the rat, $P2Y_1$ receptor mRNA is expressed likewise in the brain and skeletal muscle, but is also found in the heart (strongly), liver and kidney (unlike in the chicken) (Tokuyama et al 1995). It could, therefore, be the suramin-sensitive P2Y receptor which is inferred (with other P2Y receptors) to be on rat hepatocytes (Tomura et al 1992, Keppens & De Wulf 1993). In the mammal, $P2Y_1$ receptor mRNA is also expressed strongly in arterial endothelial cells (Henderson et al 1995, Boeynaems et al 1996, this volume).

Receptors of the $P2Y_1$ type are very abundant in the brain. Specific binding of dATPαS is of similar high affinity to that with recombinant $P2Y_1$ and these sites are found at the density, exceptional for a brain receptor, of 35–40 pmoles per mg brain membrane protein (Simon et al 1995b). The *in situ* abundance of $P2Y_1$ mRNA and of the binding sites, noted above, bears this out. These sites have the pharmacology (in ligand binding competition) of the $P2Y_1$ receptor, with similar K_i values thereto (Table 1). These statements apply equally to the chicken and the rat brain (Simon et al 1995b).

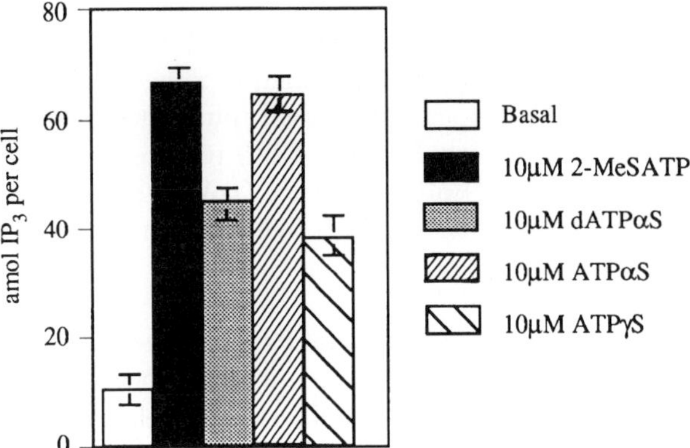

FIG. 1. P2Y agonists induce inositol-1,4,5-trisphosphate ($InsP_3$) accumulation in intact COS-7 cells transiently expressing the recombinant purinoceptor. The amount of $InsP_3$ synthesized is expressed (at a saturating agonist concentration) as attomoles per cell ± SEM (three independent determinations, each in triplicate). (From Simon et al 1995b.)

TABLE 1 Potencies of purinoceptor-active ligands at the chick brain native and the recombinant purinoceptors

Ligand	*Expressed in COS-7 cells K_i (nM)*	*Expressed in Xenopus oocytes EC_{50} (nM)*	*Full (F) or partial (P) agonist*[a]	*Native brain P2Y purinoceptors*[b] *K_i (nM)*
2-MeSATP	69 ± 23	10 ± 1	F	34 ± 8
dATPαS	23 ± 7	ND	P(67%)	17 ± 2
ATPαS	63 ± 22	ND	P	32.0 ± 9
ATP	48 ± 13	155 ± 50	P(70%)	47.8 ± 4
ADP	171 ± 19	258 ± 40	P	530 ± 183
2-MeSADP	326 ± 176	ND	F	170 ± 38
ATPγS	52 ± 17	ND	P	ND
ADPβS	91 ± 18	~150	P	ND
Suramin	1592 ± 206	230 ± 80[c]		1052 ± 244
Reactive blue 2	944 ± 201	580 ± 130[c]		1472 ± 278
UTP	6077 ± 327	> 10 000		> 10 000
Adenosine	> 10 000	> 10 000		> 10 000
L-β,γ-meATP	> 10 000	> 10 000		> 10 000

All values are expressed as the mean ± SEM of at least three independent determinations.
[a]The behaviour in expression, relative to 2-MeSATP.
[b]For comparison, values obtained on brain membranes from the newly-hatched chick (Simon et al 1995) are also shown.
[c]Represents an IC_{50} value for antagonists, suramin and Reactive blue 2.
ND, not determined. (From Simon et al 1995a,b.)

We cannot state that all of the P2Y receptors revealed so far in the brain are $P2Y_1$ sites, since in theory those binding sites could include also one or more other subtypes which are closely similar in their binding properties. If so, these could be yet uncloned subtypes, since their pharmacology (Table 1, last column) does not fit any known P2Y (or P2X) receptor (e.g. because of the complete insensitivity to UTP) other than $P2Y_1$. We conclude that the brain contains a relatively high number of $P2Y_1$-like receptors.

Native occurrence of the $P2Y_1$ receptors

There is one case so far where the functional $P2Y_1$ receptor in a *native* tissue can be identified as being expressed alone, i.e. with no other P2 purinoceptors present. This is in cells isolated from the capillary endothelium from the rat brain, a tissue which *in situ* shows a mixed P2 purinoceptor responsiveness, suggesting that $P2Y_1$ and $P2Y_2$ receptors are present (Frelin et al 1993). From those cells, C. Frelin and co-workers in Sophia Antipolis (France) have isolated an endothelial cell line (B10), also ATP-responsive. These cells have lost the $P2Y_2$ receptor mRNA (Feolde et al 1995) and the only P2Y-like receptor mRNA detectable by low stringency PCR priming is that of the rat $P2Y_1$, which is abundant and whose full-length cDNA was cloned from there (Webb et al 1996b).

Parallel to this, the B10 cells bind nucleotides, to give the selectivity and relative affinities which are characteristic of the recombinant $P2Y_1$ receptor (Webb et al 1996b), confirming that these endothelial cells possess that receptor subtype on their surface. However, while the nucleotides mobilize Ca^{2+} in these cells, this is not associated with an increase in $InsP_3$ but *is* with a cAMP increase (Webb et al 1996b). This is in sharp contrast to the transduction mediated by avian $P2Y_1$ receptors, where $InsP_3$ formation is clearly induced (Filtz et al 1994, Simon et al 1995a). Moreover, we can compare a rat glioma cell line (C6-2B), which has been shown to contain a P2Y purinoceptor (along with other purinoceptor(s)) which likewise does not link to $InsP_3$ formation, but was shown to couple to adenylate cyclase inhibition (Boyer et al 1993). From this C6-2B cell line, also, we have amplified the entire rat $P2Y_1$ receptor cDNA sequence (Webb et al 1996b). In contrast, from cultured brain astrocytes (Neary & Norenberg 1992), where a P2Y receptor activation produces $InsP_3$, the same rat $P2Y_1$ receptor entire sequence was also amplified (Webb et al 1996b). While a strict proof cannot be claimed for the identity of the $P2Y_1$ receptor seen functionally in these latter two cases and that recognized by its mRNA sequence, we cannot associate any other type of response there with that sequence. We therefore conclude that the $P2Y_1$ receptor, while commonly coupling to phospholipase C, can in some native cells couple instead through another transductional pathway.

Other recombinant P2Y receptors

The P2U (now $P2Y_2$) receptor cDNA was isolated independently by Lustig and co-workers (Lustig et al 1993) and its structure and properties have been described fully by Lustig et al (1996, this volume). In that receptor, UTP is equipotent with ATP and 2-MeSATP is inactive.

A further P2Y receptor cDNA was cloned, again starting from chick brain (Barnard et al 1994), with a protein sequence which is about equidistant from the $P2Y_1$ and $P2Y_2$ sequences. This was designated as $P2Y_3$. It responded to ATP and other nucleotides both in oocyte expression and by Ca^{2+} mobilization in transfected mammalian cells (Webb et al 1996a). This receptor, which also occurs in various peripheral tissues, has a nucleotide selectivity that does not match any detected previously in those tissues, presumably because of overlap with other P2 purinoceptors present there. This shows a preference for nucleoside diphosphates, with ATP and 2-MeSATP relatively weakly active. It is expressed most strongly in the spleen.

A potential fourth P2Y receptor of this series has been recognized by cDNA cloning and characterized by us from the HEL cell line. The HEL cell line is derived from the megakaryocyte lineage, responds to platelet-activating agents and can be induced to express platelet-like characteristics (Schwaner et al 1992, Vittet et al 1992). It is known that HEL cells respond to ADP with Ca^{2+} transients (Kalambakas et al 1992, Vittet et al 1992). The ATP response to ADP is antagonized by 10 μM ATP (Akbar et al 1996a), effects which occur also at the P2T purinoceptor of platelets (Hourani & Hall 1996, this volume). HEL cells contain cell membrane high affinity binding sites for ADP, which resemble the P2T receptor sites of platelets (Kalambakas et al 1992). Therefore HEL cells are a potential source of the P2T receptor. However, the $P2Y_1$ and the $P2Y_2$ receptors are also present, as shown both by the nucleotide responses seen and by isolating their cDNA clones from HEL cells (Ayyanathan et al 1996, Akbar et al 1996a). Moreover, differences were found between the platelet and HEL cell responses to ADP and in the protein labelled in each by a photo-affinity derivative of ADP (Akbar et al 1996a); hence, the identity of an ADP receptor in the HEL cell with the platelet P2T receptor is not clear.

In addition to $P2Y_1$ and $P2Y_2$, another P2Y receptor cDNA (HP212) was indeed found from HEL cells, which shares sequence identity with the $P2Y_1$ and $P2Y_2$ receptors. In transfected cell expression it gives the following binding affinity profile: 2-Cl-ATP > ATP > 2-MeSATP > UTP. Its properties will be described fully elsewhere (Akbar et al 1996b). The mRNA of this receptor is particularly abundant in the human heart and this is likely to be a P2Y subtype which has been found to be prominent in cardiac myocytes.

We have cloned a further cDNA clone (R5) encoding another protein of the seven-transmembrane-domain class which shares a high degree of sequence identity with the $P2Y_1$–$P2Y_3$ receptors. The source used was the glioma cell line C6-2B, which is of interest because its P2Y-type receptors have been well characterized in the native state (by T. K. Harden and co-workers). They include a subtype which is negatively coupled to adenylate cyclase (Boyer et al 1993) and a receptor which is apparently specific for pyrimidine nucleotides and couples to $InsP_3$ formation (Lazarowski & Harden 1994). From these cells, starting with PCR using primers based on known P2Y receptor sequences, we have cloned cDNAs which correspond to rat $P2Y_1$ (as in the B10 cells, and noted above), to rat P2U and to the full-length R5 sequence. Despite the use of consensus primers covering sequences common to the known P2Y receptors and lowered stringency, no other P2Y-type receptor could be detected. It seems likely, therefore, that the R5 sequence is of the C6-2B 'pyrimidine-selective' receptor: this currently awaits confirmation by expression and functional testing. Certainly, the R5 sequence should be, by its sequence relationship and by the conservation in it of the motifs of the $P2Y_1$ receptor (discussed below), a further receptor of the P2Y series.

Yet another P2Y receptor subtype has been discovered indirectly. Kaplan et al (1993) reported an mRNA which becomes expressed in T lymphocytes upon their activation (in the chicken), which encodes a 308-residue protein (6H1) with the 7TM structure. This is an 'orphan receptor', whose mRNA was absent in tissues except T cells and then only when activated. We (with Duncan Henderson) noted the homology of this to the chicken $P2Y_3$ receptor (Fig. 2). Accordingly, we have expressed 6H1 in COS-7 cells and found a very high level of expressed binding sites for [^{35}S]dATPαS (Fig. 3). Displacement of this binding showed the affinity order: ATP $>$2-MeSATP $>>\alpha,\beta$-meATP, UTP (Fig. 3). We have concluded that 6H1 is a new P2Y receptor subtype (Webb et al 1996c). As is discussed by Di Virgilio et al (1996, this volume), human T lymphocytes express P2Z receptors but may express also P2Y receptors upon activation. It will be interesting to see if these are the human homologue of 6H1.

The sequences HP212, 6H1 and R5 are here given their arbitrary clone names, but will need to be put into the numerical series of $P2Y_4$ onwards. This is not done here, pending agreement on the chronology of these and any other simultaneously described receptors.

Protein sequences of the P2Y receptors

The six P2Y receptors described above, which must arise from six independent genes, share sufficient protein sequence identity (Table 2) to constitute a family within the rhodopsin superfamily of G protein-coupled receptors. This is a new family which is separate from all other G protein-coupled receptors: for example, the most similar sequences to the $P2Y_1$ receptor are several diverse receptors from different families, none more that $\sim$25% identical to $P2Y_1$ (Barnard et al 1994).

Within this family, the members are more diverse, in sum, than in any other of the known G protein-coupled receptor families (Table 2). The relationships between them are displayed in the tree of Fig. 2. This high degree of divergence suggests that intermediate members are still to be found and that the family is

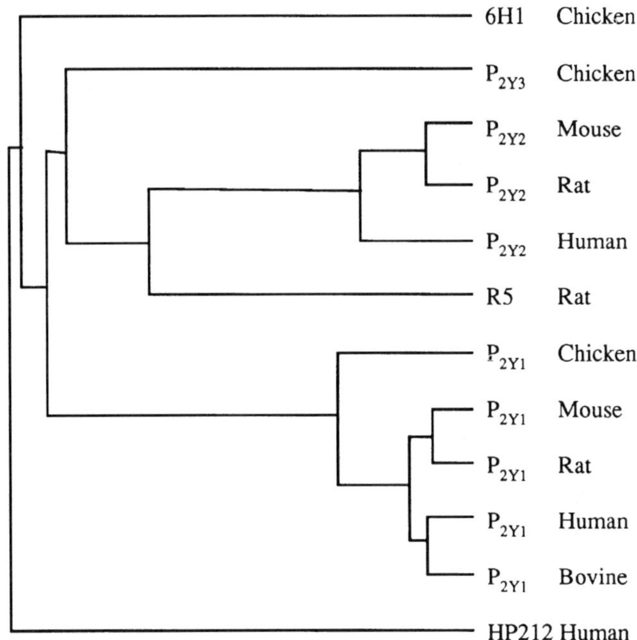

FIG. 2. A tree for minimum distances in the multiple sequence alignment of recombinant P2 purinoceptor amino acid sequences. A $P2Y_1$ sequence isolated from turkey (Filtz et al 1994) is not included as there is only one amino acid difference between this and the chicken sequence. For the other references see the text.

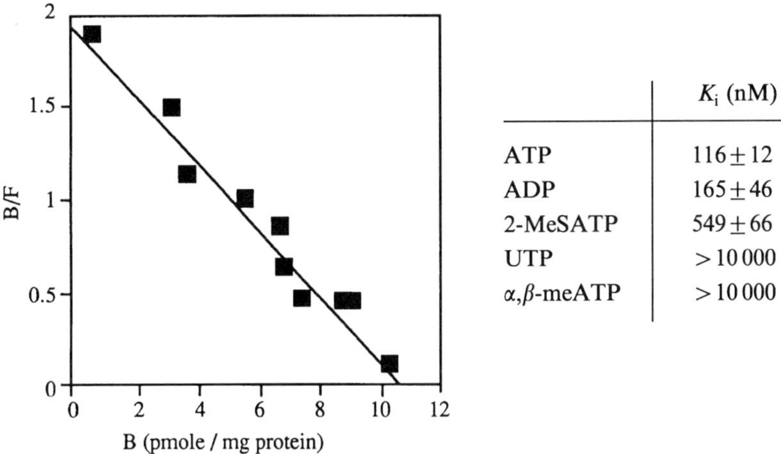

	K_i (nM)
ATP	116 ± 12
ADP	165 ± 46
2-MeSATP	549 ± 66
UTP	> 10 000
α,β-meATP	> 10 000

FIG. 3. Scatchard plot for binding of [^{35}S]dATPαS to membranes of COS-7 cells transfected with the 6H1 cDNA. K_i values indicated were computed from competition of this binding. (From Webb et al 1996c.)

TABLE 2 Percentage identity of amino acid sequences of P2Y receptor subtypes

	P2Y₁		*P2Y₂*		*P2Y₃*	*HP212*	*R5*	*6H1*
Subtype	*C*	*H*	*M*	*H*	*C*	*H*	*R*	*C*
$P2Y_1$ (C)	100	86	34	37	36	27	36	32
$P2Y_1$ (H)		100	35	38	—	—	—	—
$P2Y_2$ (M)			100	89	40	28	50	33
$P2Y_3$ (C)					100	24	39	36
HP212 (H)						100	22	23
R5 (R)							100	34
6H1 (C)								100

Species: C, chicken; H, human: M, mouse; R, rat. For references see text.

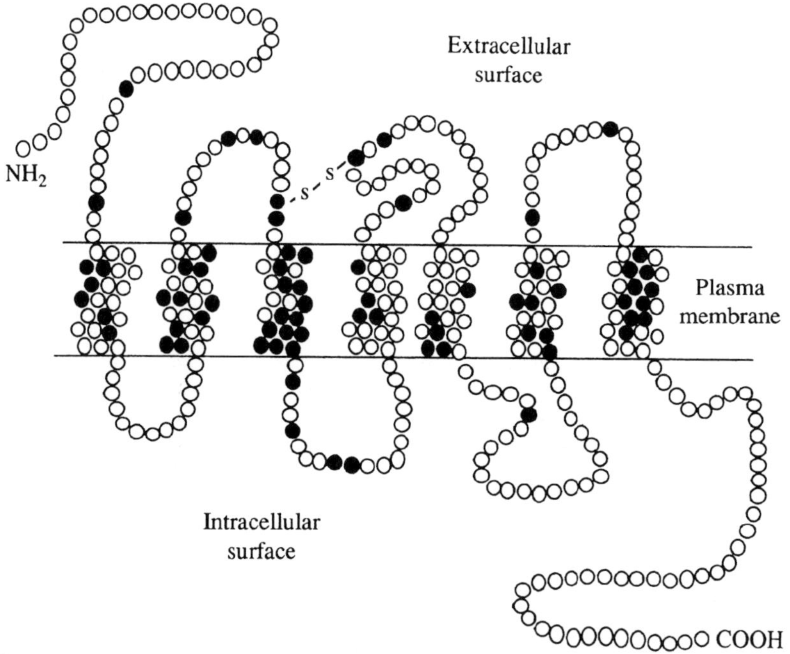

FIG. 4. Schematic diagram of the sequence of the chick $P2Y_1$ purinoceptor and its identity with the mouse $P2Y_2$, chicken $P2Y_3$ and rat R5 receptors. The $P2Y_1$ purinoceptor sequence is shown. Residues conserved between all four receptor sequences are indicated by black circles; variant residues are denoted by open circles. Note that the lengths of the receptors differ.

likely to be as numerous as the highest-populated of the transmitter G protein-coupled receptor families, namely the serotonin receptors.

Some of the fractional identity shown in Table 2 represents a conserved set of common residues. This is illustrated in Fig. 4, where the residues which are common to the $P2Y_1$ (chicken *or* human), $P2Y_2$, $P2Y_3$ and R5 receptors are shown as filled circles. Note the high conservation of TM3, which is general in this family, as is the pair of cysteines presumed to be disulfide bonded. Note also that the third intracellular loop and the tail, regions implicated in G protein specificity in other G protein-coupled receptors, have nothing in common here. Despite the use of different species for the comparisons of Table 2, those relationships seem general: where other species can be compared, e.g. chicken to human in $P2Y_1$, the same receptor is strongly conservative in sequence (86%, and 95% rat to human). However, when we consider 6H1 and HP212, we see that these are more distantly related to the four sequences compared in Fig. 4, and to each other. The divergence in structural features in HP212 is so great that a transduction mechanism different to all of the others here may be suspected.

Conclusion

At least six independent P2Y receptors have been identified at the molecular level and there will surely be others described from other laboratories. None of these has any resemblance in sequence at all to the adenosine receptors, confirming that P2 and P1 receptors are fundamentally unrelated. The number of the structures and of the pharmacologies seen in the P2Y G protein-coupled receptor family, and their wide distribution in peripheral tissues and in brain regions, denote a variety of important roles in intercellular signalling for these receptors.

Acknowledgements

We thank Dr P. Leff (Astra, UK) for his interest and support. Dr C. Léon kindly sent the abstract of the paper cited, when in press. A part of this work was supported by Astra Research, UK. We benefited from collaborations on various topics with G. Burnstock and B. King (University College, London), B. J. Krishek and T. G. Smart (School of Pharmacy, London), I. A. Dainty, D. G. Elliott, D. J. Henderson, P. Leff and G. M. Smith (Astra, UK), C. Frelin (CNRS, Sophia Antipolis, France), M. Matsumoto and R. Shingai (Morioka, Japan), J. T. Neary (Miami, USA) and R. S. Sundick (Detroit, USA), which are shown in our cited publications.

References

Abbracchio MP, Burnstock G 1994 Purinoceptors: are there families of P_{2X} and P_{2Y} purinoceptors? Pharmacol Ther 64:445–475

Akbar GKM, Dasari VR, Sheth SB, Ashby B, Mills DCB, Kunapuli SP 1996a Characterization of P2 purinergic receptors on human erythroleukemia cells. J Receptor Signal Transd Res, in press

Akbar GKM, Webb TE, Dasari VR et al 1996b Molecular cloning of a novel cardiac P2 purinoceptor from human erythroleukemia cells, in press

Allsup DJ, Boarder MR 1990 Comparison of P_2 purinergic receptors of aortic endothelial cells with those of adrenal medulla: evidence for heterogeneity of receptor subtype and of inositol phosphate response. Mol Pharmacol 38:84–91

Ayyanathan K, Webb TE, Sandhu AK, Athwal RS, Barnard EA, Kunapuli SP 1996 Cloning and chromosomal localization of the human P_{2Y1} purinoceptor. Biochem Biophys Res Commun 218:783–788

Barnard EA 1992a Receptor classes and the transmitter-gated ion channels. Trends Biochem Sci 17:368–374

Barnard EA 1992b Classes of receptor subunits, analysis and reconstitution. In: Burgen AV, Barnard EA, Roberts GCK (eds) Receptor subunits and complexes. Cambridge University Press, Cambridge, p 97–117

Barnard EA, Burnstock G, Webb TE 1994 The G protein-coupled receptors for ATP and other nucleotides: a new receptor family. Trends Pharmacol Sci 15:67–70

Boarder MR, Weisman GA, Turner JT, Wilkinson GF 1995 G protein-coupled P_2 purinoceptors: from molecular biology to functional responses. Trends Pharmacol Sci 16:133–139

Boeynaems J-M, Communi D, Pirotton S, Motte S, Parmentier M 1996 Involvement of distinct receptors in the actions of extracellular uridine nucleotides. In: P2 purinoceptors: localization, function and transduction mechanisms. Wiley, Chichester (Ciba Found Symp 198) p 266–277

Boyer JL, Lazarowski ER, Chen X, Harden TK 1993 Identification of a P_{2Y}-purinergic receptor that inhibits adenylate cyclase. J Pharm Exp Ther 267:1140–1146

Burnstock G, Kennedy C 1985 Is there a basis for distinguishing two types of P_2 purinoceptor? Gen Pharmacol 5:433–440

Di Virgilio F, Ferrari D, Falzoni S et al 1996 P2 purinoceptors in the immune system. In: P2 purinoceptors: localization, function and transduction mechanisms. Wiley, Chichester (Ciba Found Symp 198) p 290–305

Edwards FA, Gibb AJ 1993 ATP: a fast neurotransmitter. FEBS Lett 325:86–89

Feolde E, Vigne P, Breittmayer JP, Frelin C 1995 ATP is a partial agonist of atypical P_{2Y} purinergic receptors in rat brain microvascular endothelial cells. Br J Pharmacol 115:1199–1203

Filtz TM, Li Q, Boyer JL, Nicholas RA, Harden TK 1994 Expression of a cloned P2Y purinergic receptor that couples to phospholipase C. Mol Pharmacol 46:8–15

Fredholm BB, Abbracchio MP, Burnstock G et al 1994 Nomenclature and classification of purinoceptors. Pharmacol Rev 46:143–156

Frelin C, Breittmayer JP, Vigne P 1993 ADP induces inositol phosphate-independent intracellular Ca^{2+} mobilization in brain capillary endothelial cells. J Biol Chem 268:8787–8792

Henderson DJ, Elliot DG, Webb TE, Dainty IA 1995 Cloning and characterisation of a bovine P_{2Y} receptor. Biochem Biophys Res Commun 212:648–656

Hourani SMO, Hall DA 1996 P2T purinoceptors: ADP receptors on platelets. In: P2 purinoceptors: localization, function and transduction mechanisms. Wiley, Chichester (Ciba Found Symp 198) p 53–70

Kalambakas SA, Robertson FM, O'Connell SM, Sinha S, Vishnupad K, Karp GI 1992 Adenosine diphosphate stimulation of cultured hematopoietic cell lines. Blood 81:2652–2657

Kaplan MH, Smith DI, Sundick RS 1993 Identification of a G protein coupled receptor induced in activated T cells. J Immunol 151:628–636

Kennedy C, Leff P 1995 How should P_{2X} purinoceptors be classified pharmacologically? Trends Pharmacol Sci 16:168–174

Keppens S, De Wulf H 1993 The complex interaction of ATP and UTP with isolated hepatocytes. How many receptors? Gen Pharmacol 24:283–289

Lazarowski ER, Harden TK 1994 Identification of a uridine nucleotide-selective G-protein-linked receptor that activates phospholipase C. J Biol Cem 269:11830–11836

Léon C, Vial C, Cazenave J-P, Gachet C 1996 Cloning and sequencing of a human cDNA encoding endothelial P_{2Y1} purinoceptor. Gene, in press

Lustig KD, Shiau AK, Brake AJ, Julius D 1993 Expression cloning of an ATP receptor from mouse neuroblastoma cells. Proc Natl Acad Sci USA 90:5113–5117

Lustig KD, Weisman GA, Turner JT, Garrad R, Shiau AK, Erb L 1996 P2U purinoceptors: cDNA cloning, signal transduction mechanisms and structure–function analysis. In: P2 purinoceptors: localization, function and transduction mechanisms. Wiley, Chichester (Ciba Found Symp 198) p 193–207

Miras-Portugal MT, Castro E, Mateo J, Pintor J 1996 The diadenosine polyphosphate receptors: P2D purinoceptors. In: P2 purinoceptors: localization, function and transduction mechanisms. Wiley, Chichester (Ciba Found Symp 198) p 35–52

Neary JT, Norenberg MD 1992 Signaling by extracellular ATP: physiological and pathological considerations in neuronal–astrocytic interactions. Prog Brain Res 94:145–151

North RA 1996 P2X receptors: a third major class of ligand-gated ion channels. In: P2 purinoceptors: localization, function and transduction mechanisms. Wiley, Chichester (Ciba Found Symp 198) p 91–110

O'Connor SE, Dainty IA, Leff P 1991 Further subclassification of ATP receptors based on agonist studies. Trends Pharmacol Sci 12:137–141

Parr CE, Sullivan DM, Paradiso AM et al 1994 Cloning and expression of a human P_{2U} nucleotide receptor, a target for cystic fibrosis pharmacotherapy. Proc Natl Acad Sci US 91:3275–3279

Schwaner I, Seifert R, Schultz G 1992 Receptor mediated increases in cytosolic Ca^{2+} in the human erythroleukemia cell line involve pertussis toxin-sensitive and -insensitive pathways. Biochem J 281:301–307

Simon J, Webb TE, King BJ, Burnstock G, Barnard EA 1995a Pharmacological characterisation of a recombinant P_{2Y} purinoceptor. Eur J Pharmacol 291:281–289

Simon J, Webb TE, Barnard EA 1995b Characterization of a P_{2Y} purinoceptor in the brain. Pharmacol Toxicol 76:302–307

Surprenant A 1996 Functional properties of native and cloned P2X receptors. In: P2 purinoceptors: localization, function and transduction mechanisms. Wiley, Chichester (Ciba Found Symp 198) p 208–222

Tokuyama Y, Hara M, Jones EMC, Fan Z, Bell GI 1995 Cloning of rat and mouse P_{2Y} purinoceptors. Biochem Biophys Res Commun 211:211–218

Tomura H, Okajima F, Kondo Y 1992 Discrimination between two types of P_2 purinoceptors by suramin in rat hepatocytes. Eur J Pharmacol 226:363–365

Vittet D, Mathieu M-N, Launay J-M, Cevillard C 1992 Platelet receptor expression on three human megakaryoblast-like cell lines. Exp Hematol 20:1129–1134

Webb TE, Simon J, Krishek BJ et al 1993 Cloning and functional expression of a brain G-protein-coupled ATP receptor. FEBS Lett 324:219–225

Webb TE, Simon J, Bateson AN, Barnard EA 1994 Transient expression of the recombinant chick brain P_{2Y1} purinoceptor and localization of the corresponding mRNA. Cell Mol Biol 40:437–442

Webb TE, Henderson D, King BF et al 1996a A novel G protein-coupled P_2 purinoceptor (P_{2Y3}) activated preferentially by nucleoside diphosphates. Mol Pharmacol, in press

Webb TE, Feolde E, Vigne P et al 1996b The P_{2Y} purinoceptor in rat brain microvascular endothelial cells couples to inhibition of adenylate cyclase, in press

Webb TE, Kaplan MG, Barnard EA 1996c Identification of 6H1 as a P_{2Y} purinoceptor: $P2Y_5$. Biochem Biophys Res Commun 219:105–110

Ziganshin AU, Ziganshin LE, King BF, Burnstock GB 1995 Characteristics of ecto-ATPase of Xenopus oocytes and the inhibitory actions of suramin on ATP breakdown. Eur J Physiol 429:412–418

DISCUSSION

Leff: We have done a full characterization of clone 803 ($P2Y_1$). With the availability of an ectoATPase inhibitor, we have been able to check whether ectoATPase is a problem in cloning and expression systems. We found that it wasn't, which is good news. The interesting thing is that we were able then to make comparisons between 803 as our standard with the previously defined P2Y receptors from functional studies going back over the last 10–15 years. Using the ectoATPase inhibitor, we were able to show that the endothelial-dependent responses in the guinea-pig aorta mapped very closely to the cloned receptor, once the ectoATPase was taken into account. When you do this and then go back through the literature, using potency orders fairly roughly at this point, you can start to do a bit of pattern recognition. You do see classes of receptors which in general look like 803 and others which do not. We haven't completed that sort of literature survey, but I think that when we do, we'll end up with a similar argument to the one that Charles Kennedy and I arrived at with the P2Xs. When one takes everything into account, a new definition of $P2Y_1$ might be required. Then you can relate the earlier data to it, which is a nice sort of conservative exercise to go through.

One thing that does give me a problem though (and this isn't a personal criticism of you—it is something that most pharmacologists accept that molecular biologists tend to do), is that you use one-point dose–response curves in characterizing the clone. When you want to go through the exercise I was just talking about it's impossible to fit such data in: you have to get the full curve. I'd be reticent to say that ADP is more active, for instance, without the full binding curves when you say that clone 103 is or isn't similar to existing UTP receptors.

Barnard: I agree entirely. Of course, one ought to have full dose–response curves when describing P2Y receptor subtypes. For several I have described, the definition *is* firmly based on dose–response curves from the Ca^{2+} release or electrophysiology or ligand binding studies, but unfortunately this cannot be the case for the $P2Y_3$ receptor (clone 103). That is because the $P2Y_3$ receptor is lethal to the cells we have expressed it in. Using several hundred oocytes, Brian King was able to construct such a plot for the one case of ADP (Webb et al 1996) but this was not practicable for other cases. Consequently, only a small time window can be found for the response before the cell dies. It would be interesting to know why this is, but it is a reproducible finding. It is not a technical problem because it is not found in any other of the P2Y clones studied under similar conditions. For the mammalian cells transfected with that $P2Y_3$

cDNA, this effect can be overcome, since in the selection process for a stably-expressing cell line the lethality would add to the selection pressure and, indeed, the stable P2Y_3-expressing line finally isolated recently by Duncan Henderson of your laboratory appears to express only a relatively low level of the receptor. With that line, therefore, dose–response curves (for intracellular Ca^{2+} release with P2Y_3 agonists) could recently be constructed (Webb et al 1996).

Brändle: Have you tried coexpression of the P2Y_1 with the P2Y_3 receptor to avoid the lethal effects on oocytes?

Barnard: Why would you think that that would be a good thing to do?

Brändle: In some cases ligand-gated ion channels express heteromers. Isn't it possible that the Y_3 chain needs a second chain for expression otherwise the Y_3 homomer is lethal for the cell?

Barnard: This is not known to me for any G protein-coupled receptor. Although for one or two types there has been evidence reported for dimers of them in the cell surface, this is not accepted as a general phenomenon and certainly they are not thought to form heteromers. There is no reason why they should.

Abbracchio: I am particularly intrigued by the P2Y_3 purinoceptor that you found expressed in large amounts in chick brain. Do you have data on the cell distribution in the developing brain? Is this receptor expressed by neurons and/or by other non-neuronal cells? Since you have found that it is lethal when expressed, does this receptor play any physiological role in development-associated programmed cell death?

Barnard: The P2Y_3 receptor *is* expressed in the brain, and indeed was cloned from brain mRNA. However, it was at a low level in the Northern blot of brain, and this level has proved too low for regional mapping of the mRNA *in situ*. It is abundant in the spleen, and the cell death relationship might be relevant in the immune system cells there.

Boeynaems: You said that the expression of the rat P2Y_1 receptor leads to a Ca^{2+} response without a rise of $InsP_3$. Is this due to an influx of Ca^{2+}, or to a mobilization of intracellular Ca^{2+}?

Barnard: It's a Ca^{2+} transient—a release of internal Ca^{2+}. There is no evidence for an influx of Ca^{2+}, since it can also be obtained by Dr C. Frelin in a Ca^{2+}-free medium.

North: What is the function of a seven-transmembrane receptor (P2Y_3) which, when it binds its agonist, kills the cell?

Barnard: We have pondered upon this. Firstly, we can't be sure that this is not simply an artefact of expression. The P2Y_3 receptor has a high lethality in expression systems, but it may be that the native cells carrying these receptors do not show lethality. It is extremely unusual—I don't know of any other transmitter receptor gene transfection that causes this effect, so I cannot give a fuller answer.

Di Virgilio: If Peter Krammer had asked the same question when he isolated the Fas/APO-1 antigens, we would be lagging behind in our knowledge of T cell-mediated cytotoxicity and apoptosis. We now know that so-called suicide receptors are expressed on the plasma membranes of most cells. When these are activated under the proper conditions they kill the cell.

Burnstock: 'The proper condition' is the key phrase here: in the whole body, these things are programmed to work at the right time.

Abbracchio: Going back to the issue of a seven-transmembrane receptor(s) mediating cell death: it may be not particularly surprising—there are indeed other examples. Adenosine induces apoptosis of thymocytes and lymphocytes. All adenosine receptors are G protein-coupled. Moreover, the involvement of a G protein-coupled receptor in adenosine-induced death of human thymocytes has been recently supported by data showing that in these cells apoptosis by adenosine is dependent upon an early and sustained inositol-1,4,5-trisphosphate release (Szondy 1994).

Barnard: But I spoke only of observations made with transfection or oocyte injection. Adenosine receptors give no lethality in those systems.

Burnstock: I don't find it particularly surprising that the actions of purines are sometimes lethal. Once we accept the involvement of ATP in programmed cell death (for references see Burnstock 1996, this volume), then it is easy to accept the idea of purinoceptor involvement in certain stages of development.

North: Is $P2Y_3$ expressed in T lymphocytes?

Barnard: No.

North: So it's not likely to be important in T cell selection.

Di Virgilio: No, but it may turn out that other purinergic receptors are involved in T cell selection.

Burnstock: I think we sometimes make the mistake in biology of focusing too tightly on one area. Many of us here are a bit biased towards ATP. If you went to a NO meeting you might think that NO was responsible for everything. Another focused area concerns growth factors. However, what one often finds is that there are synergistic actions in biology, certainly when it comes to complex processes like regeneration. For example, even if you don't get an action of ATP, when you combine it with actions of growth factors or NO, you might get a huge response. Don't always assume because you don't see ATP doing something on its own that it isn't vital to some synergistic process. We have an example—we have been transplanting the enteric nervous system into the brain. To cut a long story short, it turns on massive growth of central neurons, involving a growth factor released from enteric glial cells working together with NO and purines released from non-adrenergic, non-cholinergic inhibitory nerves. Each on their own doesn't produce neurite outgrowth, but put them together and there is remarkable growth of CNS neurons. With ATP, we may only have part of the story. A general principle that I would recommend is that no matter how excited we are about ATP, we should always try to interface with other areas of interest.

Surprenant: Four years ago I went to a meeting that was all about K^+ channels and every person was showing yet another mutagenesis experiment with little apparent regard for the function of these channels. I had the question: are the ever more detailed mutagenesis studies actually telling us anything? I was impressed by Chris Miller, who said that the molecular biology work was probably not at that time telling us anything about their functional roles, but that we were in a new information-gathering period. His opinion was that it was going to be a few years before the importance of the functional aspects was realized. This made me think that all this minute mutagenesis might be worth doing.

Wiley: When it comes to function, there's no doubt that these receptors are part of the story, but then the other inseparable part of this story concerns which cells in the body are releasing nucleotides that act on those receptors. At the moment we don't even know the precise mechanism by which ATP or UTP exits from some cells. We know that these nucleotides are released from endothelial cells, for example, under conditions of increased flow. We've looked at certain tumour cell lines and shown that they also release significant amounts of ATP which is easily measured by luminometry. But we are still in this information-gathering phase, and until we know from which cells and by what mechanisms the nucleotides are released, it's going to be hard to put all this information together.

Westfall: The release of nucleotides from endothelial cells is not only induced by shear stress, which we have known about for a while, but also by naturally occurring endogenous vasoactive substances such as noradrenaline, bradykinin and 5-HT. The release of ATP from endothelial cells by noradrenaline is mediated by α_1 receptors. Curiously, endothelial cells in various blood vessels have varying proportions of the α_1 receptor subtype that mediates ATP release. The physiological importance of this is as yet unknown.

Burnstock: In nerves ATP is almost certainly vesicle bound, but in endothelial cells release may not involve vesicles.

Westfall: The process of release of ATP isn't known, but it certainly does happen. I think that's important.

Harden: The turkey erythrocyte membrane is a very useful model for the study of the P2Y purinoceptor. This receptor activates phospholipase C (PLC) in a fashion that is entirely dependent on the presence of guanine nucleotides. Little basal enzyme activity is observed with these membranes, but addition of ATP, ADP or nucleotide analogues in the presence of GTP results in a marked increase in inositol phosphate formation. The selectivity of this receptor for activation by adenine nucleotides is consistent with that initially defined for P2Y purinoceptors by Burnstock & Kennedy. Since this model system provides a membrane assay for a P2Y purinoceptor-promoted functional response and provides a source of highly purified plasma membranes, it has the potential advantage of providing a preparation in which binding of a radioligand to a P2Y purinoceptor can be studied under conditions similar to those in which a P2Y purinoceptor-promoted functional response also can be measured.

We have used a number of radioligands, including [^{35}S]ADPβS, in attempts to label the P2Y purinergic receptor on turkey erythrocyte plasma membranes.

[^{35}S]ADPβS binds to a site(s) on these membranes which, with a selected group of competing agonist ligands, fits pharmacological properties observed in functional assays of P2Y purinoceptors. For example, 2-methylthioATP potently inhibits radioligand binding and α,β-methylene ATP is relatively impotent. There is a good correlation between the K_i values of these and those of 10 other agonists for inhibition of [^{35}S]ADPβS binding to turkey erythrocytes and their respective EC_{50} values for stimulation of PLC. In addition, a number of the P2Y purinoceptor agonists synthesized by Ken Jacobson that show high potencies for activation of PLC in the turkey erythrocyte membrane system are also similarly potent inhibitors of radioligand binding. Saturation analyses of [^{35}S]ADPβS binding using 2-methylthioATP to define 'specific binding' generated B_{max} values of approximately 3 pmol/mg protein, which is approximately 10-fold higher than the density of β-adrenergic receptors in the same membranes. Although these published results (Cooper et al 1989) were consistent with the conclusion that [^{35}S]ADPβS could be used in a radioligand binding assay for P2Y purinoceptors on turkey erythrocyte plasma membranes, subsequent work has cast doubt on this view. For example, 2′-deoxyATP ($EC_{50} = 2 \times 10^{-5}$ M) and 3′-deoxyATP ($EC_{50} > 1 \times 10^{-4}$ M) are both very weak agonists (and are without antagonist activities) at the PLC-linked P2Y purinoceptor on turkey erythrocyte membranes. However, both potently inhibit [^{35}S]ADPβS binding in these membranes (K_i of 3′-deoxyATP $= 8 \times 10^{-9}$ M and K_i of 2′-deoxyATP $= 2 \times 10^{-7}$ M). Other adenine nucleotide analogues have also been identified that activate PLC and inhibit radioligand binding over very different concentration ranges. Many assay conditions have been tested without success in attempts to circumvent these large differences in relative orders of potency of agonists for inhibition of radioligand binding and activation of the P2Y purinoceptor-mediated response. The discrepancies also remain if binding and activity measurements are made under initial rate conditions. None of our data are consistent with the existence of more than one P2Y purinoceptor in turkey erythrocyte membranes that might explain some of the lack of correlation of binding and activity measurements. We have also radiolabelled several of the high potency 2-thioether derivatives that have been synthesized by Ken Jacobson and colleagues. Binding data with each of these additional radioligands are essentially undistinguishable from results obtained with [^{35}S]ADPβS. Finally, [^{35}S]deoxyATPαS has been used as a radioligand by others to label putative P2 purinoceptor binding sites. In our hands, this molecule is a very poor agonist (and is not an antagonist) at the PLC-activating P2Y purinoceptor of turkey erythrocyte membranes and at the cloned turkey ($P2Y_1$) receptor or its human $P2Y_1$ homologue stably expressed in 1321N1 human astrocytoma cells.

Although we remain hopeful that a high affinity agonist radioligand can be developed to radiolabel P2 purinoceptors, we are not convinced that any of the available radiolabelled agonists have sufficient binding selectivity to be used as a radioligand. High affinity non-nucleotide antagonists remain the most likely useful molecules for unambiguous radiolabelling of P2Y and other P2 purinoceptors.

Jacobson: In turkey erythrocytes we even get antagonists competing with the expected K_i values. The binding of deoxyATPαS in these membranes has many characteristics of P2Y receptors.

Barnard: As I had detailed in our presentation, Josef Simon and Tania Webb in our laboratory have used [^{35}S]dATPαS to label several of the P2Y receptor subtypes after expression in COS-7 cells, obtaining self-consistent results. That labelling is displaced by ATP and other nucleotides with a rank order and potencies which are different, and are characteristic, for each receptor subtype expressed. There is no receptor binding site for [^{35}S]dATPαS on native cells or cells transfected with the vector alone. For the P2Y$_1$ receptor the pharmacology at the dATPαS binding site has been compared with that obtained functionally with the expressed receptor. As was shown in our Table 1, the affinities of 4 agonists measured by means of the dATPαS binding at the chicken P2Y$_1$ receptor are quite similar to their functional EC$_{50}$ values (for ATP and two others, within a factor of 3, and for one, 2-MeSATP, 6.9-fold). In making such comparisons it must be borne in mind that the functional response with such agonists is not measured on the receptor in the COS cells themselves, since the latter show interference from native P2U-like receptors (although these do not bind dATPαS); *Xenopus* oocytes were used here and the coupled G proteins and other factors which could affect both affinity and EC$_{50}$ may differ from those of the COS cells so that a closer correspondence would not necessarily be expected. Moreover, the functional measurements on a G protein-coupled receptor, even when they can be performed on the same cell as the binding, are reflecting a complex set of equilibria involving components additional to the receptor and ligand, and these can vary with the individual agonist and the receptor expression level, as reviewed very pertinently by Kenakin (1995). Hence, it is theoretically possible that discrepancies with binding occur with some agonists and not others.

On the question whether dATPαS and ADPβS are abnormal in being very weak agonists but binding very strongly, this is not observed in the conditions we have employed. With the chicken P2Y$_1$ receptor ADPβS was equipotent with ATP in the oocyte and so was dATPαS in COS-7 cells for InsP$_3$ formation, although it was a partial agonist relative to ATP (Simon et al 1995). Likewise, on the chicken P2Y$_3$ receptor in the oocyte dATPαS and ADPβS are both agonists, roughly equipotent with ATP (Webb et al 1996). dATPαS was also tested in the Ca^{2+} release system in Jurkat cells and is an agonist. Again, their efficacies are lower and exact comparisons could therefore be difficult.

In general, the measurement of equilibrium binding affinities on expressed P2Y-series receptors is, in fact, very difficult in most cases with the presently-available tools. Most host cell lines have intrinsic P2U-like receptors which bind radiolabelled ATP and UTP. The ligand we discussed, [^{35}S]dATPαS, is useful because it does not bind to those receptors; however among many commonly-used cell lines tested (including the 1321N1 astrocytoma cells successful in functional measurements), all except COS-7 cells have too many of the low-affinity, high-abundance binding sites for that and other nucleotides, i.e. they show binding at an interfering level which does not saturate up to 1 μM ligand. These are presumed to be nucleotidase or other non-receptor low-affinity binding sites. It is also feasible to make these receptor binding measurements on native endothelial cells in cases where they express very abundantly only one P2Y subtype, as in the microvascular endothelial B10 cells I have discussed. In summary, difficulties which may be found on measuring the binding of P2Y receptors in a given cell type do not mean that the ligand in question cannot be applied in other cases.

Leff: Experimentally, why are you assuming that the 15 compounds that don't fit the pattern don't interact in the same way with the same receptor? How do you know they are the same type of agonist as the others? Have you been able to antagonize all the agonists with the same ligand?

Harden: We know whether they are P2Y agonists and whether they block the receptor or not. We have seen no evidence of multiple receptors from additivity or cross desensitization experiments. We have not determined antagonist K_i values with some of the agonists whose binding properties don't fit.

Leff: Is there a blocker that you feel able to use to block those agonists?

Harden: That is a good experiment. What we should do now is determine a K_i with suramin and PPADS at that receptor. If we determine antagonist K_i values with an agonist that fits and an agonist that doesn't fit and get the same result, then we can assume it's the same receptor. We've seen no data that suggest that they are not activating the same receptor, but that's a good point.

Burnstock: For years the drug companies have had this problem. For rapid and mass analysis of new compounds they go in for binding. Unfortunately, sometimes when they take their best compounds based on binding and try them in the body, they are relatively ineffective.

Williams: A common reason for lack of activity when going from *in vitro* binding to *in vivo* functional models relates to pharmacokinetics, specifically plasma half-life. I remember working on a cholecystokinin project where peptide ligands active in the very low nanomolar range consistently showed no activity in traditional animal models of schizophrenia. It came as no surprise that the half-life of many of these supposedly stable peptides was less than one minute while their potential effects were measured 14 minutes later!

Leff: With Ken Harden's data, you've got a set of compounds which clearly do correlate, which are agonists and are suffering in principle all the criticisms that have been mentioned. There's no reason in my mind (assuming you're talking about the same generic class of agonist with the same types of properties and the same intrinsic abilities to desensitize) to expect the binding and functional data not to correlate.

Edwards: But different agonists do have different desensitization characteristics.

Leff: But that's a question of whether you are arguing that desensitization follows activation or is parallel to it: if it follows it, then they should correlate; if it is parallel, perhaps not. That is what you have got to invoke. You have to say that there are 500 compounds which operate in a mechanistically similar way, and then some structurally closely related ones which are mechanistically dissimilar. Regarding the comparison of binding and functional data, there is a theoretical basis whereby now you would expect correlation. For years, when people were doing binding studies on G protein-linked systems and trying to do them on ion channel-linked systems, they thought that as long as they used things like high levels of GTP or stable nucleotides, they could measure what pharmacologists call affinity. If you believe anything that's been said recently about inverse agonism and two-state models, then the pharmacological data and the binding data will both be expected to contain the elements of affinity, then conformational change, that is, efficacy—and to roughly the same degree. So what you've shown there is actually the theoretical expectation for 500 compounds. As Frances Edwards is saying, there is an intrinsic difference in one set of agonists—the 15—and the other 500.

Harden: In my experience with other G protein-linked receptors, we wouldn't have had to go to such extremes. We wouldn't have had to explain away spurious binding data on the basis of this and that with the β receptor, serotonin receptor and the muscarinic receptor, all of which again makes me question the validity of all P2 receptor binding assays. I think this idea of a desensitized state is wrong and does not explain lack of correlation of potencies determined in binding assays with those determined in assays of activity. We're not talking about an ion channel type receptor in which desensitization is a rapid consequence of receptor activation—we're talking about a G protein-linked receptor studied *in membranes* under conditions in which there is no evidence for desensitization occurring.

We can generate the same data if, instead of doing equilibrium binding assays, we do initial rate competition curves. So we can generate those same curves at 15 s of binding rather than at 15 min. I don't think something is happening during time that moves the curve one way or another.

Williams: Your concerns about the B_{max} need to be viewed in the context of Eric Barnard's finding of a B_{max} of 37–39 pmol/mg protein for the $P2Y_1$ receptor in the CNS. This may argue for a very important functional role for this receptor.

Harden: Can someone explain to me how you can have an agonist that is not an antagonist, has an apparent potency of 10 μM in an activity assay, and in a binding assay now has a potency of 2 nM, whereas 2-MeSATP which in the activation assay has a potency of 10 nM is also 10 nM in a binding assay? It doesn't make sense.

Leff: It would be alright if all the agonists did that, but I think Eric was really saying that, depending on the system, you could expect affinity estimates that you get by binding to be to the left of an EC_{50}. According to traditional receptor theory, that would be impossible: the affinity is going to sit to the right of the functional curve. But there are reasons why you could expect that to go wrong. Again, it's because you've got a discorrelation that you have the problem.

Williams: The muscarinic agonists synthesized at Merck by John Saunders (Freedman et al 1990) may be of relevance here. These were a series of 'super' agonists represented by L-670207 that were more efficacious than carbachol in stimulating muscarinic receptor-linked transduction systems. It appeared that these molecules had the unique property of undergoing a receptor-associated change in conformation once they were bound. This change, a reflection of the intrinsic flexibility of these azanorbornane oxadiazoles was the key to efficacy. They were good binders as well as being efficacious. The synthesis of more rigid molecules which is what a lot of companies were doing at the time to get high affinity ligands resulted in molecules that bound but lacked efficacy. This explained to molecular modellers why there was a discrepancy in binding versus function.

References

Burnstock G 1996 P2 purinoceptors: historical perspective and classification. In: P2 purinoceptors: localization, function and transduction mechanisms. Wiley, Chichester (Ciba Found Symp 198) p 1–34

Cooper CL, Morris AJ, Harden TK 1989 Guanine nucleotide-sensitive interaction of radiolabeled agonist with a phospholipase C-linked P2Y purinergic receptor. J Biol Chem 264:6202–6206

Freedman SB, Harley EA, Patel S et al 1990 A novel series of non-quaternary oxadiazoles acting as full agonists at muscarinic receptors. Br J Pharmacol 101:575–580

Kenakin T 1995 Agonist–receptor efficacy. II. Agonist trafficking of receptor signals. Trends Pharmacol Sci 16:232–238

Simon J, Webb TE, King BJ, Burnstock G, Barnard EA 1995 Pharmacological characterization of a recombinant P_{2Y} purinoceptor. Eur J Pharmacol 291:281–289

Szondy Z 1994 Adenosine stimulates DNA fragmentation in human thymocytes by Ca^{2+}-mediated mechanisms. Biochem J 304:877–885

Webb TE, Henderson D, King BF et al 1996 A novel G protein-coupled P_2 purinoceptor ($P2Y_3$) activated preferentially by nucleoside diphosphates, in press

General discussion I

Molecular modelling of P2Y purinoceptors

Jacobson: We've been carrying out molecular modelling on a number of G protein-coupled receptors, principally adenosine and ATP receptors. I want to show you a little bit about the methodology and take you on a brief tour of the putative binding site.

Michiel van Rhee in my lab and I have compared two different approaches to G protein-coupled receptors, one of them based on the template of bacteriorhodopsin. Although we, Ad IJzerman and a number of others have made receptor models based on bacteriorhodopsin, we recently became aware of a better approach, based on the low-resolution structure of rhodopsin. There are three advantages here to using rhodopsin over bacteriorhodopsin. One is that you have a sequence homology with rhodopsin, which is a receptor in the same superfamily. Also, you can build up the helices individually before you combine them in a bundle, and you can customize them: that is, you can include the particular features of proline residues. Proline residues cause a deviation from linearity of the helix and they also cause a twist in the periodicity of the helix. You cannot account for these major structural changes if you are simply overlaying a new sequence onto the bacteriorhodopsin template. You can also put the helices together in a way that makes sense for each particular receptor. One of the important structural differences between bacteriorhodopsin and rhodopsin is the third transmembrane helix. In rhodopsin this helix is more internalized than in the bacteriorhodopsin structure. The portion of the receptor encompassing helices 5, 6 and 7 and helix 3 is where most of the small molecule ligand binding occurs. We have shown this by mutagenesis of A_{2A} adenosine receptors, and this is the overall conclusion of our P2Y model as well. Fig. 1 (*Jacobson*) shows the $P2Y_1$ receptor model, indicating the putative binding site for ATP. After docking of ATP, the receptor complex has been energetically minimized. We see that binding occurs in the top third of the helices with respect to the exofacial side. Thus the ATP molecule extends from helix 7 to helices 6 and 5. The triphosphate group according to this model is coordinated by a number of positively charged residues in the sixth and seventh helices that have already been alluded to today. The α-phosphate is coordinated to arginine 299, the γ-phosphate to lysine 269, and also to histidine 266. These three residues are commonly associated with phosphate binding to proteins in general. So it makes sense in relation to what's known about triphosphate binding in

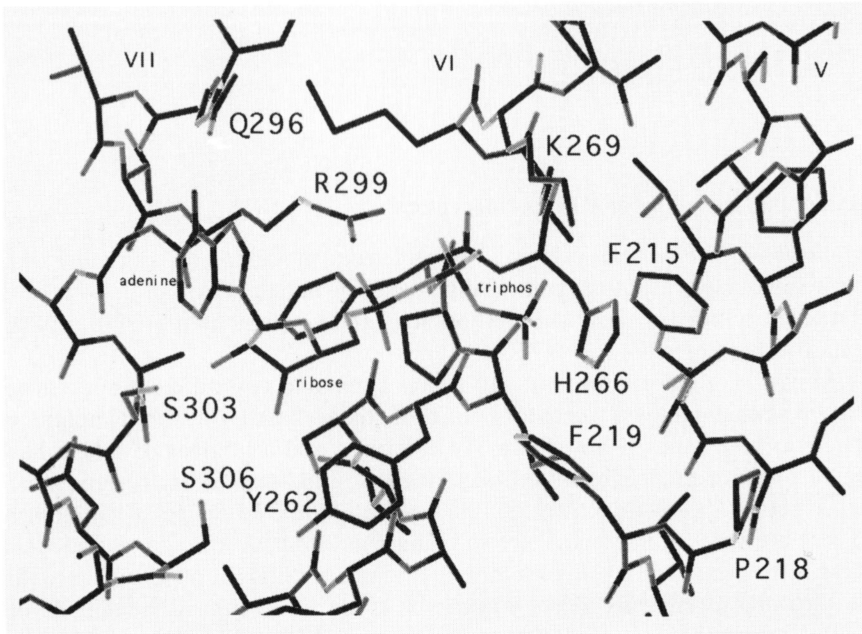

FIG. 1 (*Jacobson*) P2Y$_1$ receptor model from the plane of the membrane, showing the putative binding site for ATP. Helices are designated by roman numerals.

structures determined by X-ray crystallography. The distances measured 4 Å to arginine 299, 3–4 Å to lysine 269 and 2–3 Å to histidine 266. According to our model, the histidine can coordinate doubly with two of the oxygens and the γ-phosphate.

Table 1 *(Jacobson)* summarizes the residues we think are involved in coordination of the ATP. First of all, the triphosphate coordinates with lysine, arginine and histidine residues, as I have pointed out. The three corresponding residues in the P2U receptor were mutated by Gary Weisman and John Turner. In the P2U receptor you have arginine in that position instead of the lysine: this conservation of positive charge suggests that this is also important for the P2Y receptor. The ribose is in proximity to serines 303 and 306 which could hydrogen bond to the hydroxyl groups. The adenine moiety, according to our model, hydrogen bonds to a glutamine residue via the exocyclic NH. This is very similar to the model that we proposed for the adenosine receptors in which we have an asparagine residue hydrogen bonding at the same point at the adenine ring (Kim et al 1995). Let me stress that there is a big difference between our models for adenosine binding to A_2 receptors and the P2Y model, and that is the adenosine is almost flipped over with respect to the orientation of ATP. In the adenosine case, it's the ribose that's binding to the seventh helix,

TABLE 1 (*Jacobson*) Residues of the $P2Y_1$ receptor in proximity ($\leqslant 5$ Å) to the docked ligand (ATP) according to the molecular model of van Rhee et al (1995)

Adenine	*Ribose*	*Triphosphate*
Q296	S303	K269
	S306	R299
Y47	A302	H121
F51		H266
F55		Y125
Y100		
F120		F215
		P218
		F219

in the ATP case it's the adenine moiety that's binding to the seventh helix. We could not dock ATP in a way that was consistent with the P1 putative binding model, so P1 and P2 are really different entities structurally.

IJzerman: I'd like to make a general comment about the 3D models of G protein-coupled receptors. I think both Ken and I, having worked together in this area intensively, agree that these models are approximate and best regarded as a visualization of existing—but not integrated—biochemical and pharmacological data. As such, the models may provide vital clues for further experimentation.

Starke: In your last remark you indicated that ATP is not fitting onto the adenosine receptor. Doesn't that contrast with the idea that ATP or other nucleotides do activate certain adenosine receptors?

Jacobson: I'm acutely aware of the challenge of the so-called P3 receptor to these models, and whether this hybrid type of receptor will resemble more of the P1 or P2. It would be a good exercise to model it when it's cloned.

Starke: There is also the idea that nucleotides activate A_1 receptors.

Jacobson: I should add that there are some positively charged residues even in the P1 receptors in the loop regions. I would like to consider the possibility that phosphate of nucleotides could be pointing upwards towards the exofacial side, and in that weak binding that may be observed in certain cases to A_1 receptors there may be some interaction with the extracellular loops.

Barnard: In the range of sequences that we now know, your four basic residues for binding the phosphate are conserved in all of them, except in subtypes 6H1 and HP212, in which only one of them is. This would be a good way of sifting through the models, because as we get more and more sequences there should be a constant feature at an interacting position.

North: Which ones are not conserved?

Barnard: The histidine/tyrosine in TM3 (H121/T125) and the histidine in TM6 (H266) are conserved, except for one change in 6H1 or HP212.

North: Is the histidine important by virtue of its charge or its aromaticity?

Jacobson: We haven't done mutagenesis in that position so I can't really answer that question, but I would think it's due to its hydrogen bonding capabilities.

References

Kim J, Wess J, van Rhee AM, Shöneberg T, Jacobsen KA 1995 Site-directed mutagenesis identifies residues involved in ligand recognition in the human A_{2a} adenosine receptor. J Biol Chem 270:13987–13997

van Rhee AM, Fischer B, van Galen PJM, Jacobsen KA 1995 Modelling the P2Y purinoceptor using rhodopsin as a template. Drug Design Discov 13:133–154

P2U purinoceptors: cDNA cloning, signal transduction mechanisms and structure–function analysis

Kevin D. Lustig*, Gary A. Weisman†, John T. Turner§, Richard Garrad†§, Andrew K. Shiau‡ and Laurie Erb†

**Department of Cell Biology, Harvard Medical School, 25 Shattuck Street, Boston, MA 02115, Departments of †Biochemistry and §Pharmacology, University of Missouri, Columbia, MO 65212 and ‡Department of Biochemistry and Biophysics, University of California, San Francisco, CA 94143, USA*

Abstract. The cloning of a P2U purinoceptor cDNA has made it possible to use molecular biological approaches to investigate P2U purinoceptor function. Expression of recombinant P2U purinoceptors in mammalian cells lacking endogenous P2U purinoceptors has enabled us to characterize the receptor protein and its downstream effectors, and has allowed a partial analysis of the role of certain amino acid residues in ligand binding. These approaches have placed the pharmacological classification of the P2U purinoceptor on a firm molecular footing and have generated model systems that can be used to investigate receptor–ligand binding, regulation and signal transduction.

1996 P2 purinoceptors: localization, function and transduction mechanisms. Wiley, Chichester (Ciba Foundation Symposium 198) p 193–207

More than twenty years have passed since Geoffrey Burnstock first suggested that the reported effects of extracellular nucleotides in mammalian cells were due to activation of plasma membrane receptors termed P2 purinergic receptors (Burnstock 1972, 1978). Today, molecular genetic and pharmacological evidence suggests that P2 purinergic receptors (now called P2 purinoceptors) form a diverse receptor family consisting of both G protein-coupled receptors and ligand-gated ion channels (Dubyak & El-Moatassim 1993, Dalziel & Westfall 1994, Fredholm et al 1994). One G protein-coupled receptor—the P2U purinoceptor—was initially distinguished from other P2 purinoceptor subtypes by its ability to be activated by the pyrimidine nucleotide UTP in addition to ATP (O'Connor et al 1991). cDNAs encoding this P2U purinoceptor have recently been cloned from several species and sequence analysis confirms that the receptors are members of a new subfamily of the seven-transmembrane-domain receptor superfamily (Lustig et al 1993, Parr et al 1994). In this review, we describe progress made towards the characterization of the structure and function of recombinant P2U purinoceptors.

cDNA cloning

We used an expression cloning approach to isolate the first cDNA encoding a P2U purinoceptor. A plasmid expression library was constructed from NG108-15 mouse neuroblastoma X rat glioma hybrid cells, which express a receptor with the agonist specificity and signalling properties of a P2U purinoceptor (Lin et al 1993). The expression library was subdivided into pools of plasmids, which were screened using a *Xenopus laevis* oocyte cloning strategy (Masu et al 1987). RNA pools were microinjected into *Xenopus laevis* oocytes and functional expression of the receptor was assayed electrophysiologically. A positive pool was iteratively subdivided until a single active cDNA encoding a P2U purinoceptor was obtained (Lustig et al 1993).

The 2.4 kb cDNA had an approximately 1.1 kb open reading frame that encoded a putative 42 kDa protein with many characteristic sequence motifs of a G protein-coupled receptor (GPCR). Sequence analysis suggested the presence of seven hydrophobic regions within the protein that presumably serve as transmembrane domains. As in many GPCRs, several potential *N*-linked glycosylation sites are found near the N-terminus and several potential phosphorylation sites are found near the C-terminus (Fig. 1). A number of amino acids that are conserved in most GPCRs—including Asn^{51}, Asp^{79}, Leu^{81}, Arg^{131}, Pro^{167} and Pro^{303}—are also found in the predicted sequence of the receptor. There is one sequence near the N-terminus of the predicted protein that resembles a consensus ATP-binding motif, but these residues could be deleted with no apparent effect on receptor function (Lustig et al 1993).

The low amino acid sequence identity ($<26\%$) between the P2U purinoceptor and other GPCRs suggests that the P2U receptor is a member of a new subfamily within the GPCR superfamily (Lustig et al 1993). The P2U purinoceptor is highly conserved across species and among tissues; homologues from human airway epithelia and intestinal epithelia share approximately 90% amino acid sequence identity with the mouse neuronal P2U receptor (Parr et al 1994). This new subfamily already has multiple members; concurrently with the report of the cloning of the P2U purinoceptor gene, Webb et al (1993) reported the cloning of a gene encoding a P2Y purinoceptor that is approximately 40% identical at the amino acid level to the mouse P2U purinoceptor, with striking identity in the third putative transmembrane domain (Barnard et al 1994). Unexpectedly, both P2U and P2Y purinoceptors are more distantly related to receptors for the structurally related ligands adenosine and cAMP than to receptors for thrombin, complement factor 5a (C5a), f-Met-Leu-Phe, bradykinin and angiotensin (Lustig et al 1993, Webb et al 1993). The similarity

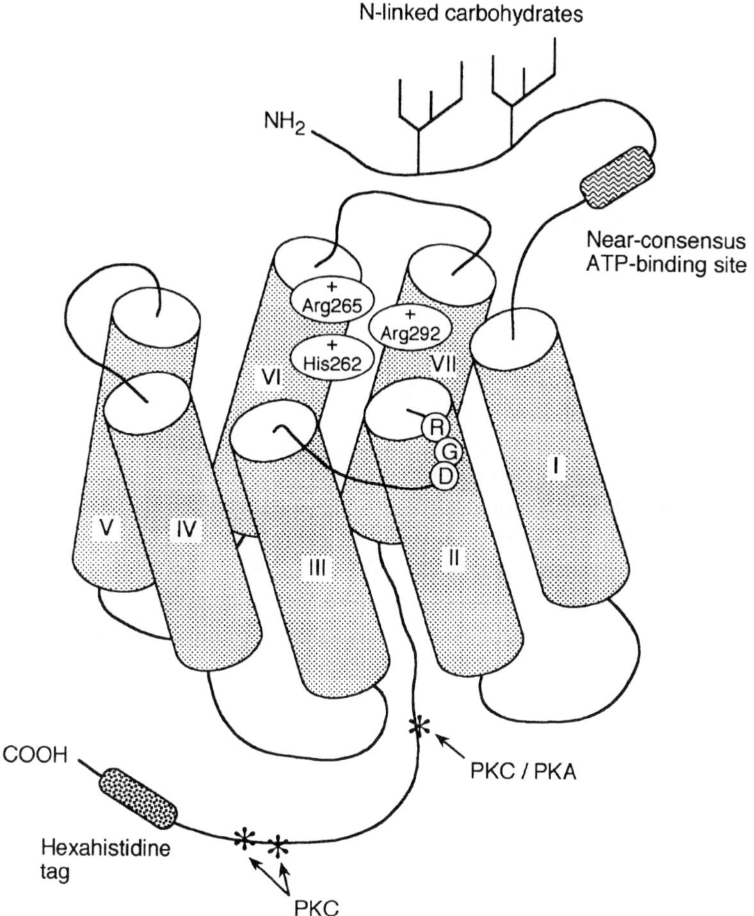

FIG. 1. Schematic of the P2U purinoceptor. The cylinders represent the seven putative transmembrane helices (I–VII) and the black lines connecting them represent the extracellular (top) and intracellular (bottom) loops of the receptor. Three positively charged residues (His^{262}, Arg^{265} and Arg^{292}) that are important for ligand binding and/or receptor activation are shown in transmembrane helices VI and VII, and a potential integrin binding motif (RGD) is shown in the extracellular loop between transmembrane domains II and III. Stars denote the location of several potential phosphorylation sites near the intracellular C-terminus. The filled box at the extreme C-terminus indicates the position of a hexahistidine tag that was added to the receptor to facilitate its biochemical isolation. The two branched figures and the filled box near the N-terminus indicate the location of two consensus *N*-linked glycosylation sites and a near consensus ATP-binding site, respectively.

between P2 purinoceptors and these peptide receptors points to the intriguing but as yet untested possibility that there are endogenous peptides capable of independently activating the P2U and P2Y purinoceptors or modifying their responses to extracellular nucleotides.

Pharmacology and signal transduction

Owing to the unavailability of cultured mammalian cell lines lacking endogenous P2U receptors, it was initially necessary for us to examine the pharmacology of the cloned receptor in *Xenopus laevis* oocytes injected with P2U receptor RNA. In voltage-clamped oocytes expressing the cloned P2U purinoceptor, inward currents were elicited by perfused ATP, UTP or ATPγS, but not by 2-methylthioATP, β,γ-methylene ATP, ADP, α,β-methylene ATP or adenosine (Lustig et al 1993). Such inward currents are presumably due to P2U purinoceptor-mediated increases in Ca^{2+} that activate endogenous Cl^- channels in the oocyte plasma membrane. Consistent with the idea that receptor activation leads to an increase in intracellular Ca^{2+}, stimulation of the cloned P2U purinoceptor by ATP, UTP or ATPγS also led to an increase in the rate of $^{45}Ca^{2+}$ efflux from the oocyte (Lustig et al 1993).

Erb et al (1993) and Parr et al (1994) have now stably expressed the recombinant P2U purinoceptor in K562 human leukaemia cells and 1321N1 human astrocytoma cells, respectively, making possible the characterization of the properties of the cloned P2U purinoceptor in mammalian cell systems. K562 and 1321N1 cells are two of only a small number of mammalian cell lines known to lack an endogenous P2U purinoceptor. In stable transfectants of both cell lines, the agonist specificity and signalling properties of the recombinant P2U purinoceptor were similar to those of the cloned P2U purinoceptor expressed in oocytes (Lustig et al 1993) as well as the endogenous P2U purinoceptor found in NG108-15 cells (Lin et al 1993), the original source of the cloned receptor. Low micromolar concentrations of ATP, UTP or ATPγS appear to activate the recombinant receptor, resulting in an increase in the level of cytoplasmic free Ca^{2+} (Erb et al 1993, Parr et al 1994). These transfectants were not responsive to the P2Y receptor agonist 2-methylthioATP or the P2X receptor agonists β,γ-methylene ATP and α,β-methylene ATP. Also without effect were the naturally occurring nucleotides GTP, TTP, CTP and AMP, and the nucleoside adenosine. Taken together, the results in amphibian oocytes and transfected mammalian cells suggest that the product of a single P2U purinoceptor gene is sufficient to confer sensitivity to both ATP and UTP. This is a surprising finding, considering that ATP and UTP are structurally less similar to each other than to other nucleotides that do not activate the P2U purinoceptor. It is of course still possible that distinct proteins generated by alternative splicing or post-translational processing could only be activated by ATP or by UTP, but not by both.

In transfected K562 cells, the P2U purinoceptor-mediated increase in cytoplasmic Ca^{2+} levels was not significantly affected by removal of extracellular Ca^{2+}, indicating that the Ca^{2+} rise was largely due to the release of Ca^{2+} from intracellular storage sites and not the influx of extracellular Ca^{2+} (Erb et al 1993). Thus the recombinant P2U purinoceptor, like endogenous P2U purinoceptors in many cell lines (Gordon 1986, Dubyak & El-Moatassim 1993), can couple to a signal transduction pathway involving the generation of Ca^{2+}-mobilizing second messengers. The second messenger responsible for Ca^{2+} mobilization is likely to be inositol-1,4,5-trisphosphate ($InsP_3$), an inositol phospholipid derivative generated after agonist stimulation of many GPCRs. 1321N1 cells stably expressing the human P2U purinoceptor respond to ATP or UTP with an increase in $InsP_3$ levels that occurs coincident with the increase in cytoplasmic Ca^{2+} levels (Parr et al 1994).

Members of the G_q and G_i families of heterotrimeric G proteins are both capable of activating phospholipase C (PLC)β, thereby generating $InsP_3$, but do so by two different mechanisms; G_q proteins signal via their GTP-bound α subunit, whereas G_i proteins signal via a complex of their β and γ subunits. The recombinant P2U purinoceptor may be able to couple to members of both G protein families since pertussis toxin, an irreversible inhibitor of G_i proteins (Simon et al 1991), inhibits by only 40% the P2U purinoceptor-mediated Ca^{2+} increase (Erb et al 1993). Pertussis toxin partially inhibits the increase in Ca^{2+} levels elicited by agonist stimulation of endogenous P2U purinoceptors in a variety of cultured cell lines (Dubyak & El-Moatassim 1993).

These results suggest a signalling model in which agonist stimulation of P2U purinoceptors leads to the G protein-dependent activation of PLC (Fig. 2). Many aspects of the signal transduction process remain to be clarified. It is not known whether other components of the signalling pathway, such as the GTP-bound α subunit of G_i proteins or the $\beta\gamma$ complex of G_q proteins, activate PLC or other downstream effectors. The identity of relevant substrates of protein kinase C, which is presumably activated by diacylglycerol, remain to be determined. Whether receptor activation leads to changes in gene expression remains to be explored. Since few biological roles for P2U purinoceptors have been delineated to date, it is difficult to predict which arms of these signalling pathways are important for receptor function.

Photoaffinity labelling and receptor isolation

The biochemical characterization of the P2U purinoceptor protein has proved difficult owing to the presence of other endogenous nucleotide-binding proteins and to the absence of selective high-affinity ligands. To circumvent both problems, we constructed a mutant P2U purinoceptor containing six consecutive histidine residues at its C-terminus (Erb et al 1993). These six histidine residues serve to 'tag' the receptor protein, enabling its rapid isolation

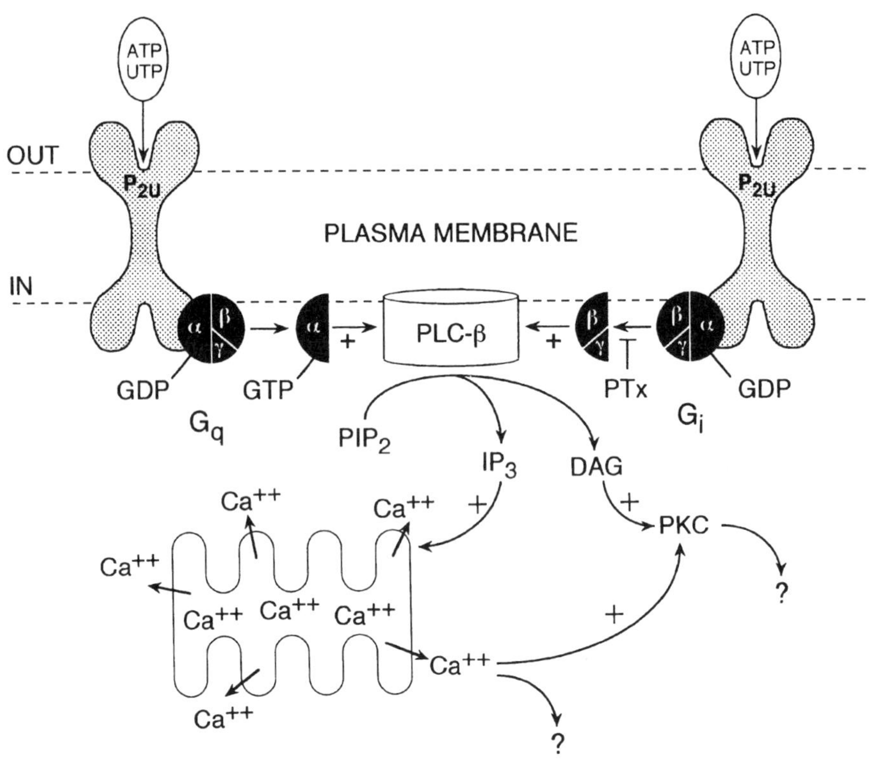

FIG. 2. Signal transduction by recombinant P2U purinoceptors. The schematic shows a model in which nucleotide-activated P2U purinoceptors stimulate the exchange of GDP for GTP on the α subunit of G_q and G_i proteins and cause the dissociation of the GTP-bound α subunit from the $\beta\gamma$ complex (blocked by pertussis toxin, PTx). Either GTP-bound α subunits from G_q or the $\beta\gamma$ complex from G_i (or both) bind to and stimulate phospholipase Cβ (PLC-β). Activated phospholipase Cβ in turn hydrolyses phosphatidylinositol-4,5-bisphosphate (PIP_2) to yield inositol-1,4,5-trisphosphate (IP_3), which liberates Ca^{2+} from intracellular storage sites, and diacylglycerol (DAG), which activates protein kinase C (PKC). Signalling through the P2U purinoceptor is modulated by multiple mechanisms, including nucleotide hydrolysis (by cell surface ectonucleotidases), receptor desensitization or down-regulation, inactivation of G proteins, degradation of IP_3 by phosphatases or the re-uptake of Ca^{2+} into storage compartments.

by metal affinity chromatography (LeGrice & Gruninger-Leitch 1990, Ljungquist et al 1989).

3′-*O*-(4-benzoyl)benzoyl ATP (BzATP), a weak photaffinity agonist of the recombinant P2U purinoceptor, was used to photolabel the histidine-tagged receptor (Erb et al 1993). Plasma membranes from K562 cells expressing the histidine-tagged P2U purinoceptor were incubated with [α-^{32}P]BzATP and then exposed to long-wavelength ultraviolet light to cross-link the photoprobe to the receptor. Labelled proteins were solubilized in urea and resolved on a Ni^{2+}-charged Sepharose column, which strongly binds proteins containing hexahistidine tags (LeGrice & Gruninger-Leitch 1990, Ljungquist et al 1989). One photolabelled protein with a molecular mass of 53 kDa was retained on the column after elution of all other labelled proteins. Consistent with the conclusion that this protein is the tagged P2U purinoceptor, only one protein band, with an apparent molecular mass of 53 kDa, was photolabelled by [α-^{32}P]BzATP to a much greater extent in plasma membranes from K562 cells stably expressing the P2U purinoceptor than in plasma membranes from untransfected K562 cells (Erb et al 1993). Since the protein putatively encoded by the P2U purinoceptor cDNA has a predicted molecular mass of 42 kDa, these findings suggest that the mature receptor in K562 stable transfectants migrates aberrantly in SDS-PAGE, perhaps due to glycosylation at one or both of the consensus *N*-linked glycosylation sites located in the N-terminal extracellular domain. ATP, but not UTP, inhibited photoincorporation of [α-^{32}P]BzATP into the 53 kDa protein (Erb et al 1993). Since site-directed mutagenesis studies suggest that the phosphate groups of ATP and UTP interact with similar binding determinants (Erb et al 1995), this may indicate that BzATP labels the P2U purinoceptor at an ATP-binding site that plays no role in ligand binding or receptor activation.

Structure–function analysis

The ligand-binding site of the P2U purinoceptor, like that of most GPCRs (Probst et al 1992, Strader et al 1994), appears to be formed by the juxtaposition of several transmembrane domains. Sequence comparisons indicated that five positively charged amino acids, in the third, sixth and seventh transmembrane helices, are conserved in both the cloned P2U and P2Y purinoceptors. Since the fully ionized form of nucleotides has been shown in at least some cell lines to activate both subtypes (Dubyak & El-Moatassim 1993), it seemed feasible that these conserved positively charged amino acids (Lys^{107}, Arg^{110}, His^{262}, Arg^{265} and Arg^{292}) were involved in binding to the negatively charged phosphate groups of ATP and UTP. To test this hypothesis, we constructed mutant P2U purinoceptor cDNAs in which one of the five positively charged residues was changed to the neutral amino acids leucine or isoleucine. These mutants were stably expressed in 1321N1 cells and the magnitude of the increase in Ca^{2+} levels in response to nucleotide treatment was used as a measure of receptor activity.

Three residues—His^{262} and Arg^{265} in transmembrane helix VI and Arg^{292} in transmembrane helix VIII—appear to play some role in ligand binding or receptor activation (Erb et al 1995). Mutation of these residues to leucine or isoleucine caused a 100- to 850-fold decrease in the potency of ATP and UTP, but did not affect their efficacy relative to the wild-type receptor. In contrast, mutation of Lys^{107} or Arg^{110} in transmembrane helix III did not alter the agonist potency or specificity of the P2U purinoceptor, indicating that these residues do not participate in ligand binding or receptor activation. Mutation of Lys^{289} in the P2U purinoceptor, which corresponds to a glutamine residue in the P2Y receptor, did not alter receptor activity, suggesting that it also is not directly involved in ligand binding. A conservative change from lysine to arginine in this position altered the rank order of potency so that the partial agonists ADP and UDP were approximately 100-fold more potent than the full agonists ATP and UTP. One possible explanation for this finding is that Lys^{289} may be positioned close enough to the ligand-binding site that its substitution by a slightly larger residue sterically interferes with ligand binding.

Molecular models of the sixth and seventh putative transmembrane domains support the hypothesis that His^{262}, Arg^{265} and Arg^{292} form a positively charged binding pocket (Erb et al 1995). Considering that the P2U receptor is most closely related to peptide GPCRs, it is intriguing that these positively charged residues lie close to residues of peptide GPCRs that have been implicated in agonist and antagonist binding by mutagenesis studies, pharmacophore mapping and molecular modelling (Strader et al 1994, Underwood et al 1994, Schambye et al 1995). His^{262} and Arg^{265} are in a region of transmembrane helix VI similar to that of His^{265} of the neurokinin receptor (implicated in non-peptide antagonist binding), Phe^{264} of the bradykinin receptor (implicated in bradykinin binding) and His^{256} of the angiotensin II receptor (implicated in antagonist binding). Arg^{292} is in a similar region of transmembrane helix VII as Tyr^{287} of the neurokinin 1 receptor (implicated in peptide and antagonist binding), Asp^{284} of the bradykinin receptor (implicated in bradykinin binding) and Asn^{295} of the angiotensin II receptor (implicated in non-peptide antagonist binding) (Fig. 3). The observation that the potency of ATP and UTP were affected equally by the mutation of the three amino acid residues suggests that the P2U receptor binding pocket interacts with the negatively charged phosphate residues of ATP or UTP and not the nucleoside moieties. Further mutagenesis studies should help to determine which residues are responsible for binding the nucleoside group of ATP and UTP.

Biological and clinical significance

A growing body of evidence suggests that the P2U purinoceptor may be the most widely distributed of the known P2 receptor subtypes. With the cloning of P2U purinoceptor cDNAs, it has become possible to use Northern analysis to determine unequivocally whether P2U purinoceptor mRNA is present in mammalian tissues. This approach was used to demonstrate the presence of P2U purinoceptor mRNA in the heart, liver, lung and kidney of both humans and mice (Lustig et al 1993, Parr et al 1994). The mRNA is also present in

```
                                       TM6

hNK1       HEQVSAKRKV VKMMIVVVCT FAICWLPFHI FFLLPYINP- DLYLK----K   281
hAT1       KNKPR-NDDI FKIIMAIVLF FFFSWIPHQI FTFLDVLIQL GIIRDCRIAD   278
hBK        KEIQT-ERRA TVLVLVVLLL FIICWLPFQI STFLDTLHRL GILSSCQDER   281
mP2U       GGLPRAKRKS VRTIALVLAV FALCPLPFHV TRILYYSFR- SLDLSCHTLN   282

Consensus .......R.. ......V... F..CWLPF.I ...L...... .....C....



                                  TM7

hNK1       FIQQVYLAIM WIAMSSTMYN PIIYCCLNDR FRLGFKHAFR      321
hAT1       IVDTAMPITI CIAYFNNCLN PLFYGFLGKK FKRYFLQLLK      318
hBK        IIDVITQIAS FMAYSNSCLN PLVYVIVGKR FRKKSWEVYQ      321
mP2U       AINMAYKITR PLASANSCLD PVLYFLAGQR LVRFARDAKP      322

Consensus .I.....I.. ..A..N.CLN P..Y...G.R F.........
```

FIG. 3. Sequence alignment of the murine P2U (mP2U), human neurokinin 1 (hNK1), human angiotensin II type 1 (hAT1) and human bradykinin (hBK) receptors. Amino acids that form the sixth and seventh transmembrane helices are shown, with boxed residues denoting amino acids that are conserved among all four receptors. Circled residues have been implicated in agonist and antagonist binding by mutagenesis studies, pharmacophore mapping and molecular modelling (see text).

human placenta and skeletal muscle and in mouse brain, spleen and testes. Additional information concerning the cellular distribution of P2U purinoceptors has been obtained from functional assays in which changes in intracellular second messenger levels or closely associated signalling events were assessed after treatment of cultured cells with P2U purinoceptor agonists. By this criterion, cell types expressing P2U purinoceptors—in at least some mammalian species—include endothelial and epithelial cells, aortic smooth muscle cells, osteoblasts, pituitary and thyroid cells, hepatocytes, lung alveolar type II cells, astrocytes and a variety of blood cells, including macrophages, monocytes and neutrophils (Dubyak & El-Moatassim 1993, Gordon 1986).

Although P2U purinoceptors appear to be ubiquitious, in only a few instances has enough evidence accumulated to suggest their physiological significance. P2U purinoceptors on the apical surface of the vascular endothelium appear to be intimately involved in the interaction between endothelial cells and platelets (and possibly other blood cells), and may act in an antithrombolytic manner (Boeynaems & Pearson 1990). Likewise, there is some evidence that P2U purinoceptors regulate glycogenolysis in hepatic tissue and participate in the regulation of skeletal and smooth muscle tone (Boarder et al 1995, Dubyak & El-Moatassim 1993).

The P2U purinoceptor has attracted interest as a target for pharmacotherapy. In airway epithelial cells from cystic fibrosis patients, P2U purinoceptors couple to a Cl^- secretory pathway distinct from the defective cystic fibrosis transmembrane conductance regulator (Brown et al 1991, Mason et al 1991). Clinically, there is evidence that inhaled UTP, acting through P2U purinoceptors on airway epithelial cells, can bypass the block to epithelial Cl^- transport and increase Cl^- secretion in cystic fibrosis patients (Knowles et al 1991). Activation of these P2U purinoceptors with more effective subtype-selective analogues may help to improve the clinical outlook for patients with cystic fibrosis.

Much remains to be learned about the function of P2U purinoceptors in both physiological and pathophysiological settings. Considering the diversity of tissues in which they are expressed, it seems likely that P2U purinoceptors will have multiple biological roles. Nevertheless, it is not yet clear why P2U purinoceptors are ubiquitously expressed and why their activation has such drastically different effects in so many different tissues. How *in vitro* effects relate to the receptor's biological roles in a living animal remains to be determined.

Acknowledgements

We are very grateful to Julie Theriot for artwork and for critically reading the manuscript. K. D. L. gratefully acknowledges the support of the Department of Biochemistry & Biophysics at the University of California, San Francisco. This work was supported in part by a predoctoral fellowship from the Howard Hughes Medical Institute (A. K. S.), and by grants from the NIH and the Cystic Fibrosis Foundation.

References

Barnard EA, Burnstock G, Webb TE 1994 G protein-coupled receptors for ATP and other nucleotides: a new receptor family. Trends Pharmacol Sci 15:67–70

Boarder MR, Weisman GA, Turner JT, Wilkinson GF 1995 G protein-coupled P_2 purinoceptors: from molecular biology to functional responses. Trends Pharmacol Sci 16:133–139

Boeynaems JM, Pearson JD 1990 P2 purinoceptors on vascular endothelial cells: physiological significance and transduction mechanisms. Trends Pharmacol Sci 11:34–37

Brown HA, Lazarowski ER, Boucher RC, Harden TK 1991 Evidence that UTP and ATP regulate phospholipase C through a common extracellular 5′-nucleotide receptor in human airway epithelial cells. Mol Pharmacol 40:648–655

Burnstock G 1972 Purinergic nerves. Pharmacol Rev 24:509–581

Burnstock G 1978 A basis for distinguishing two types of purinergic receptor. In: Straub RW, Bolis L (eds) Cell membrane receptors for drugs and hormones: a multidisciplinary approach. Raven, New York, p 107–118

Dalziel HH, Westfall DP 1994 Receptors for adenine nucleotides and nucleosides: subclassification, distribution, and molecular characterization. Pharmacol Rev 46:449–466

Dubyak GR, El-Moatassim C 1993 Signal transduction via P_2-purinergic receptors for extracellular ATP and other nucleotides. Am J Physiol 265:577C–606C

Erb L, Lustig KD, Sullivan DM, Turner JT, Weisman GA 1993 Functional expression and photoaffinity labeling of a cloned P_{2U} purinergic receptor. Proc Natl Acad Sci USA 90:10449–10453

Erb L, Garrad R, Wang Y, Quinn T, Turner JT, Weisman GA 1995 Site-directed mutagenesis of P2U purinoceptors. Positively charged amino acids in transmembrane helices 6 and 7 affect agonist potency and specificity. J Biol Chem 270:4185–4188

Fredholm BB, Abbracchio MB, Burnstock G et al 1994 Nomenclature and classification of purinoceptors. Pharmacol Rev 46:143–156

Gordon JL 1986 Extracellular ATP: effects, sources and fate. Biochem J 233:309–319

Knowles MR, Clarke LL, Boucher RC 1991 Activation by extracellular nucleotides of chloride secretion in the airway epithelia of patients with cystic fibrosis. N Engl J Med 325:533–538

LeGrice SFJ, Gruninger-Leitch F 1990 Rapid purification of homodimer and heterodimer HIV-1 reverse transcriptase by metal chelate affinity chromatography. Eur J Biochem 187:307–314

Lin TA, Lustig KD, Sportiello MG, Weisman GA, Sun GY 1993 Signal transduction pathways coupled to a P_{2U} purinoceptor in neuroblastoma X glioma (NG108-15) cells. J Neurochem 60:1115–1125

Ljungquist C, Breitholtz A, Brink-Nilsson H, Moks T, Uhlen M, Nilsson B 1989 Immobilization and affinity purification of recombinant proteins using histidine peptide fusions. Eur J Biochem 186:563–569

Lustig KD, Shiau AK, Brake AJ, Julius D 1993 Expression cloning of an ATP receptor from mouse neuroblastoma cells. Proc Natl Acad Sci USA 90:5113–5117

Mason SJ, Paradiso AM, Boucher RC 1991 Regulation of transepithelial ion transport and intracellular calcium by extracellular ATP in human normal and cystic fibrosis airway epithelium. Br J Pharmacol 103:1649–1656

Masu Y, Nakayama K, Tamaki H, Harada Y, Kuno M, Nakanishi S 1987 cDNA cloning of bovine substance-K receptor through oocyte expression system. Nature 329:836–838

O'Connor SE, Dainty IA, Leff P 1991 Further subclassification of ATP receptors based on agonist studies. Trends Pharmacol Sci 12:137–141

Parr CE, Sullivan DM, Paradiso AM et al 1994 Cloning and expression of a human P_{2U} nucleotide receptor, a target for cystic fibrosis pharmacotherapy. Proc Natl Acad Sci USA 91:3275–3279

Probst WC, Snyder LA, Schuster DI, Brosius J, Sealfon SC 1992 Sequence alignment of the G-protein coupled receptor superfamily. DNA Cell Biol 11:1–20

Schambye HT, Hjorth SA, Weinstock J, Schwartz TW 1995 Interaction between the nonpeptide angiotensin antagonist SKF-108,566 and histidine 256 (HisVI-16) of the angiotensin type I receptor. Mol Pharmacol 47:425–431

Simon MI, Strathmann MP, Gautam N 1991 Diversity of G proteins in signal transduction. Science 252:802–808

Strader CD, Fong TM, Tota MR, Underwood D, Dixon RAF 1994 Structure and function of G protein-coupled receptors. Annu Rev Biochem 63:101–132

Underwood DJ, Strader CD, Rivero R, Patchett AA, Greenlee W, Prendergast K 1994 Structural model of antagonist and agonist binding to the angiotensin II, AT1 subtype, G protein coupled receptor. Chemistry & Biology 1:211–221

Webb TE, Simon J, Krishek BJ et al 1993 Cloning and functional expression of a brain G-protein-coupled ATP receptor. FEBS Lett 324:219–225

DISCUSSION

Leff: Relating to the activation of the P2U receptor, we notice that UDP and ADP go from partial to full agonists when Lys^{289} is changed to Arg. This suggests that the conformational change involving the activation of the receptor has become more productive for those agonists. Doesn't that tell you something about the mechanism or the nature of the conformational change which constitutes receptor activation?

Lustig: We just don't know enough about the 3D structure of the transmembrane domain or any other part of the receptor to be sure. Until we get a 3D structure of agonist-bound and unoccupied receptor it is going to be very difficult to conclude anything about the nature of the conformational change underlying receptor activation.

Boucher: All your data were expressed as percentage of maximal: if you look at the absolute Ca^{2+} changes, do UDP and ADP become more active or were they the same?

Lustig: They become more active.

Barnard: The fact that the arginines which you mentioned can vary to a very different type of residue in one, two or even three receptors in this series, as now seen, could mean that they are not part of the phosphate binding site. I think that we should, for safety, only really consider at the moment residues that are conserved in them all.

Lustig: We know so little about how the 3D structures of these receptors are actually formed that we can't make very many predictions based solely on sequence similarity. It could be that there are other residues that can substitute to form the correct conformation at the receptor binding site.

Barnard: There would need to be a particularly big change in this case if you are going to go from TM6 to another TM containing an arginine (or lysine) in changing the agonist binding site between one receptor and another. You would also have to make the assumption that the binding site for ATP has a very different configuration for ATP in different members of the series. In one case it is aligned with that site in TM6 binding the phosphate, and in another case it is aligned with TM3. In TM6, in the subtypes lacking the arginine, there is no other positive residue there. I think that involvement is less likely.

Jacobson: As long as you have at least one of the positive charges preserved, you can say that there is some similarity in the phosphate binding. You are generally right that we should not consider this a rigid candidate that applies to all the subtypes, and we should consider the possibility that there may be even multiple binding modes within the same subtype for ATP. There would have to be many points of recognition for ATP, and a given area of recognition within a particular subtype may not be as significant as you would think. I'm not that troubled by it.

Barnard: The RGD sequence, which is the integrin binding site, is strangely only present in P2U. It is not present in any of the other sequences that we have. What does that mean? I suspect that, if anything, it makes it more significant.

Lustig: Since the P2U receptor is expressed in so many tissues, perhaps it points to a general role for the receptor in cell–cell or cell–matrix adhesion and communication.

Barnard: In support of that, the P2U receptor is present in almost all cell lines. There are even binding sites that have the pharmacology of P2U receptors on a line of the non-responding astrocytoma cells. Maybe those are poorly coupled to the transduction mechanisms.

Lustig: We also have had difficulty in identifying cell lines that do not express a P2U receptor. Whether the P2U receptor is ubiquitously expressed *in vivo*, however, remains to be determined, since it is possible that the P2U receptor is up-regulated after the cells are cultured.

Burnstock: There are many examples of endothelial cells that have both P2Y and P2U receptors, but there are some with only one or the other. It would be a mistake to assume that P2U receptors are universal.

Harden: Eric Barnard, how are you detecting the P2U receptor on astrocytes?

Barnard: Either by the response to UTP in the Ca^{2+} transient (reported in some astrocytic cultures and also found by my co-workers in various potential host cell lines) or, where they do not give that response, by binding with [^{3}H]UTP and an appropriate competition series with P2U agonists.

Harden: We've made many attempts to radiolabel P2U receptors on 1321N1 human astrocytoma cells and membranes from other cell lines. Although we can get binding that sort of fits, it never truly fits P2U receptor binding pharmacology. Moreover, we find no mRNA for P2U receptors in 1321N1 cells.

Barnard: I had said that it is a 'P2U-like' binding.

Harden: I wouldn't think it is the receptor, though.

Boucher: By PCR we did not detect any evidence for a P2U or $P2Y_2$-type receptor in 132 cells. One of the issues, however, is that there are different clones of 132s: unless we are all using the same clone, we can't be sure.

Barnard: We have applied PCR to some of the cell lines which we have looked at and found P2U receptor mRNA present. That might go along with the integrin binding motif: maybe the protein, even if not coupled, has a universal role in an attachment of the cell surface.

Wiley: We've just completed a study showing that when you have lymphocytes adherent to endothelial cells, agonists for the P2U purinoceptor can produce a Ca^{2+} transient not only in the endothelial cell but also in the adherent lymphocyte. But this signal transmission doesn't occur if the lymphocyte is simply adjacent to the endothelial cell; it has to be tightly adherent. We know that P2U is involved because P2Y agonists don't produce this transmission, but P2U agonists do (Wiley et al 1995). It could be that the P2U purinoceptor, through its RGD sequence, adheres to a counter receptor on the lymphocyte surface which actually transmits a Ca^{2+} signal from one cell to another.

Abbracchio: You mentioned the presence of these consensus phosphorylation sites. Do you have any data on their functional significance? For example, are they substrates for protein kinase A, C or maybe β-adrenergic receptor kinase? What happens in receptors carrying mutations at these sites? Do the mutated receptors show different sensitivities to agonists?

Lustig: Fernando Gonzalez' lab has been working with Gary Weisman's lab to try to answer those questions. They have made mutants in many of the consensus phosphorylation sites and are currently determining whether these mutants affect receptor activity or receptor down-regulation. Unfortunately, the results are not out yet.

IJzerman: Going back to Paul Leff's question about which residues are important for binding and which are important for activation of this receptor: we've used all of the G protein-coupled receptor sequences (about 700) to do statistical analysis on which residues seem to be important for either activation or binding. It is possible to isolate selected classes of receptors, such as adenosine receptor sequences, and put them against, for instance, all the amine receptor sequences or perhaps even all the receptor sequences that are present in this huge database. What we see then is that so-called correlated mutations often appear, which means that a pair of residues tends to stay conserved, or mutates in tandem. This suggests some functional correlation between the two residues. For P2 receptors we would need about 10–12 really different sequences to do such an analysis. It might be that if you did the analysis you would find that the arginine seems to be important as you have shown and, additionally, that is as correlated to another residue that may be very important for activation or binding. So far, we have observed such correlated mutations in many other receptor classes.

Di Virgilio: I have a comment on the up-regulation of the P2Y/U receptors. We observed that under some conditions of activation, you can induce human lymphocytes to show P2Y/U-linked responses which they don't show under resting conditions.

Lustig: What conditions are those?

Di Virgilio: When they are mitogenically activated.

I also have a comment on your very interesting Lys^{289} to Arg mutation. Perhaps this explains why Eric Barnard was unable to find an ADP receptor in the megakaryocytic cell line he was looking at. He started looking for an ADP receptor and ended up with an ATP receptor. The conformation in the native cell might be such that ADP is the lethal agonist. Just a small switch in the conformation converts this receptor into an ATP receptor.

Lustig: Alternatively, perhaps there is no receptor that is specific for ADP. Instead, there may be an accessory factor that can bind to an ATP receptor and 'convert' it to a receptor that can be activated by ADP.

Barnard: I agree that one cannot rule out an exotic system of that type, but I still tend to think that this is likely to be due to another receptor gene.

Reference

Wiley JS, Chen JR, Jamieson GP, Thurlow PJ 1995 Agonists for endothelial P_2 purinoceptors trigger a signaling pathway producing Ca^{2+} responses in lymphocytes adherent to endothelial cells. Biochem J 311:589–594

Functional properties of native and cloned P2X receptors

Annmarie Surprenant

Glaxo Institute for Molecular Biology, 14 Chemin des Aulx, 1228 Plan-les-Ouates, CH-1211 Geneva, Switzerland

Abstract. Electrophysiological experiments on dissociated smooth muscle and neurons have revealed three distinct phenotypes of P2X receptor: (1) a rapidly desensitizing, α,β-methylene ATP-sensitive response typical of most smooth muscle; (2) a non-desensitizing, α,β-methylene ATP-insensitive response characteristic of PC12 phaeochromocytoma cells and rat superior cervical ganglion neurons; and (3) a non-desensitizing, α,β-methylene ATP-sensitive response observed in sensory neurons. All of these purinoceptors share a similar cationic and high Ca^{2+} permeability and sensitivity to blockade by suramin, Cibacron blue, oxidized ATP, pyridoxal-5-phosphate and pyridoxalphosphate-6-azophenyl-2′,4′-disulfonic acid. Heterologous expression of two forms of cloned P2X receptors (from rat vas deferens and PC12 cells) reveals that each cloned receptor can reconstitute native responses with remarkable fidelity. Such results suggest that homo-oligomeric channels may be formed from single subunits of the P2X receptor in smooth muscle, PC12 cells and some neurons. The third phenotype observed in native cells might result from co-assembly of subunits of the cloned receptors. However, co-expression studies show that these two forms of the P2X receptor do not heteropolymerize. Therefore, the non-desensitizing, α,β-methylene ATP-sensitive response observed in sensory neurons may result from a distinct P2X receptor or from heteropolymerization of more than one distinct P2X purinoceptor.

1996 P2 purinoceptors: localization, function and transduction mechanisms. Wiley, Chichester (Ciba Foundation Symposium 198) p 208–222

ATP-gated cationic currents were first described in sensory neurons (Jahr & Jessell 1983, Krishtal et al 1983); they have now been described in numerous smooth muscle, neuronal and glial cells as well as in cochlear hair cells (see Surprenant et al 1995). These ATP-induced currents result from the activation of P2X purinoceptors. Although P2X purinoceptors share many functional similarities across this wide variety of cell types, differences in specific physiological and pharmacological properties allow delineation of three distinct receptor types (Fig. 1A). Recently, genes encoding two forms of the P2X receptor ($P2X_1$, $P2X_2$) have been isolated (Valera et al 1994, Brake et al

1994, North 1996, this volume). When expressed in oocytes or mammalian cells, ATP and other agonists evoke currents that are strikingly similar to those observed in native cells from which the genes had been cloned (Fig. 1B). This chapter summarizes our electrophysiological studies that provide the basis for the aforementioned statements.

Experimental rationale

We have taken a comparative approach to studying the properties of ionotropic P2X receptors in native cells and when heterologously expressed in mammalian cell lines (HEK293 human kidney fibroblasts and CHO cells), in order to reduce discrepancies in results that often arise due to differences in experimental methods (e.g. tissue dissociation and culture conditions, drug delivery methods, internal and external solutions for electrophysiological recordings). In studies summarized below, whole-cell recordings were made using the same internal and external solutions throughout, agonists were applied by a fast-flow 'concentration-clamp' U-tube delivery system, neurons and smooth muscle were all obtained from adult animals and were examined 3–7 days after dissociation, and heterologously expressed P2X receptors were examined 18–48 h after lipofectin-induced transfection (see Khakh et al 1995, Evans et al 1995). Experiments were carried out on rat superior cervical ganglia (SCG), nodose and coeliac neurons, guinea-pig coeliac neurons, rat vas deferens and tail artery smooth muscle, nerve growth factor-differentiated PC12 cells, and HEK cells expressing the P2X receptor cloned from rat vas deferens or human bladder smooth muscle ($P2X_1$) or cloned from PC12 cells ($P2X_2$; cDNA provided by A. Brake and D. Julius, UCSF).

Kinetics of ATP-gated currents

Like other ionotropic receptor channels (i.e. nicotinic, glutaminergic, type A γ-aminobutyric acid [$GABA_A$]), the latency to onset of ATP-gated currents is rapid, occurring within a few milliseconds. Time to peak current is concentration-dependent, as is also the case at other ligand-gated channels, but rise times of currents evoked by maximum concentrations of ATP fall into three groups. Smooth muscle and the cloned $P2X_1$ receptor show the fastest rise times ($\tau_{onset} \sim 5$ ms), while PC12 cells, rat SCG neurons and the cloned $P2X_2$ receptor are the slowest ($\tau_{onset} \sim 25$ ms) and sensory neurons exhibit an intermediate time constant ($\tau_{onset} \sim 15$ ms). The most striking difference in kinetic properties among P2X receptors is the marked desensitization observed during continuous agonist application in dissociated smooth muscle cells and the cloned $P2X_1$ receptor compared with the relative absence of desensitization observed at almost all other P2X receptors (Fig. 1). Desensitization is remarkably similar between native P2X receptors in smooth muscle and

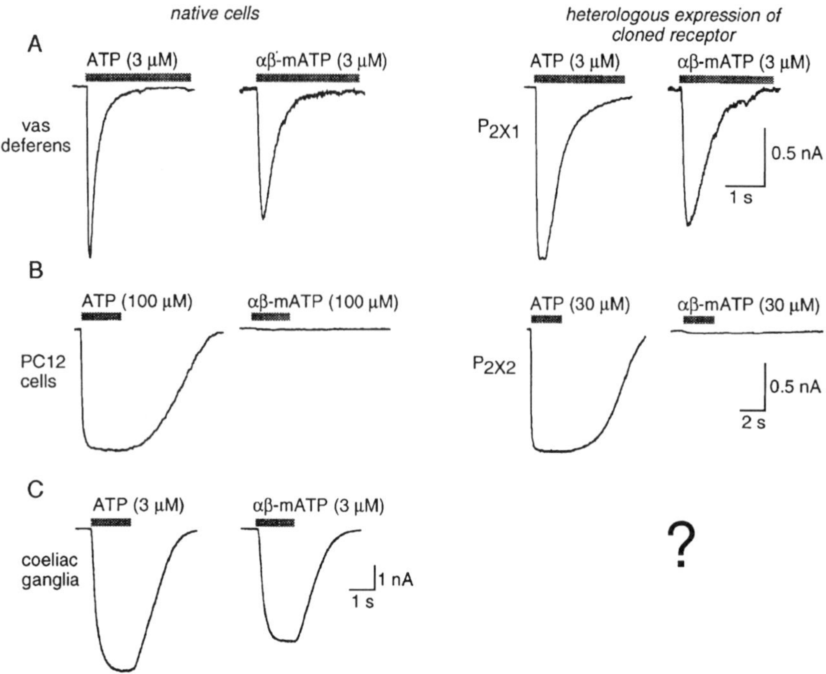

FIG. 1. Three distinct phenotypes can be distinguished in native cells (A–C, left-hand traces); these are termed (A) *α,β-meATP-sensitive, desensitizing*, (B) *α,β-meATP-insensitive, non-desensitizing* and (C) *α, β-meATP-sensitive, non-desensitizing*. When the $P2X_1$ receptor cDNA, cloned from rat vas deferens, or the $P2X_2$ receptor cDNA, cloned from PC12 cells, are transiently transfected into HEK293 cells (A and B, right-hand traces), responses are strikingly similar to endogenous P2X receptors. Traces are whole-cell currents recorded as described in text; bars above each trace indicate duration of agonist application. (Figure is adapted from review by Humphrey et al 1995.)

heterologously expressed $P2X_1$ receptors, and follows a single exponential time course (τ_d) which is about 300 ms for EC_{90} concentrations of ATP; desensitization during α,β-methylene ATP (α,β-meATP) application is significantly slower for equivalent amplitude currents (see also Evans & Kennedy 1994). Although there is little desensitization of ATP-gated currents through the $P2X_2$ receptor and other non-desensitizing receptors when agonist is applied for a few seconds, a significant desensitization can be observed at these sites during longer application periods (10–120 s; Bean 1990, Nakazawa et al 1990). Nevertheless, the distinct differences during short agonist applications make it reasonable to characterize native and cloned P2X receptors as ‘desensitizing’ and ‘non-desensitizing’.

Desensitizing responses are also associated with a dramatic rundown of the response with repeated agonist applications (Fig. 2). This rundown

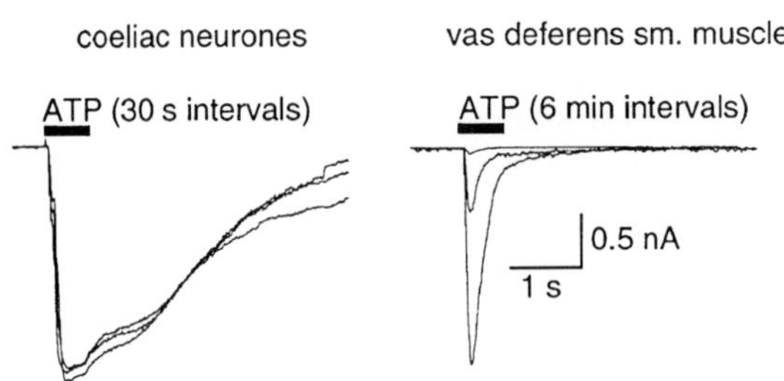

FIG. 2. Examples of rundown and lack of rundown in currents elicited by ATP applied to rat vas deferens smooth muscle or guinea-pig coeliac neurons at intervals of 30 s (coeliac neurons) or 6 min (vas deferens). (Recordings from vas deferens reproduced with permission from Khakh et al 1995.)

phenomenon has been observed in all studies in which single dissociated smooth muscle cells have been examined, as well as in studies in which the $P2X_1$ receptor has been expressed in oocytes or the mammalian cell lines, HEK293 and CHO cells (Benham 1989, Benham & Tsien 1987, Friel 1988, Inoue & Brading 1991, Evans & Kennedy 1994, Khakh et al 1995, Valera et al 1994). At first sight, this seems a puzzling finding in light of what is known about excitatory junction potentials (EJPs) in smooth muscle, which almost certainly result from synaptically released ATP activating $P2X_1$ receptors (Sneddon & Burnstock 1984, Evans & Surprenant 1992, see Surprenant et al 1995). That is, sympathetic nerve stimulation at frequencies greater than about 0.5 Hz results in strong facilitation of EJPs in most smooth muscle, particularly vas deferens and arterial smooth muscle (Burnstock & Holman 1961, Sneddon & Burnstock 1984, Evans & Surprenant 1992, see Hirst 1989). It may be that synaptically activated $P2X_1$ receptors do undergo a similar rundown process but this is not apparent in the EJP during repetitive stimulation because release from individual sympathetic terminals (varicosities) onto smooth muscle occurs only very intermittently (Cunnane & Searl 1994).

Table 1 summarizes the kinetic properties of several native P2X receptors as well as the cloned $P2X_1$ and $P2X_2$ receptors. It is clear that onset kinetics fall into three groups, whereas desensitization and rundown properties separate into two distinct categories.

Ionic permeability

It is a general finding that ATP-gated ion channels are not only permeable to monovalent cations (equally permeable to both Na^+ and K^+) but they also show a very high Ca^{2+} permeability. Ca^{2+} permeability ratios ($P_{Ca^{2+}}/P_{Na^+}$) calculated from reversal potential measurements in altered Ca^{2+} solutions are about 4 : 1 for a number of smooth muscle and neuronal preparations as well as for the cloned $P2X_1$ and $P2X_2$ receptors (Benham & Tsien 1987, Benham 1989, Nakazawa et al 1991, Furukawa et al 1994, Valera et al 1994). Direct measurements of Ca^{2+} flow through P2X channels in sympathetic neurons have shown that only about 8% of the total current evoked by ATP is carried by Ca^{2+} under normal physiological conditions (Rogers & Dani 1995), presumably because there is such an excess of external Na^+ ions relative to Ca^{2+}. Another general characteristic of native and cloned P2X receptors is the inward rectification observed in both whole-cell and single-channel recordings (Khakh et al 1995a, Clous 1994, Brake et al 1994, Valera et al 1994). However, it has been noted for endogenous responses in neurons that the degree of inward rectification for a given neuronal population shows considerable variability from cell to cell (Khakh et al 1995a); we have observed similar variability in the degree of rectification observed in HEK293 cells expressing either the $P2X_1$ or $P2X_2$ receptor. The mechanisms responsible for rectification of P2X receptor channels remains to be determined, but it does not appear to involve Mg^{2+} ions, as is the case for NMDA receptors and inwardly rectifying K^+ channels (Sprengel & Seeburg 1995, Jan & Jan 1994). In the main, whole-cell recordings of ATP-gated currents over a wide range of preparations have revealed few, if any, differences in the properties of the cationic pore of P2X receptors. There are only limited studies in which single-channel properties of P2X receptors have been examined; most estimates of unitary conductance of these channels are in the range of 12–20 pS (Benham & Tsien 1987, Bean et al 1990, Cloues 1995, Nakazawa et al 1991, Valera et al 1994, but see Fieber & Adams 1991). It is not known whether this reflects receptor subtype heterogeneity, but preliminary studies on cloned P2X receptors suggest single

TABLE 1 Kinetic properties of native and cloned P2X purinoceptors

	$P2X_1$[a]	*Vas deferens*	*$P2X_2$*	*PC12*	*SCG*	*Nodose*	*Coeliac*
τ_{onset} (ms)[b]	4.8	5.6	22	18	24	12.4	9.8
τ_d (s)[b]	0.34	0.39	43	38	49	24	29
Rundown (%)[c]	100	100	15	18	12	19	22
Recovery (min)[d]	6–8	8–12	0.5–1	0.5–1	1	1	1

[a]Values are means from 3–14 experiments carried out on cloned $P2X_1$ receptor obtained from human bladder, acutely dissociated rat vas deferns smooth muscle, cloned $P2X_2$ receptor obtained from PC12 cells, nerve growth factor-differentiated PC12 cells, rat SCG and nodose neurons and guinea-pig coeliac neurons.
[b]τ_{onset} and τ_d measured from currents evoked in response to EC_{90} concentration of ATP applied for 2 s.
[c]Rundown defined as percentage decrease from initial response occurring at the fifth response during ATP application (0.5 s duration) at 30 s intervals.
[d]Time to recover to >80% of initial response after five applications of ATP delivered at 30 s intervals.

channel properties may well show considerable differences (R. J. Evans, unpublished results).

Antagonist properties

It is widely acknowledged that purinoceptor research has suffered significantly from a lack of selective, competitive, high-affinity antagonists. Although a number of effective antagonists have become available over the last five years, particularly suramin, pyridoxalphosphate-6-azophenyl-2′, 4′-disulfonic acid (PPADS) and pyridoxal-5′-phosphate (P5P), these have not allowed adequate separation between classes of P2 purinoceptors nor among P2X receptor subtypes (see reviews by Abbracchio & Burnstock 1994, Kennedy & Leff 1995). In various studies in whole tissues, these have sometimes acted as competitive antagonists, and differences in antagonist effects occasionally have been used to distinguish subtypes. We have examined a number of compounds which have been shown to inhibit P2X-mediated responses in whole tissues (Table 2). At native and cloned P2X receptors, none of the antagonists we examined showed competitive inhibition of ATP-gated currents, with the exception of low concentrations of suramin ($\leqslant 10\,\mu$M). PPADS, P5P and DIDS (4,4′-diisothiocyanato-stilbene-2,2′-disulfonic acid) were very slowly reversible, or irreversible, while actions of suramin were extremely rapid (2–10 s for onset and offset of inhibition). It is important to note that suramin and PPADS are effective at concentrations between 1–10 μM (Table 2); it has been a consistent observation in our studies that higher concentrations ($\geqslant 100\,\mu$M) produce non-selective inhibition of these receptors as well as other ion channels (e.g. 5-HT_3, $GABA_A$, delayed rectifier K^+ and N-type Ca^{2+} channels). Currently, we consider ATP-evoked inward currents that require 100 μM or more of suramin or PPADS for inhibition to be 'antagonist-insensitive'. The inhibition by oxidized ATP, which has been reported to be an irreversible P2Z purinoceptor antagonist (Wiley et al 1992), was fully reversed within 10–15 min of removal. The covalent linking of oxidized ATP to P2Z receptors in lymphocytes (Wiley et al 1992) was observed after a 24 h incubation period and therefore cannot be compared directly to our protocol in which antagonists are applied for 8–20 min prior to agonist application. Curare (10–100 μM) has been reported to be an effective blocker of ATP-evoked currents in PC12 cells and in oocytes expressing the $P2X_2$ receptor cloned from PC12 cells (Nakazawa et al 1991, Brake et al 1994). However, we have not observed any significant curare-mediated inhibition of these currents in our studies (Table 1). The results summarized in Table 1 make clear the need for new selective antagonists: cell lines stably expressing cloned $P2X_1$ and $P2X_2$ receptors are now available which should be of assistance in developing such selective compounds.

TABLE 2 Antagonist properties of native and cloned P2X purinoceptors

	$P2X_1$[a]	*Vas deferens*	$P2X_2$	*PC12*	*SCG*	*Nodose*	*Coeliac*
Suramin[b]	1	2	8	5	4	12	3
PPADS	1	0.5	1	3	2	5	1.5
P5P	8	10	10	25	25	21	20
Oxo-ATP	25	30	20	NT	30	NT	15
DIDS	5	1	100	NT	NT	NT	NT
Curare	>300	>300	>300	>100	NT	>100	>100

[a]Preparations are the same as described in Table 1.
[b]All values are concentrations in μM producing half-maximal inhibition (IC_{50} values) of inward current evoked by EC_{90} concentration of ATP; values shown are means from 3–12 experiments. NT, not tested; PPADS, pyridoxalphosphate-6-azophenyl-2′,4′-disulfonic acid; P5P, pyridoxal-5′-phosphate; oxo-ATP, oxidized ATP; DIDS, 4,4′-diisothiocyanatostilbene-2,2′-disulfonic acid.

Agonist properties

Because of the lack of selective antagonists, pharmacological characterization of P2 purinoceptors has relied primarily on agonist potency profiles. These types of studies in multicellular tissues face three major difficulties. First, ATP and other hydrolysable analogues are subject to potent but variable degradation by ectonucleotidases. Secondly, ubiquitously present metabotropic P2Y receptors are activated by most of the same agonists. Thirdly, mixed populations of P2X receptor subtypes may be activated. Concentration-clamp methods of agonist application during whole-cell recordings from single cells in which G protein-associated pathways have been dialysed away can overcome most of the first two problems; similar recordings made from cells expressing a single subtype of a cloned P2X receptor can obviate the third problem. Comparison of agonist concentration–response curves obtained in this manner reveal two distinct patterns of agonist actions at P2X receptors (Table 3). ATP, 2-methylthioATP, 2-chloroATP and ADP are full agonists at all sites examined, with ATPγS, Ap_5A and 3′-*O*-(4-benzoyl)benzoyl ATP (BzATP) acting as partial agonists in all groups. Thus, these agonists all exhibit similar rank order of agonist potency; however, EC_{50} values for these agonists at native smooth muscle, guinea-pig coeliac and rat sensory neurons, and cloned $P2X_1$ receptors, are approximately 10-fold lower than for the cloned $P2X_2$ receptor, rat SCG, coeliac, nucleus tractus solitarius neurons and native PC12 cells. In the first group, α,β-meATP is approximately equipotent with ATP but it is virtually ineffective in the second group. Of interest are the actions of the D and L isomers of β,γ-meATP which appear to differentiate a third group: α,β-meATP-sensitive neurons (e.g. guinea-pig coeliac, rat nodose)

TABLE 3 Agonist properties of native and cloned P2X purinoceptors

	$P2X_1$[a]	*Vas deferens*	*$P2X_2$*	*PC12*	*SCG*	*Nodose*	*Coeliac*
ATP[b]	0.9	3	8	40	43	3	3.3
α,β-methylene ATP	2.2	1	>300	>300	>300	9	13
2-methylthio ATP	1	3	3	45	46	0.4	3
2-chloroATP	0.5	NT	3	32	NT	NT	2
ATPγS	3	NT	21	60	NT	10	8
ADP	73	80	220	300	NT	80	65
D-β,γ-methylene ATP	3	5	>300	>300	>300	>300	>300
Ap_5A	3	NT	90	200	NT	NT	NT
BzATP	0.7	NT	23	90	120	NT	12

[1]Preparations are as described in Table 1.
[2]Values are average EC_{50} concentrations in μM. In all preparations, ATP, 2-methylthioATP, 2-chloroATP and ADP were full agonists while ATPγS, BzATP and Ap_5A produced maximum of 50–70% of the ATP response.
NT, not tested.

are relatively insensitive to D- and L-β,γ-meATP whereas these agonists are about equipotent to ATP and α,β-meATP in native smooth muscle and the cloned $P2X_1$ receptor (Table 3, Trezise et al 1995).

Homomeric and heteromeric P2X receptor channels?

The results described above show that the physiological and pharmacological properties of native P2X receptors in smooth muscle are reconstituted with remarkable fidelity by expression of the $P2X_1$ receptor in oocytes or mammalian cells. Similarly, there are few differences in properties of native (i.e. nerve growth factor-differentiated PC12 cells) and cloned $P2X_2$ receptors. Thus, it seems likely that native receptors in these preparations are formed by association of homomeric subunits in an as yet unknown stochiometry, although it is emphasized that there is no direct evidence that P2X receptors in smooth muscle or in PC12 cells form by homopolymerization of these specific subunits.

As indicated by results presented in this chapter and illustrated in Fig. 1c, a third phenotype is observed in a number of neuronal preparations. This phenotype, *α,β-meATP-sensitive, non-desensitizing*, exhibits kinetic properties similar to the cloned $P2X_2$ receptor (Table 1) and agonist potency properties similar to the cloned $P2X_1$ receptor (Table 3). This receptor phenotype may result from a new gene product, or from heteropolymerization of $P2X_1$ and $P2X_2$ to form a molecularly distinct architecture with new functional properties. We therefore asked whether co-expression of the two P2X subtypes causes an ATP-gated cationic current which exhibits minimal desensitization but is fully activated by α,β-meATP or whether the resulting current represents the simple addition of the two independent subtypes. We also took advantage of the findings that repeated agonist applications at 30 s intervals result in >95% decrease in current produced by activation of $P2X_1$ with little decrease in response in cells expressing $P2X_2$ (Fig. 2, Table 1). These studies showed that all currents recorded from cells co-transfected with $P2X_1$ and $P2X_2$ could be described by the addition of independent subunits; thus, no evidence for heteropolymerization of these subunits was obtained. An example of one of these experiments is illustrated in Fig. 3. Nevertheless, as it seems certain that these two receptors are only the first of this new multigene family (North 1996, this volume), it remains to be seen whether ATP-gated currents observed in sensory and central neurons form from homomeric or heteromeric association of new P2X receptor subunits.

Summary

Results have been presented which support the conclusion that three functionally distinct types of P2X receptors are observed in smooth muscle

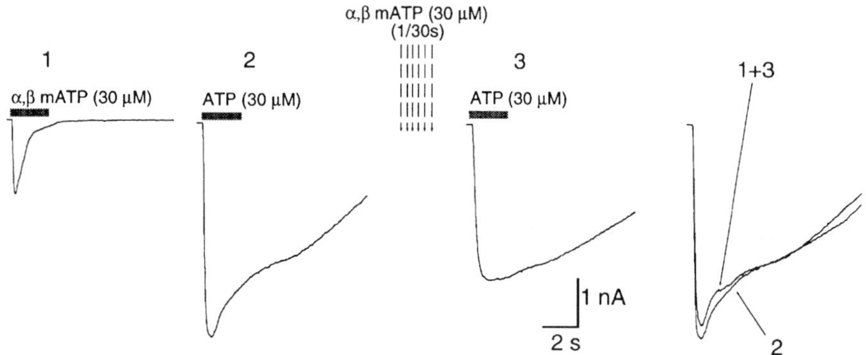

FIG. 3. Heteropolymerization does not occur when cloned $P2X_1$ and $P2X_2$ receptors are co-expressed in HEK293 cells. Recordings are from HEK cells co-transfected with cDNA encoding both $P2X_1$ and $P2X_2$ receptors. Application of α,β-methylene ATP (α,βmATP) evokes current response identical to that observed when only $P2X_1$ receptors are expressed (e.g. see Fig. 1a), but when ATP is applied to the same cell a biphasic current response occurs. After repeated applications of α,β-methylene ATP (indicated by arrows, current traces not shown but response runs down as illustrated in Fig. 2), a subsequent application of ATP evokes only a sustained component. The superimposed traces on the right are the current to the initial application of ATP (labelled 2) and a digital addition of the currents labelled 1 and 3; these results indicate that the response is due to the simple addition of two individual channels.

and neurons. These three types of P2X receptor responses share similar antagonist potency profiles and ionic permeability properties; however, they can be distinguished by differences in channel kinetics and agonist potency. The first type can be termed *α,β-meATP-sensitive, desensitizing*; this is observed in smooth muscle and is represented by homomeric forms of the cloned $P2X_1$ receptor. This receptor shows marked desensitization during brief applications of agonist and dramatic rundown of the response with repeated agonist application; agonist EC_{50} concentrations are in the low micromolar range. The second type is *α,β-meATP-insensitive, non-desensitizing*; it is observed in PC12 cells, rat SCG and nucleus tractus solitarius neurons and is represented by homomeric $P2X_2$ receptors. This receptor shows little desensitization or rundown; agonist EC_{50} values are 10-fold higher than α,β-meATP-sensitive, desensitizing receptors. The third type is *α,β-meATP-sensitive, non-desensitizing*; it is seen in guinea-pig coeliac and rat sensory neurons. No gene product encoding this type of receptor has been published and it does not form by heteropolymerization of the $P2X_1$ and $P2X_2$ receptors. Continued electrophysiological and molecular biological approaches will determine whether these three types will suffice to describe new endogenous and cloned P2X receptors.

References

Abbracchio MP, Burnstock G 1994 Purinoceptors: are there families of P_{2X} and P_{2Y} purinoceptors? Pharmacol Ther 64:445–475

Bean BP 1990 ATP-activated channels in rat and bullfrog sensory neurons: concentration dependence and kinetics. J Neurosci 10:1–10

Bean BP, Williams CA, Ceelen PW 1990 ATP-activated channels in rat and bullfrog sensory neurons: current–voltage relation and single channel behavior. J Neurosci 10:11–19

Benham CD 1989 ATP-activated channels gate calcium entry in single smooth muscle cells dissociated from rabbit ear artery. J Physiol 419:689–701

Benham CD, Tsien RW 1987 A novel receptor-operated Ca^{2+}-permeable channel activated by ATP in smooth muscle. Nature 328:275–278

Brake AJ, Wagenbach MJ, Julius D 1994 New structural motif for ligand-gated ion channels defined by an ionotropic ATP receptor. Nature 371:519–523

Burnstock G, Holman ME 1961 The transmission of excitation from autonomic nerve to smooth muscle. J Physiol 155:115–133

Cloues R 1995 Properties of ATP-gated channels recorded from sympathetic neurons: voltage dependence and regulation by zinc ions. J Neurophysiol 73:312–319

Cunnane TC, Searl TJ 1994 Neurotransmitter release mechanisms in autonomic nerve terminals. Adv Second Messenger Phosphoprotein Res 29:425–459

Evans RJ, Kennedy C 1994 Characterization of P_2 purinoceptors in the smooth muscle of the rat tail artery: a comparison between contractile and electrophysiological responses. Br J Pharmacol 113:853–860

Evans RJ, Surprenant A 1992 Vasoconstriction of guinea-pig submucosal arterioles following sympathetic nerve stimulation is mediated by the release of ATP. Br J Pharmacol 106:242–249

Evans RJ, Lewis C, Buell G, Valera S, North RA, Surprenant A 1995 Pharmacological characterization of heterologously expressed ATP-gated cation channels (P_{2X} purinoceptors). Mol Pharmacol 48:178–183

Fieber LA, Adams DJ 1991 Adenosine triphosphate-evoked currents in cultured neurons dissociated from rat parasympathetic cardiac ganglia. J Physiol 434:239–256

Friel DD 1988 An ATP-sensitive conductance in single smooth muscle cells from the rat vas deferens. J Physiol (Lond) 401:361–380

Furukawa K, Ishibachi H, Akaike N 1994 ATP-induced currents in neurons freshly dissociated from the tuberomammillary nucleus. J Neurophysiol 71:868–873

Hirst GDS 1989 Neuromuscular transmission in intramural blood vessels. In: Schultz SG, Wood JD, Rauner BB (eds) Handbook of physiology, section 6: The gastrointestinal system. American Physiological Society, Bethesda, MD, p 1635–1666

Humphrey PPA, Buell G, Kennedy I et al 1995 New insights on P_{2X} purinoceptors. Naunyn-Schmiedeberg's Arch Pharmacol 35:1–12

Inoue R, Brading AF 1991 Human, pig and guinea-pig bladder smooth muscle cells generate similar inward currents in response to purinoceptor activation. Br J Pharmacol 103:1840–1846

Jahr CE, Jessell TM 1983 ATP excites a subpopulation of rat dorsal horn neurons. Nature 304:730–733

Jan LY, Jan YN 1994 Potassium channels and their evolving gates. Nature 371:119–122

Kennedy C, Leff P 1995 How should P_{2X} purinoceptors be classified pharmacologically? Trends Pharmacol Sci 16:168–174

Khakh BS, Humphrey PPA, Surprenant A 1995 Electrophysiological properties of P_{2X} purinoceptors in rat superior cervical nodose and guinea-pig coeliac neurons. J Physiol 484:385–395

Krishtal OA, Marchenko SM, Obukhov AG 1983 Receptor for ATP in the membrane of mammalian sensory neurons. Neurosci Lett 35:41–45

Nakazawa K, Fujimori K, Takanaka A, Inoue K 1990 An ATP-activated conductance in pheochromocytoma cells and its suppression by extracellular calcium. J Physiol 428:257–272

Nakazawa K, Inoue K, Fujimori K, Takanaka A 1991 ATP activated single channel currents recorded from cell free patches of pheochromocytoma cells. Neurosci Lett 119:5–8

North RA 1996 P2X receptors: a third major class of ligand-gated ion channels. In: P2 purinoceptors: localization, function and transduction mechanisms. Wiley, Chichester (Ciba Found Symp 198) p 91–109

Rogers M, Dani JA 1995 Comparison of quantitative calcium flux through NMDA, ATP and ACh receptor channels. Biophys J 68:501–506

Sneddon P, Burnstock G 1984 ATP as cotransmitter in rat tail artery. Eur J Pharmacol 106:149–152

Sprengel R, Seeburg PH 1995 Ionotropic glutamate receptors. In: North RA (ed) Ligand- and voltage-gated ion channels. CRC Press, London, p 213–264

Surprenant A, Buell G, North RA 1995 P_{2X} receptors bring new structure to ligand-gated ion channels. Trends Neurosci 18:224–229

Trezise DJ, Michel AD, Grahames CBA, Khakh BJ, Surprenant A, Humphrey PPA 1995 The selective P_{2X} purinoceptor agonist, β,γ-methylene-L-adenosine 5′-trisphosphate, discriminates between smooth muscle and neuronal P_{2X} receptors. Naunyn-Schmiedeberg's Arch Pharmacol 351:603–609

Valera S, Hussy N, Evans RJ et al 1994 A new class of ligand-gated ion channel defined by P_{2X} receptor for extracellular ATP. Nature 371:516–519

Valera S et al 1995 Characterization and chromosomal localization of a human P_{2X} receptor from urinary bladder. Recept Channels 3:283–290
Wiley JS, Chen JR, Snook MB, Jamieson GP 1992 The P_{2Z} purinoceptor of human lymphocytes: actions of nucleotide agonists and irreversible inhibition by oxidized ATP. Br J Pharmacol 112:946–950

DISCUSSION

Edwards: It would be nice if the ATP receptor/channel classifications were so simple, but I fear that the ATP receptor from the medial habenula is not going to fit into any of the classes very well. It doesn't seem to rectify, it is α,β-meATP sensitive and it's highly Ca^{2+} permeable.

Surprenant: With one exception (rat dorsal root ganglia; Bean 1990) studies to date on endogenous P2X receptors have found rather high $Ca^{2+}:Na^{+}$ permeability ratios (in the order of 4:1; see Surprenant et al 1995).

Concerning rectification of P2X channels, we continue to be puzzled by the remarkable variability in the degree of rectification from cell to cell, or oocyte to oocyte, that we see in heterologously expressed $P2X_1$ and $P2X_2$ receptors or in endogenous receptors (e.g. Khakh et al 1995).

With regard to desensitization, it would appear that the P2X receptor in medial habenula shows a desensitization pattern that is intermediate between the smooth muscle $P2X_1$ and the PC12 $P2X_2$ receptor.

Edwards: Although it is hard to study under our experimental conditions, I would say that they're probably highly desensitized.

Kennedy: You see the rundown of the $P2X_1$ receptors *in situ* in the vas deferens smooth muscle cells, and you see it in the HEK cells: do you see it also in the oocytes?

Surprenant: Yes, it's as dramatic.

Kennedy: And when you're looking at the Ca^{2+} permeability, say in HEK cells, if you are at very low or zero Ca^{2+}, does it still run down?

Surprenant: The rundown of the $P2X_1$ receptor is not altered significantly by Ca^{2+} concentrations between 0.1 mM and 100 mM, nor by Mg^{2+} from 0 to 10 mM.

Miras-Portugal: What effect does Ap_5A have on the P2X receptors?

Surprenant: Ap_5A is a partial agonist in that it produces currents which are about 60% of the maximum ATP response. The EC_{50} for the Ap_5A response is in the 3–5 μM range.

Miras-Portugal: But only for the $P2X_1$?

Surprenant: No, Ap_5A works for all of these receptors.

Miras-Portugal: With the same efficacy?

Surprenant: It looks the same for all the receptors.

Miras-Portugal: The reason I ask is because in bovine noradrenergic chromaffin cells, where a P2X receptor similar to that cloned from PC12 cells exists, Ap_4A and Ap_5A were unable to induce the extracellular Ca^{2+} entry, nor could they imitate the secretory response induced by ATP on noradrenaline release that would be expected if they were agonists.

Surprenant: There are two clones for which we haven't looked at Ap_5A. It may be that these don't show the same response as the others.

Harden: Have you quantified the amount of P2X receptor protein that is expressed? Could this have anything to do with your capacity to see heteropolymerization?

Surprenant: Our calculations and our results indicate that we can detect expression of individual subunits as long as the ratio is at least 10 : 1—that is, if one receptor type is expressed at less than 10% of the other, we would be unable to detect the presence of the low expressing receptor. Our results suggest that the receptor types we have co-expressed do express at approximately the same rate. We base this on the finding that maximum currents are similar in cells expressing each of the four P2X receptors we have cloned so far—$P2X_1$, $P2X_2$, $P2X_3$ (Lewis et al 1995) and $P2X_4$ (C. Lewis, R. A. North, G. Buell & A. Surprenant, unpublished results).

Boucher: Have you any single-channel data on these receptors?

Surprenant: No, but we really need those data. Richard Evans has recently begun working on that. So far we have very little information, but the unitary conductance looks to be about the same, although some stay open longer and others are very 'flickery'.

Boucher: Do they really rectify if you have symmetric solutions?

Surprenant: We don't have sufficient data under symmetrical ion solutions to answer that question.

Westfall: Your PC12 cells are all nerve growth factor treated (differentiated). Have you ever looked at the non-differentiated PC12 cells? What happens?

Surprenant: Yes, we have. The only thing that happens is that the responses are 20–100-fold bigger. The dose–response curve is still the same.

Westfall: There are a lot of changes in PC12 cells when you treat them with nerve growth factor. That is a big magnification of the response.

Surprenant: As far as we can see and from what is published, that seems to be all that happens.

Leff: The key difference between the nodose and the PC12 cell types concerns the agonists. They all move rightward on the concentration axis going from one to the other and the response to α,β-meATP collapses. If this wasn't a direct channel-linked system, that behaviour would have resembled a change in receptor reserve in terms of classical receptor theory, but it would be argued that this can't apply because you haven't got the post-receptor coupling to allow a reserve to exist. But what you can say (as we understand from David Colquhoun's [1973] work and my own recent work [Leff 1995]) is that if you have different relative ratios of an inactive and active receptor conformation, you can get these different profiles of agonist activity, partial agonists becoming antagonists and so on. If we didn't know these were physically different entities then, pharmacologically speaking, we could invoke such an explanation. Is there an explanation which would allow you to say that, for example, the nodose has become tipped towards the inactive conformation by co-expression with the other type? Does that make any sort of sense?

Surprenant: That may be possible, but one of the striking things we saw when we did the co-expression studies and saw that we did have a new channel (Lewis et al 1995) was that the agonist properties were identical to the one that was α,β-meATP sensitive. So we didn't have anything in between—it was either one or the other, at least within a plus or minus twofold range.

Barnard: This is a very unusual system for ion channel receptors. In almost all other cases where a heteromer can be formed, the single subunits give no or poor responses. There are a few exceptions, for instance the α-subunit of the glycine receptor. Can you get any clues about the stoichiometry by varying the amount of DNA?

Surprenant: We have made only very preliminary experiments in which we have varied cDNA concentrations in order to see whether we can gain any information concerning the stoichiometry of P2X subunit assembly. To date, the results are all inconclusive—or undecipherable.

Barnard: Another feature which is often diagnostic of a heteromer is the cooperativity—heteromers tend to have a different Hill number. Do you find a change in Hill slope from that of the homomer?

Surprenant: I was hoping you wouldn't ask me that! We are just now obtaining sufficient data that will allow us to compare dose–response curves in this kind of detail.

North: We are searching for our keys under the streetlight when we think about the possible forms of these molecules coming together. The problem at the moment is that we have only really two phenotypes: that is, there are only two things that you can measure—α,β-meATP sensitivity and desensitization—so that limits your ability to ask the question whether or not they will actually heteropolymerize. We need many more clear phenotypic differences to be able to answer the question that Eric Barnard is raising: do three or four different clones contribute to a channel? Some of these subunits are in fact insensitive to antagonists, so that would give us a further way to separate their contributions to a multimeric channel. But what we would really like are major differences in single channel properties.

Barnard: Could you get any clue from looking at antibody localization? Do you find co-localization of receptor types?

North: Yes, in some cells.

Barnard: Of those which will co-polymerize?

North: No, we haven't done that because we have antibodies only for $P2X_1$ and $P2X_2$ receptors. The critical experiment to do is co-immunoprecipitation from native tissues.

Surprenant: As Eric Barnard just alluded to, it is an unusual channel in that two subunits (e.g. $P2X_2$ and $P2X_3$) that express very well on their own, also heteropolymerize to express a new channel with a distinct phenotype (Lewis et al 1995). I would like to speculate that a cell may express one type of receptor in response to a certain stimulus or a certain environment, another one in response to a different environment, and when they are both being made at the same time they heteropolymerize to form a functionally distinct receptor type.

References

Bean BP 1990 ATP-activated channels in rat and bullfrog sensory neurons: concentration dependence and kinetics. J Neurosci 10:1–10

Colquhoun D 1973 The relation between classical and cooperative models for drug action. In: Rang HP (ed) Drug receptors. MacMillan, London, p149–181

Khakh BS, Humphrey PPA, Surprenant A 1995 Electrophysiological properties of P_{2X} purinoceptors in rat superior cervical nodose and guinea-pig coeliac neurons. J Physiol (Lond) 484:385–395

Leff P 1995 The two-state model of receptor activation. Trends Pharmacol Sci 16:89–97

Lewis C, Neidhart S, Holy C, North RA, Buell G, Surprenant A 1995 Coexpression of $P2X_2$ and $P2X_3$ receptor subunits can account for ATP-gated currents in sensory neurons. Nature 377:423–435

Surprenant A, Buell G, North RA 1995 P_{2X} receptors bring new structure to ligand-gated ion channels. Trends Neurosci 18:224–229

ATP as a co-transmitter with noradrenaline in sympathetic nerves—function and fate

Charles Kennedy, Gerald J. McLaren, Tim D. Westfall and Peter Sneddon

Department of Physiology and Pharmacology, University of Strathclyde, Royal College, 204 George Street, Glasgow G1 1XW, UK

Abstract. ATP and noradrenaline are co-stored in synaptic vesicles in sympathetic nerves and when co-released act postjunctionally to evoke contraction of visceral and vascular smooth muscle. In the original purinergic nerve hypothesis it was proposed that ATP would then be sequentially broken down to ADP, AMP and adenosine. Although such breakdown can be measured, it is not clear how the time-scale of breakdown compares with the time-course of the postjunctional actions of ATP. We have investigated the role of ectoATPase in modulating purinergic neurotransmission in the guinea-pig vas deferens using ARL67156 (formerly FPL67516), a recently developed inhibitor of ectoATPase. ARL67156 (1–100 μM) potentiated neurogenic contractions in a concentration-dependent manner. Onset of potentiation was rapid and the effect reversed rapidly on washout of the drug. The effect was also frequency dependent, being greater at lower frequencies. The purinergic component of the neurogenic contraction was isolated using the α_1 antagonist prazosin (100 nM) and ARL67156 caused a similar potentiation. ARL67156 also potentiated contractions evoked by exogenous ATP (100 μM), but had no effect on those of the stable analogue α,β-methylene ATP (500 nM). In the presence of the P2 purinoceptor antagonist PPADS (100 μM), ARL67156 also had no effect on contractions evoked by noradrenaline (10 μM) or KCl (40 mM). These results are consistent with an inhibitory action of ARL67156 on ectoATPase and suggest that ectoATPase modulates purinergic transmission in the guinea-pig vas deferens. When released from sympathetic nerves, ATP acts at the P2X purinoceptor, a ligand-gated cation channel, to evoke depolarization and contraction. In single acutely dissociated smooth muscle cells of the rat tail artery, studied under voltage-clamp conditions, ATP and its analogues evoke an inward current, with a rank order potency of 2-methylthioATP = ATP > α,β-methylene ATP. This is very different from the order of potency for evoking contraction in whole vessel rings, which is α,β-methylene ATP $\gg$2-methylthioATP$\geq$ATP. This discrepancy can be explained by a previously unrecognized attenuation of the action of ATP and 2-methylthioATP, but not α,β-methylene ATP, by ectoATPase in whole tissues.

1996 P2 purinoceptors: localization, function and transduction mechanisms. Wiley, Chichester (Ciba Foundation Symposium 198) p 223–238

For many years noradrenaline was thought to be the sole neurotransmitter released from sympathetic nerves. Acting mainly at α_1-adrenoceptors, noradrenaline induces contraction of visceral and vascular smooth muscle via the inositol-1,4,5-trisphosphate ($InsP_3$) second messenger system and release of intracellular Ca^{2+} stores (Fig. 1), or in some cases by depolarizing the smooth muscle cells (Bolton & Large 1986, Walsh 1994). However, in many studies a nonadrenergic component to the sympathetic response in tissues such as blood vessels and the vas deferens was reported and it is now clear that ATP is co-stored with nonadrenaline and when released can evoke smooth muscle contraction (Fig. 1). Neuropeptide Y is also co-stored in sympathetic nerves and in some tissues potentiates contractions evoked by noradrenaline and ATP (see Kennedy 1993b). Thus, sympathetic neurotransmission is not the simple, single neurotransmitter process previously envisaged. Rather, multiple transmitters act and interact, in varying degrees depending on the tissue, to produce an integrated control of smooth muscle tone (for reviews see Burnstock 1990, von Kügelgen & Starke 1991, Kennedy 1993a). Here we will concentrate on recent advances in our knowledge of how ATP acts postjunctionally and how these actions are terminated by breakdown of ATP by ectoATPase.

Electrophysiological actions of endogenously released ATP

The initial postjunctional response of the vas deferens and most blood vessels to sympathetic nerve stimulation is a rapid, transient excitatory junction potential (EJP). Although EJPs are abolished by tetrodotoxin and guanethidine, they are resistant to α-adrenoceptor blockade (see von Kügelgen & Starke 1991). EJPs were first shown to be mediated by ATP in the early 1980s in the guinea-pig vas deferens, where they were inhibited by the P2 purinoceptor antagonist 3-*O*-3[*N*-(4-azido-2-nitrophenyl)amino] proprionyl ATP ($ANAPP_3$, Sneddon et al 1982, Sneddon & Westfall 1984). Subsequently, they have been shown to be inhibited by other P2 purinoceptor antagonists, such as suramin (Sneddon 1992) and pyridoxalphosphate-6-azophenyl-2′,4′-disulfonic acid (PPADS, McLaren et al 1994) and by desensitization of the P2 purinoceptor by the potent agonist α,β-methylene ATP (α,β-meATP, Sneddon & Burnstock 1984a). Likewise, EJPs in blood vessels such as the rat tail artery are inhibited by α,β-meATP (Sneddon & Burnstock 1984b) and suramin (McLaren et al 1995) (Fig. 2).

In most tissues, noradrenaline released from sympathetic nerves has no effect on the membrane potential of smooth muscle cells. However, in the rat tail artery, stimulation of the sympathetic nerves with a train of pulses produces an electrical response with two distinct phases (Sneddon & Burnstock 1984b, McLaren et al 1995). Each stimulus evokes a rapid EJP and as the train of pulses progresses a slow depolarization develops (Fig. 2). The slow depolarization, which is smaller

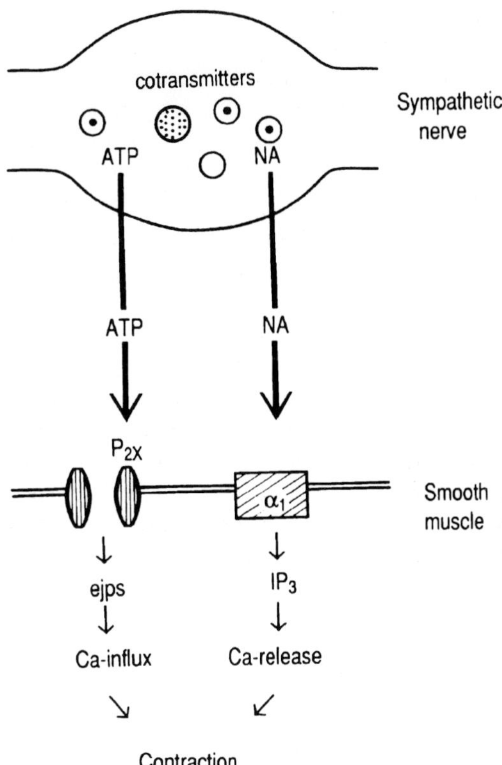

FIG. 1. Schematic representation of co-transmission by ATP and noradrenaline in sympathetic nerves. NA, noradrenaline; α_1, α_1-adrenoceptor; IP_3, inositol-1,4,5-trisphosphate; ejps, excitatory junction potentials. Neuropeptide Y can also function as a co-transmitter, mainly by potentiating contractions to ATP and noradrenaline, but has been omitted here for clarity.

than the EJPs, reaches a peak 5–20 s after the last EJP and decays back to resting membrane potential within 60 s. The magnitude of the slow depolarization increases with the frequency of stimulation and with the number of pulses in the train. Suramin abolishes the EJPs and increases the amplitude of the slow depolarization by an unknown mechanism. On the other hand, phentolamine has no effect on the EJPs, but abolishes the slow depolarization. Thus, in this vessel both purinergic and noradrenergic components to the electrical response to sympathetic nerve stimulation can be seen.

Mechanical actions of endogenously released ATP

In the guinea-pig vas deferens, EJPs summate until the depolarization is sufficient to trigger action potentials and Ca^{2+} ion influx, whereas

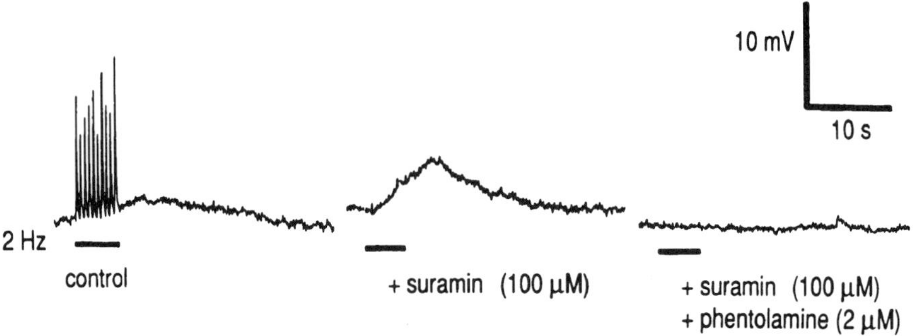

FIG. 2. The effects of suramin and phentolamine on EJPs and slow depolarization in the rat tail artery. The traces show abolition of EJPs and potentiation of the slow depolarization by suramin (100 μM). Subsequently, phentolamine (2 μM) abolished the slow depolarization. Solid bars indicate electrical field stimulation for 5 s at 2 Hz. All data were obtained in a single cell of stable membrane potential.

noradrenaline initiates synthesis of $InsP_3$ and release of internal Ca^{2+} stores. Together, ATP and noradrenaline produce a characteristic biphasic response (Fig. 3). The initial phasic component of the contraction is predominantly purinergic, as it is virtually abolished by $ANAPP_3$ (Sneddon et al 1982, Sneddon & Westfall 1984), α,β-meATP (Meldrum & Burnstock 1983), suramin (Bailey & Hourani 1995) and PPADS (McLaren et al 1994, Fig. 3). The smaller, tonic phase of the contraction is predominantly noradrenergic, as it is greatly inhibited by α-adrenoceptor antagonists (Sneddon et al 1982, Westfall et al 1996).

Unlike the vas deferens, neurogenic contractions of most blood vessels tend to be monophasic. That a component of this response is resistant to α-adrenoceptor antagonists and mediated by ATP was first shown in the mid 1980s (see von Kügelgen & Starke 1991, Kennedy 1993a). Subsequently, in most vessels studied a purinergic component of the neurogenic contraction has been demonstrated. However, the relative contribution of ATP and noradrenaline varies greatly between vessels. For example, in the rat tail artery ATP only contributes up to about 10% of the peak contraction (Bao et al 1993), whereas in the rabbit mesenteric artery ATP is the sole excitatory neurotransmitter, although noradrenaline is released and contributes to feedback-inhibition (Ramme et al 1987). The relative contribution of ATP and noradrenaline within a given vessel can also depend on the frequency of nerve stimulation (Kennedy et al 1986, Bao et al 1993). Thus, in the rabbit ear artery the action of ATP is much more prominent at low frequencies of nerve stimulation and, as the frequency increases, so the contribution of ATP decreases and that of noradrenaline increases (Kennedy et al 1986). This has been suggested to be due to differences in the time-course of action of ATP and noradrenaline. Exogenous ATP evokes rapid transient contractions which fade in the

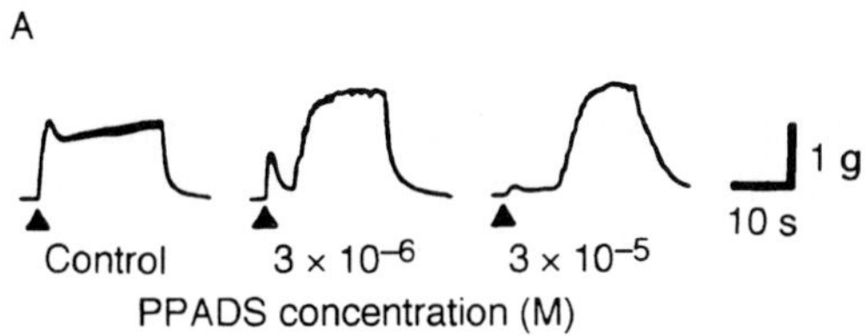

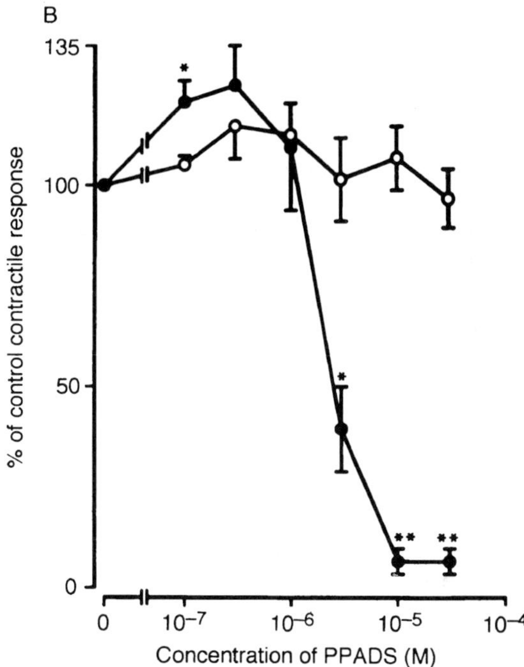

FIG. 3. The effect of PPADS on biphasic contractions of the guinea-pig vas deferens to sympathetic nerve stimulation. (A) Contractions elicited by a train of pulses for 20 s at 4 Hz (▲, start of stimulation) in the absence and presence of PPADS (3 μM and 30 μM). (B) The mean data ($n = 6$) for the effect of PPADS (100 nM–3 μM) on the initial phasic response (●) and the secondary tonic response (○). Vertical lines show SEM; $*P < 0.05$; $**P < 0.01$. (From McLaren et al 1994.)

continued presence of agonist, i.e. a postjunctional mechanism appears to limit the effects of ATP. Exogenous noradrenaline on the other hand evokes slower, more maintained contractions that are not subject to desensitization over the same time scale. If the postjunctional responses to endogenously released ATP and noradrenaline are similar to those of the exogenously administered drug, then increasing the number of pulses applied to the nerve is likely to lead to a progressively smaller action of ATP, but an increasing action of noradrenaline.

Mechanisms of action of ATP

The patch-clamp technique has proved very useful for studying the actions of ATP at the level of the single vascular and visceral smooth muscle cell, and a number of novel interactions of ATP with ionic currents have been recorded. In some tissues ATP activates a slowly developing cation current and a Ca^{2+}-dependent Cl^- current (see Kennedy 1993b), but by far the most common response is activation of a rapidly developing cation current (I_{ATP}) (Fig. 4a). This current is inward at negative membrane potentials and so depolarizes the cell. At physiological concentrations, Ca^{2+} ions can then enter the cell via two routes: (a) through the ATP-sensitive cation channels; and (b) via dihydropyridine-sensitive Ca^{2+} channels which are opened by the ATP-induced depolarization. Similar to the EJPs seen in the vas deferens and most blood vessels, I_{ATP} and the depolarization develop rapidly and are not maintained during continued application of ATP, i.e. desensitization is seen. Thus, it is considered that ATP acts at the P2X purinoceptor, a ligand-gated cation channel, to evoke I_{ATP}, the current underlying EJPs.

The P2X purinoceptor was originally defined by a rank order of agonist potency of α,β-meATP$\gg$2-methylthioATP (2-MeSATP)$\geq$ATP, with α,β-meATP approximately three orders of magnitude more potent than ATP (Burnstock & Kennedy 1985, Kennedy 1990). However, it is now clear that the influence of ectoATPase enzymes on agonist potency is much greater than anticipated, leading to an erroneous characterization of the P2X purinoceptor. Thus, the potencies of ATP and 2-MeSATP are decreased 100–1000-fold by breakdown; when this is prevented, ATP and 2-MeSATP are more potent than α,β-meATP as agonists at the P2X purinoceptor. This conclusion is based on results obtained using several approaches to study the actions of agonists at P2X purinoceptors in the absence of ectoATPase activity (see Kennedy & Leff 1995). Our approach was to determine the ability of ATP, 2-MeSATP and α,β-meATP to evoke inward currents in single, acutely dissociated smooth muscle cells of the rat tail artery using the voltage-clamp technique (Evans & Kennedy 1994). Agonists were applied rapidly to cells (equilibration time <20 ms) using a concentration-clamp system. Under these conditions the influence of ectoATPase on agonist action was likely to be substantially reduced or abolished.

In single cells, ATP activated concentration-dependent, transient inward currents, with a latency to onset of less than 3 ms, consistent with an action at a ligand-gated cation channel (Fig. 4a). 2-MeSATP and α,β-meATP evoked similar transient inward currents at the same concentrations. The order of potency was ATP$=$2-MeSATP$\geq\alpha,\beta$-meATP. All responses were abolished by the P2X purinoceptor antagonist suramin. Full concentration–response curves could not be obtained, but the partial curves for ATP and 2-MeSATP were less than an order of magnitude to the left of that for α,β-meATP.

ATP (Fig. 4b) and its analogues also produced suramin-sensitive contractions of rat isolated tail artery rings, with α,β-meATP ≫2-MeSATP>ATP. α,β-meATP was almost a thousand times more potent than ATP, and 2-MeSATP was eight times more potent than ATP. This is the 'classical' profile of a P2X purinoceptor, but is very different from that seen in single cells. ATP evoked inward currents at concentrations three orders of magnitude lower than those required to elicit contraction (Fig. 4b). Likewise, 2-MeSATP was two orders of magnitude more potent in the single cells. In contrast, α,β-meATP evoked both responses at similar concentrations.

The great difference in the pharmacology of the receptor seen in single cells and in artery rings was surprising, as opening of a ligand-gated cation channel is the initial step by which P2X purinoceptor agonists contract smooth muscle. The simplest explanation is that ATP, 2-MeSATP and α,β-meATP act at the same site (the P2X purinoceptor) to evoke inward currents and contraction, but their relative potencies are determined by differences in their breakdown in

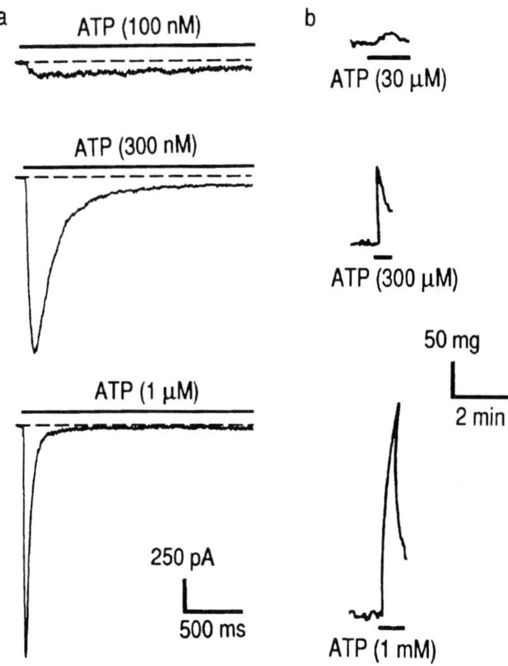

FIG. 4. Excitatory effects of ATP in the rat tail artery. (a) Inward currents evoked by ATP when applied rapidly for 2 s using a U-tube superfusion system as indicated by the solid bars. The dashed lines represent zero current levels. The holding potential was −60 mV. Each record is from a different cell. (b) Contractions evoked by ATP in rat isolated tail artery smooth muscle rings. All records are from the same preparation. (Traces in (a) are from Evans & Kennedy 1994.)

intact muscle and single cells. ATP and 2-MeSATP are broken down by ectonucleotidases, but α,β-meATP is relatively resistant.

Breakdown of ATP following release from nerves

The original purinergic nerve hypothesis envisaged that after being released from nerves, ATP would be sequentially broken down to ADP, AMP and adenosine, which would in turn be taken up by a specific transport system into the prejunctional nerve bouton and the postjunctional tissue, or would diffuse out into the general circulation (Burnstock 1972). While such a pattern of breakdown can certainly be measured in a wide range of tissues where ATP acts as a neurotransmitter, it is unclear how much of an influence these enzymes have on purinergic neurotransmission. If breakdown is slower than the action of ATP, then it is likely that the ectoATPase will have minimal influence on neurotransmission, simply 'mopping up' ATP after it has acted. If, on the other hand, breakdown and the action of ATP occur at comparable rates, then ectoATPase will functionally modulate purinergic neurotransmission, analogous to the action of acetylcholinesterase at the neuromuscular junction.

The main difficulty in determining the time course of breakdown of ATP has been the lack of selective inhibitors of ectoATPase. A wide range of compounds have been tested as inhibitors, but although some are effective, their selectivity and potency are generally very poor (see Ziganshin et al 1994). However, a recently developed analogue of ATP—6,*N*,*N*-diethyl-D-β,γ dibromomethylene ATP (ARL67156, formerly known as FPL67156)—may be a major breakthrough in this field. ARL67156 inhibits ectoATPase with a pIC_{50} of 4.62 in human blood cells (Crack et al 1995) and 5.1 in the rat vas deferens (Khakh et al 1995). It enhances contractions of the rabbit ear artery evoked by exogenous ATP, but not responses to the stable analogue α,β-meATP (Crack et al 1995). This suggests that its enhancing effect on contractions is due to inhibition of ATP breakdown and not to non-selective effects on the smooth muscle, since if this were the case then it would be expected that responses to α,β-meATP would also have been enhanced. ARL67156 is also a weak P2X purinoceptor antagonist, with a pA2 of 3.3 (Crack et al 1995), which implies that if used at 10–100 μM, then effective inhibition of ectoATPase should be achieved with relatively little receptor antagonism. We have studied the effects of ARL67156 on neurotransmission in the guinea-pig vas deferens (Westfall et al 1996).

Co-transmission in the guinea-pig vas deferens

As described above, stimulation of the sympathetic nerves evokes a biphasic contraction of the guinea-pig vas deferens. The initial phasic component is predominantly purinergic, while the secondary tonic component is largely noradrenergic. ARL67156 produces a rapid and reversible potentiation of the neurogenic contractions (Fig. 5). This was maximal within 10 min, well maintained over 30 min, and reversed rapidly on washout. The effect of ARL67156 was also concentration dependent. No effect was seen at 1 μM, but from 5–100 μM the peak amplitude of the contraction progressively increased. This is very similar to the potency with which ARL67156 inhibits ectoATPase activity in biochemical assays in human blood cells (Crack et al 1995) and the rat vas deferens (Khakh et al 1995). At 100 μM, ARL67156 inhibited ectoATPase activity in human blood cells by approximately 70%. Higher concentrations were not used in our studies because of limited availability of the compound and because of the antagonistic action of ARL67156 at P2X purinoceptors at concentrations above 100 μM. However, it is clear that if the ectoATPase activity in the guinea-pig vas deferens could be inhibited completely, without concomitant antagonism of P2X purinoceptors, then even greater potentiation of the neurogenic contractions is possible.

The amplitude of the potentiation was inversely related to the frequency of stimulation. In the presence of 100 μM ARL67156, the peak contraction evoked at 1 Hz was approximately doubled, whereas at 8 Hz only a 38%

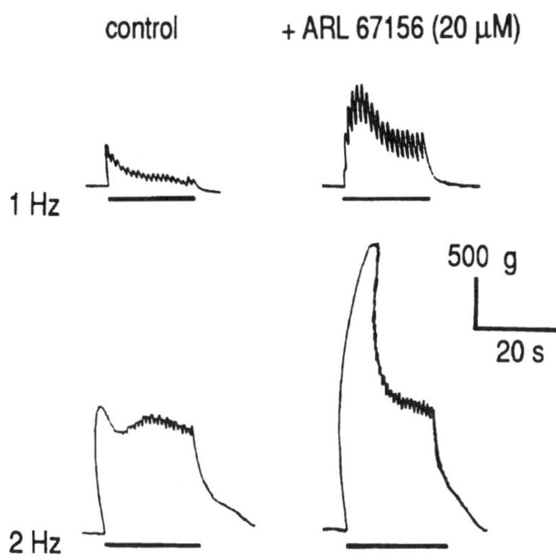

FIG. 5. The effect of ARL67156 on neurogenic contractions of the guinea-pig vas deferens. The traces on the left show control responses to stimulation of the sympathetic nerves for 20 s at 1 and 2 Hz. Traces on the right show responses in the same tissue after 10 min in the presence of ARL67156 (100 μM).

increase was seen. Thus, contractions evoked by low stimulation frequencies appear to be influenced by ectoATPase more than contractions evoked at high frequencies of stimulation. Interestingly, although ARL67156 also potentiates neurogenic contractions in the guinea-pig urinary bladder, it does so in a frequency-independent manner. The reason for this difference is not known.

Although the initial phasic, neurogenic response is predominantly purinergic, it is partially mediated by noradrenaline. Therefore, in order to study the effects of ARL67156 on a purely purinergic response, contractions were evoked under conditions in which the noradrenergic component was blocked. The α_1-adrenoceptor antagonist prazosin (100 nM) abolished contractions evoked by 10 μM noradrenaline and depressed the peak amplitude of the contraction evoked by nerve stimulation at 4 Hz by 21%. The neurogenic response remaining in the presence of prazosin was abolished by the P2X purinoceptor antagonist PPADS, indicating that it was mediated entirely by ATP. ARL67156 (100 μM) potentiated both phases of the remaining response and the peak amplitude was almost doubled, similar to the potentiated seen in the absence of prazosin. Thus, these results are consistent with ARL67156 increasing neurogenic contractions by potentiating the purinergic component.

Effect of ARL67156 on exogenous agonists

In order to support this mode of action of ARL67156 we looked at its effect on contractions evoked by submaximal concentrations of exogenous ATP and several other agonists. ATP (100 μM) and its stable analogue α,β-meATP (0.5 μM) both produced rapid transient contractions which reached a peak in about 5 s and then subsided rapidly, even in the continued presence of the agonist. This time-course is similar to that of the rapid component of neurogenic contractions. In the presence of ARL67156 (100 μM) the peak response to ATP was increased and, although the response was still transient, tension remained elevated for longer. The potentiation reversed rapidly on washout of ARL67156. In contrast, neither the peak amplitude nor the time course of the response to α,β-meATP were affected by ARL67156, consistent with ARL67156 acting to inhibit the breakdown of ATP.

Next we investigated the effects of ARL67156 on contractions evoked by noradrenaline (10 μM) and KCl (40 mM) and were surprised to find that both were potentiated. Indeed the potentiation of noradrenaline was greater than that of ATP. These results were unexpected and could suggest that while ARL67156 may well inhibit ectoATPase, it also causes a non-selective sensitization of the smooth muscle to contractile agents. However, it was not obvious how noradrenaline, KCl and ATP could be potentiated by ARL67156, but α,β-meATP be unaffected. This led us to consider a synergistic model which is still consistent with a selective action of ARL67156. That is, while the levels of ATP in the extracellular space in the guinea-pig vas deferens in the absence of nerve stimulation are normally low because of the activity of ectoATPase, in the prolonged presence of the ectoATPase inhibitor ATP levels rise such that, although subthreshold for contraction, they are high enough to act at P2 purinoceptors to sensitize the smooth muscle to other agonists which do not act through P2 purinoceptors. This appeared to be the case, because when these experiments were repeated in the presence of the P2 purinoceptor antagonist PPADS, ARL67156 no longer potentiated contractions evoked by exogenous noradrenaline and KCl. This is consistent with a number of previous reports that ATP and noradrenaline can interact synergistically in the guinea-pig vas deferens (Holck & Marks 1978, Sakai et al 1979, Kazic & Milosavljevic 1980). The intracellular pathways which mediate this receptor cross-talk remain to be determined.

Properties of ectoATPase

Numerous studies using whole tissue or purified enzyme preparations have shown that ectoATPase is highly dependent upon Ca^{2+} and/or Mg^{2+} ions, that in most tissues the optimum pH for activity is about 7.5, and that the enzyme has a broad selectivity for hydrolysing nucleotides (see Ziganshin et al 1994). Subsequent cloning studies have isolated cDNA sequences encoding ectoATPase and these were found to be identical to those of members of the immunoglobulin superfamily of cell adhesion molecules (Lin & Giuidotti 1989, McGuaig et al 1993, Najjar et al 1993, Sippel et al 1994). Most of these large glycoproteins are located outside the cell and the region that encodes ectoATPase activity is found, as expected, in the extracellular portion. Cell adhesion molecules are essential for physical interactions between cells, so this may explain the widespread distribution of ectoATPase. Interestingly, cell adhesion molecule function can be modified by agents such as insulin and growth factors which act via receptor tyrosine kinases (Najjar et al 1993), suggesting the intriguing possibility that ectoATPase activity may also be regulated by these factors.

Future studies

Although we now have a much greater understanding of how ATP acts as a co-transmitter with noradrenaline from sympathetic nerves, a number of points are still unclear. For example, even though sympathetic nerve stimulation evokes purinergic EJPs in most blood vessels studied, it is not clear why this leads to a large purinergic component to the contraction in some vessels, but only a small component in others. Furthermore, ATP and noradrenaline have been shown to act synergistically in many tissues, but little is known about the pathways underlying this interaction. Similarly, the number of different ectoATPase molecules that exist and their distribution remains uncertain, and the effect of agents which modulate cell adhesion on ectoATPase activity remains unknown. Clarification of these other uncertainties will give us a better understanding of these systems and so could lead to novel therapeutic applications of drugs which interact with purinergic neurotransmission.

References

Bailey SJ, Hourani SMO 1995 Effects of suramin on contractions of the guinea-pig vas deferens induced by analogues of adenosine 5′-triphosphate. Br J Pharmacol 114: 1125–1132

Bao J-X, Gonon F, Stjärne L 1993 Frequency- and train length-dependent variation in the roles of postjunctional α_1- and α_2-adrenoceptors for the field stimulation-induced neurogenic contraction of the rat tail artery. Naunyn-Schmiedebergs Arch Pharmakol 347:601–616

Bolton TB, Large WA 1986 Are junction potentials essential? Dual mechanism of smooth muscle cell activation by transmitter released from autonomic nerves. Q J Exp Physiol 71:1–28

Burnstock G 1972 Purinergic nerves. Pharmacol Rev 24:509–581

Burnstock G 1990 Purinergic mechanisms. Ann N Y Acad Sci 603:1–17

Burnstock G, Kennedy C 1985 Is there a basis for distinguishing two types of P_2-purinoceptor? Gen Pharmacol 16:433–440

Crack BE, Pollard CE, Beukers MW et al 1995 Pharmacological and biochemical analysis of FPL 67156, a novel, selective inhibitor of ecto-ATPase. Br J Pharmacol 114:475–481

Evans RJ, Kennedy C 1994 Characterisation of P_2-purinoceptors in the smooth muscle of the rat tail artery: a comparison between contractile and electrophysiological responses. Br J Pharmacol 113:853–860

Holck MI, Marks BH 1978 Purine nucleoside and nucleotide interactions on normal and subsensitive alpha adrenoceptor responsiveness in guinea-pig vas deferens. J Pharmacol Exp Ther 205:104–117

Kazic T, Milosavljevic D 1980 Interaction between adenosine triphosphate and noradrenaline in the isolated vas deferens of the guinea-pig. Br J Pharmacol 71:93–98

Kennedy C 1990 P_1 purinoceptor and P_2 purinoceptor subtypes—an update. Arch Int Pharmacodyn Ther 303:30–50

Kennedy C 1993a ATP as a cotransmitter with noradrenaline in sympathetic perivascular nerves. In: Edvinsson L, Uddman R (eds) Vascular innervation and receptor mechanisms: new perspectives. Academic Press, Orlando, FL, p 187–199

Kennedy C 1993b Cellular mechanisms underlying the excitatory actions of ATP in vascular smooth muscle. Drug Dev Res 28:423–427

Kennedy C, Leff P 1995 How should P_{2X} purinoceptors be classified pharmacologically? Trends Pharmacol Sci 16:168–174

Kennedy C, Saville V, Burnstock G 1986 The contributions of noradrenaline and ATP to the responses of the rabbit central ear artery to sympathetic nerve stimulation depend on the parameters of stimulation. Eur J Pharmacol 122:291–300

Khakh BS, Michel AD, Humphrey PPA 1995 Inhibition of ectoATPase and Ca-ATPase in rat vas deferens by P_2-purinoceptor antagonists. Br J Pharmacol 115:2P

Lin SH, Guidotti G 1989 Cloning and expression of a cDNA coding for a rat liver plasma membrane ecto-ATPase: the primary structure of the ecto-ATPase is similar to that of the human biliary glycoprotein I. J Biol Chem 264:4408–4414

McCuaig K, Rosenberg M, Nedellec P, Turbide C, Beauchemin N 1993 Expression of the *Bgp* gene and characterization of mouse colon biliary glycoprotein isoforms. Gene 127:173–183

McLaren GJ, Lambrecht G, Mutschler E, Bäumert HG, Sneddon P, Kennedy C 1994 Investigation of the actions of PPADS, a novel P_{2X}-purinoceptor antagonist in the guinea-pig isolated vas deferens. Br J Pharmacol 111:913–917

McLaren GJ, Kennedy C, Sneddon P, 1995 The effects of suramin on purinergic and noradrenergic neurotransmission in the rat isolated tail artery. Eur J Pharmacol 277:57–61

Meldrum LA, Burnstock G 1983 Evidence that ATP acts as a co-transmitter with noradrenaline in sympathetic nerves supplying the guinea-pig vas deferens. Eur J Pharmacol 92:161–163

Najjar SM, Accili D, Phillipe N, Jernberg J, Margolis R, Taylor SI 1993 pp120/ecto-ATPase, an endogenous substrate of the insulin receptor tyrosine kinase, is expressed as two variably spliced isoforms. J Biol Chem 268:1201–1206

Ramme D, Regenold JT, Starke K, Busse R, Illes P 1987 Identification of the neuroeffector transmitter in jejunal branches of the rabbit mesenteric artery. Naunyn-Schmiedebergs Arch Pharmakol 336:267–273

Sakai KK, Hymson DL, Shapiro R 1979 The effects of adenosine and adenine nucleotides in the guinea-pig vas deferens: evidence for existence of purinergic receptors. Life Sci 24:1299–1308

Sippel CJ, McCollum MJ, Perlmutter DH 1994 Bile acid transport by the rat liver canalicular bile acid transport/ecto-ATPase protein is dependent on ATP but not its own ecto-ATPase activity. J Biol Chem 269:2800–2826

Sneddon P 1992 Suramin inhibits excitatory junction potentials in guinea-pig vas deferens. Br J Pharmacol 107:101–103

Sneddon P, Burnstock G 1984a Inhibition of excitatory junction potentials in guinea-pig vas deferens by α,β-methylene-ATP: further evidence for ATP and noradrenaline as cotransmitters. Eur J Pharmacol 100:85–90

Sneddon P, Burnstock G 1984b ATP as a co-transmitter in rat tail artery. Eur J Pharmacol 106:149–152

Sneddon P, Westfall DP 1984 Pharmacological evidence that adenosine triphosphate and noradrenaline are co-transmitters in the guinea-pig vas deferens. J Physiol 347: 561–580

Sneddon P, Westfall DP, Fedan JS 1982 Cotransmitters in the motor nerves of the guinea-pig vas deferens: electro-physiological evidence. Science 218:693–695

von Kügelgen I, Starke K 1991 Noradrenaline–ATP co-transmission in the sympathetic nervous system. Trends Pharmacol Sci 12:319–324

Walsh MP 1994 Regulation of vascular smooth muscle tone. Can J Physiol Pharmacol 72:919–936 (erratum: 1994 Can J Physiol Pharmacol 72:1257)

Westfall TD, Kennedy C, Sneddon P 1996 Enhancement of sympathetic purinergic neurotransmission in the guinea-pig isolated vas deferens by the novel ecto-ATPase inhibitor ARL 67156. Br J Pharmacol, in press

Ziganshin AU, Hoyle CHV, Burnstock G 1994 Ectoenzymes and metabolism of extracellular ATP. Drug Dev Res 32:134–146

DISCUSSION

Illes: I have a question relating to the contraction measurements of the vas deferens. You explain your data by receptor synergism or receptor cross-talk—

wouldn't another explanation also be possible? Various authors have suggested that noradrenaline initiates a so-called 'cascade transmission' in smooth muscle organs (e.g. von Kügelgen & Starke 1991). This means that the activation of α_1-adrenoceptors by endogenously released noradrenaline releases ATP from the smooth muscle. This ATP may activate muscular P2 receptors and amplify the contractile effect of noradrenaline. So, if you block the degradation of ATP by ARL67156, you may get a larger response to noradrenaline by increasing the synaptic concentration of ATP, even without the need to hypothesize about a cross-talk between α_1-adrenoceptors and P2 receptors.

Kennedy: You're suggesting that ARL67156 potentiates noradrenaline because it releases ATP from the smooth muscle. However, whenever release of ATP by noradrenaline has been reported in the guinea-pig vas deferens, its time-course has been much slower than the time-course of the response to noradrenaline seen in our experiments. So it is unlikely that this is the mechanism responsible.

Westfall: There is some evidence for ATP release from smooth muscle, but in guinea-pig vas deferens it's really pretty difficult to show. Consequently, I don't think it's a major factor.

Burnstock: Postjunctional release of ATP is relatively easy to demonstrate in the rat vas deferens, but it is hard to show it in the guinea-pig vas deferens.

Westfall: But the timing is off. In terms of this very fast event that they see, it is difficult to envisage this feed-forward type of amplification.

Starke: When we stimulate sympathetic nerves in guinea-pig vas deferens, we find a decrease in ATP release by prazosin and a further decrease by suramin. We get release of ATP by noradrenaline acting through α_1-adrenoceptors and by α,β-meATP acting apparently through P2X purinoceptors, so we do think that ATP release from postjunctional sites occurs. This might of course then contribute to your potentiation.

Illes: In 1987 we published a paper with Klaus Starke showing that vasoconstriction in small branches of the rabbit mesenteric artery was exclusively due to the release of ATP from postganglionic sympathetic nerve terminals (Ramme et al 1987). Noradrenaline was only a kind of presynaptic modulator. A few years later, we found that when a higher frequency of stimulation was used, noradrenaline also became a neuroeffector transmitter (A. Raether & P. Illes, unpublished results). Do you see a noradrenergic component in your experiments if the frequency of stimulation is increased?

Kennedy: You are referring to blood vessels of course. Regarding the guinea-pig vas deferens, there is a noradrenaline component in both phases. We blocked it with prazosin and showed that noradrenaline contributes approximately 20% of the phasic component. The contribution of noradrenaline to the tonic component is more difficult to quantify—because the tonic component is very uneven, there is not always a well defined maximum response. It really depends on the tissue you're looking at, and the guinea-pig vas deferens is well characterized. The finding that both noradrenaline and ATP contribute to both components is not a novel one.

Starke: Have you done any experiments on the potentiation of neurogenic contractions by ARL67156 in the presence of theophylline? Theophylline will remove presynaptic adenosine-mediated inhibitory modulation whilst leaving intact excitatory purinergic transmitters. Such experiments would help determine whether inhibition of adenosine production by ARL67156 modulates neurotransmission *per se*.

Kennedy: That is the next series of experiments we intend doing. However, Sneddon et al (1984) did show in the guinea-pig vas deferens that the P1 purinoceptor antagonist 8-phenyltheophylline had no effect on neurogenic contractions. Consequently, the involvement of endogenous adenosine in the potentiating action of ARL67156 is uncertain.

Edwards: I don't understand why potentiation is greatest at low frequencies. Do you think this indicates that the ectoATPase is saturated very quickly?

Kennedy: This is a difficult question to answer. I said at the start that it could be analogous to cholinesterase at neuromuscular junctions. So I spoke to the local neuromuscular junction experts and asked them if there was frequency dependence at the neuromuscular junction if you inhibit cholinesterase. The summary of the 25 min reply was that they didn't know. The cholinesterase inhibitors that they use tend to have other actions as well; there are presynaptic nicotinic receptors which can be inhibitory or excitatory. The situation at the synapses that we are studying may be as complicated.

Edwards: On the same subject, I don't know how physically restricted that synapse is.

Kennedy: In the guinea-pig vas deferens the synaptic junction is approximately 20 nm across, which makes it one of the smallest junctions in the autonomic nervous system.

Burnstock: But we've got every reason to suppose that there is release of transmitter from nerve varicosities up to a micron away from the nearest smooth muscle membrane which is still effective. There's no clear evidence that it is only the closest junctions at 20 nm that release ATP.

Hourani: I wanted to pick up on the point of the formation of adenosine from ATP, and whether that has pre-synaptic or indeed post-synaptic inhibitory effects. The difference you see between the potentiation by ARL67156 of nerve-stimulated contractions of the vas deferens and the bladder could be related to the fact that in the bladder the major breakdown product is inosine rather than adenosine, so you might get less potentiation in the bladder because you've got less adenosine (Bailey & Hourani 1994). Is ATP itself potentiated in the bladder?

Kennedy: Yes.

Hourani: To the same extent as in the vas deferens?

Kennedy: Yes.

Leff: Are the differences that you have observed attributable to the way you've measured responses? You're measuring peak amplitudes: if you measured areas, would you come to a different conclusion about the comparison between the two tissues?

Kennedy: We haven't measured the area, so we can't say. But it doesn't look like it.

Westfall: If you prevent the formation of adenosine by preventing the breakdown of ATP, you would of course reduce the potential presynaptic action of adenosine to decrease transmitter release. However, ATP itself also has presynaptic modulatory effects. If the breakdown of ATP is retarded, then this results in a potentiation of the presynaptic effect. Thus the loss of presynaptic modulation by adenosine may be counterbalanced by enhancing the presynaptic effect of ATP.

References

Bailey SJ, Hourani SMO 1994 Differential effect of suramin on P_2-purinoceptors mediating contraction of the guinea-pig vas deferens and urinary bladder. Br J Pharmacol 112:219–225

Ramme D, Regenold JT, Starke K, Brusse R, Illes P 1987 Identification of the neuroeffector transmitter in jejunal branches of the rabbit mesenteric artery. Naunyn-Schmiedeberg's Arch Pharmacol 336:267–273

Sneddon P, Meldrum LA, Burnstock G 1984 Control of transmitter release in guinea pig vas deferens by prejuctional P1 purinoceptors. Eur J Pharmacol 105:293–299

von Kügelgen I, Starke K 1991 Noradrenaline–ATP co-transmission in the sympathetic nervous system. Trends Pharmacol Sci 12:319–324

ATP release and its prejunctional modulation

Klaus Starke, Ivar von Kügelgen, Bernd Driessen and Ralph Bültmann

Pharmakologisches Institut, Albert Ludwigs-Universität Freiburg, Hermann-Herder-Strasse 5, D-79104 Freiburg im Breisgau, Germany

Abstract. We studied some properties of the release of noradrenaline and ATP in isolated sympathetically innervated tissues. Release was elicited by electric stimulation and assessed as overflow of tritiated compounds (after labelling with [^{3}H]noradrenaline) and enzymically measured ATP, respectively. Evans blue, which inhibits ectonucleotidases, greatly increased the evoked overflow of ATP, indicating that a major part of the ATP was metabolized after release. Much of the ATP was postjunctional in origin. The neural fraction was isolated when postjunctional release was suppressed by prazosin (α_1-adrenoceptor antagonist) and suramin (P2 purinoceptor antagonist). Comparison of neural ATP and [^{3}H]-noradrenaline release showed that prostaglandin E_2 reduced the release of both co-transmitters to a similar extent. Activation of prejunctional α_2-adrenoceptors, however, preferentially reduced the release of [^{3}H]noradrenaline, and activation of prejunctional A_1 purinoceptors reduced preferentially the release of ATP. Nucleotides such as ATP depressed the release of [^{3}H]noradrenaline through two receptors: the well-known prejunctional A_1 receptors and a separate group of prejunctional P2 purinoceptors. P2 antagonists increased the release of [^{3}H]-noradrenaline. Overall, the results indicate differential storage, release and modulation of release of the two sympathetic co-transmitters. They also indicate that postganglionic sympathetic axons possess receptors for both co-transmitters: α_2 and P2 autoreceptors.

1996 P2 purinoceptors: localization, function and transduction mechanisms. Wiley, Chichester (Ciba Foundation Symposium 198) p 239–259

When a substance is proposed to be a neurotransmitter, two questions, among others, arise. First, can the release of the substance from the respective axons be demonstrated chemically, as 'overflow', and if so, what does the overflow tell us about the neural release? Second, do the respective axons possess prejunctional, release-modulating autoreceptors for the substance, the situation regularly encountered in the case of known transmitters?

ATP, in all likelihood, is a co-transmitter of noradrenaline in postganglionic sympathetic neurons (see Burnstock 1990, von Kügelgen & Starke 1991a). Regarding the two questions mentioned, sympathetic nerve stimulation elicits an overflow of ATP from the innervated tissue (e.g. Lew & White 1987), and it has been suggested that postganglionic sympathetic axons possess release-inhibiting P2 purinoceptors, i.e. receptors for nucleotides such as ATP (e.g. Fujioka & Cheung 1987). Here we describe some properties of the evoked overflow of ATP and compare it with the evoked overflow of noradrenaline, showing that there may be a differential release of the two transmitters. We also present evidence for prejunctional P2 purinoceptors and their operation as autoreceptors.

Methods

All experiments were carried out on isolated tissues superfused with a physiological electrolyte solution. Release of noradrenaline was determined as overflow of total tritiated compounds after preincubation of the tissue with (-)[^{3}H]noradrenaline. ATP was determined with the luciferin–luciferase technique. Sympathetic axons in the tissues were stimulated by an electrical field. All responses to field stimulation were abolished by tetrodotoxin. Details are described in the original publications.

Results and discussion

Transmitter overflow

The stimulation-evoked overflow of ATP from sympathetically innervated tissues is accompanied by an overflow of ADP, AMP and adenosine (Westfall et al 1978, Sedaa et al 1990), and it seems likely that the latter compounds are at least partly formed from released ATP. We have recently shown that Evans blue, which blocks P2X purinoceptor-mediated contractions in rat vas deferens, also blocks the degradation of ATP relatively effectively (Bültmann & Starke 1993, Bültmann et al 1995). We therefore used Evans blue to examine the hypothesis of the degradation of ATP after its release. Electrical stimulation of the rat vas deferens preincubated with [^{3}H]noradrenaline elicited contraction and an overflow of tritium (in the following interpreted as 'release of [^{3}H]noradrenaline') and ATP (Fig. 1). Evans blue caused a very large (24-fold) increase in the stimulation-evoked overflow of ATP. This was not accompanied by any increase in the evoked contraction and the release of [^{3}H]noradrenaline. The increase in ATP overflow, therefore, probably was not due to an increase in neural or smooth muscle (see below) ATP release. Rather, it was due to inhibition of the breakdown of released ATP. Figure 1, hence, supports the idea that a large part of the ATP released from cells is normally degraded, presumably by ectonucleotidases, in accord with the mode of inactivation of many neurotransmitters. The contraction probably was not

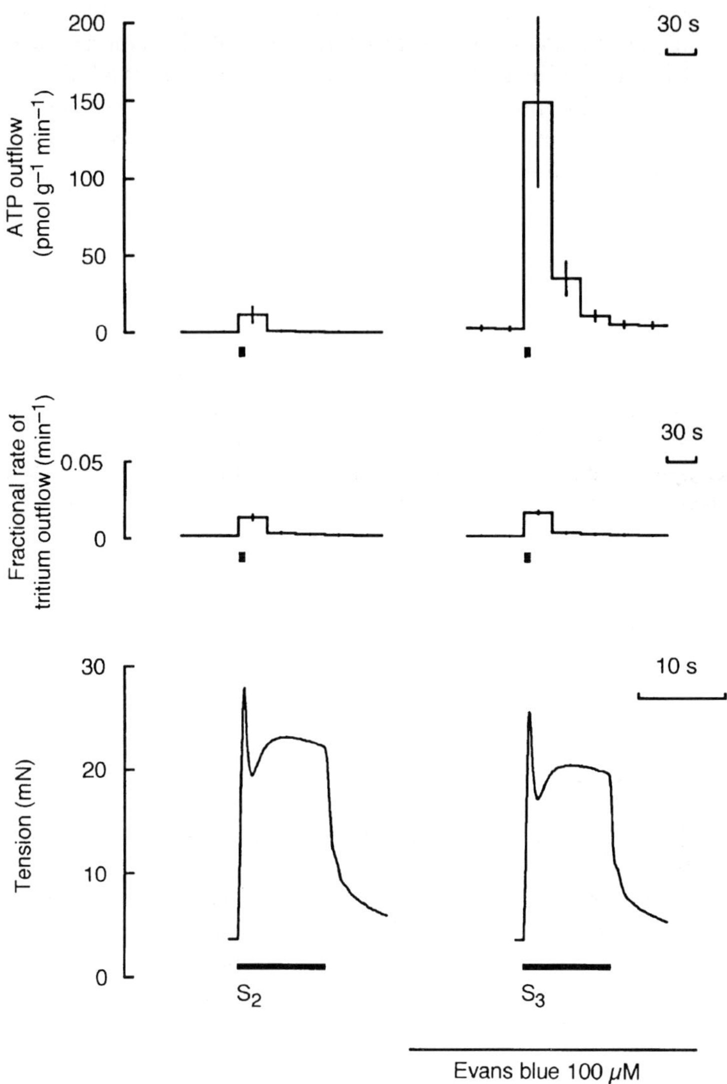

FIG. 1. Effect of Evans blue on mechanical tension, tritium outflow, ATP outflow and the response to electrical stimulation in rat vas deferens preincubated with [^{3}H]noradrenaline. There were three periods of electrical stimulation (100 pulses, 10 Hz) of which only S_2 and S_3 were evaluated. Evans blue was added 30 min before S_3. Note different time scales for tension and outflow values. (From Bültmann et al 1995.)

changed (Fig. 1) because the increase in the biophase concentration of ATP balanced the blockade, by Evans blue, of smooth muscle P2 purinoceptors.

What is the cellular source of the overflow of ATP? In contrast to other transmitters, ATP occurs in all cells. In experiments such as that of Fig. 1, ATP in fact is released not only from sympathetic axons but also from the innervated smooth muscle cells (Westfall et al 1978, Fredholm et al 1982). We have shown that activation of both α_1-adrenoceptors and P2 purinoceptors leads to postjunctional release of ATP (von Kügelgen & Starke 1991b). Blockade of these postjunctional receptors by combined administration of prazosin (0.3 μM) and suramin (300 μM), in a manner that abolishes smooth muscle contraction, therefore reduces the electrically evoked overflow of ATP by 80–90% (Driessen et al 1993). It is the remaining 10–20% that presumably reflects the purely neural release of ATP.

Modulation through prejunctional receptors is a characteristic property of transmitter release. If noradrenaline and ATP are co-transmitters, release of the latter, like that of the former, should be subject to such modulation. Prostaglandins, acting at EP_3 receptors (see Exner & Schlicker 1995), are well known to decrease the release of noradrenaline. Do they act similarly on the release of ATP? Figure 2 demonstrates the effect of prostaglandin E_2 in otherwise untreated guinea-pig vas deferens. As expected, prostaglandin E_2 reduced the release of [^{3}H]noradrenaline. The release of ATP, however, was not reduced but, at 100 nM of prostaglandin E_2, even increased (as observed by Ellis & Burnstock 1990). Are neural noradrenaline and ATP release modulated by prostaglandin E_2 in opposite directions? The contraction tracings of Fig. 2 suggest an alternative explanation: prostaglandin E_2 (100 nM) greatly increased the contraction and, therefore, could have increased the smooth muscle rather than the neural release of ATP. We repeated the experiment in tissues superfused with medium containing prazosin (0.3 μM) and suramin (300 μM). Under these conditions, contractions were abolished, the evoked overflow of ATP was much reduced, and prostaglandin E_2 (1 to 100 nM, as in Fig. 2) now caused concentration-related inhibition of both [^{3}H]noradrenaline and ATP release (Driessen & Starke 1994). Prostaglandin E_2 changes the neural release of noradrenaline and of ATP in the same, inhibitory, direction. The increase of the total release of ATP in unblocked preparations is smooth muscle in origin: prostaglandin E_2 (100 nM), while inhibiting the neural release of both noradrenaline and ATP, simultaneously potentiates their smooth muscle effects so that the contraction and the smooth muscle release of ATP are increased. This example suggests that ATP overflow experiments should be carried out in α_1-adrenoceptor- and P2 purinoceptor-blocked preparations to ascertain a purely neural origin of the ATP measured.

Effects of prejunctional receptor ligands in the guinea-pig vas deferens, in the presence of prazosin and suramin, are summarized in Table 1. Prostaglandin E_2 caused a similar percentage inhibition of the release of [^{3}H]noradrenaline

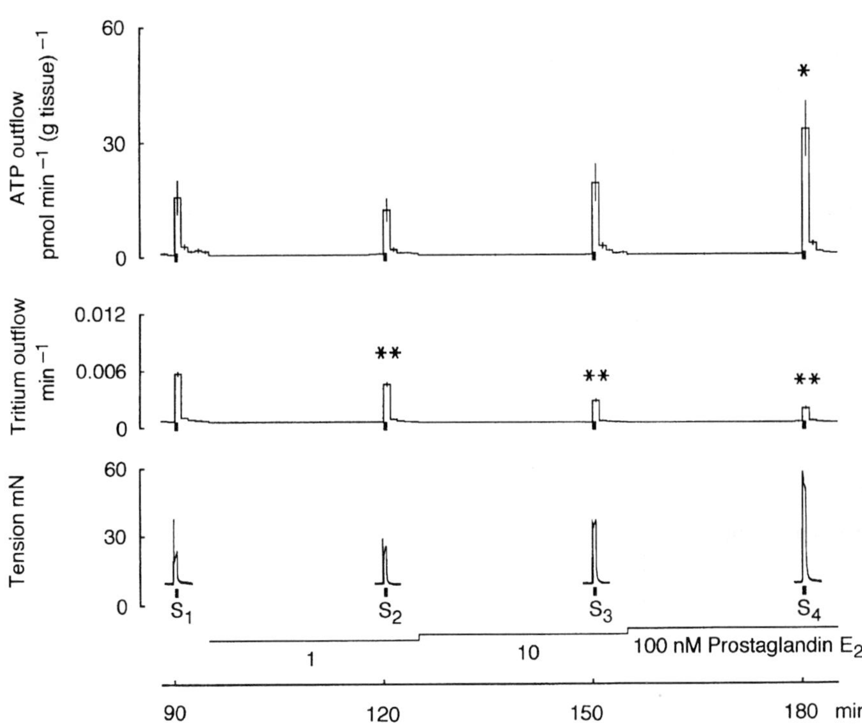

FIG. 2. Effect of prostaglandin E_2 on mechanical tension, tritium outflow, ATP outflow, and the response to electrical stimulation in guinea-pig vas deferens preincubated with [^{3}H]noradrenaline. There were four periods of electrical stimulation (210 pulses, 7 Hz; S_1 to S_4). Prostaglandin E_2 was added as indicated. Abscissae, min of superfusion. Tritium outflow is expressed as fractional rate (min^{-1}). Significant differences of S_n/S_1 overflow values from corresponding controls (solvent instead of prostaglandin E_2): *$P<0.05$; **$P<0.01$. (From Driessen & Starke 1994.)

and of ATP. Other drugs, however, caused disproportionate changes. The selective α_2-adrenoceptor agonist UK14304 (5-bromo-6-[2-imidazolin-2-ylamino]-quinoxaline) tended to produce more marked inhibition of the release of [^{3}H]noradrenaline than of ATP, and the selective α_2-adrenoceptor antagonist rauwolscine increased the release of [^{3}H]noradrenaline to a significantly greater extent than the release of ATP, indicating that activation of prejunctional α_2 autoreceptors depresses the release of noradrenaline more than the release of ATP. The opposite pattern was obtained with adenosine and the selective A_1 purinoceptor agonist CCPA (2-chloro-N^6-cyclopentyladenosine). Both reduced preferentially the evoked overflow of ATP, indicating that activation of prejunctional A_1 adenosine receptors depresses the release of

TABLE 1 Modulation of neural noradrenaline and ATP release in guinea-pig vas deferens through prejunctional receptors

Drug added before S_2 to S_4		*% Change in evoked*	
		Tritium overflow	*ATP overflow*
Prostaglandin E_2	1 nM	− 11.6 ± 1.2	−14.7 ± 10.5
Prostaglandin E_2	10 nM	− 48.2 ± 1.4	−55.0 ± 5.6
Prostaglandin E_2	100 nM	− 78.3 ± 0.8	−70.2 ± 8.0
UK14304	10 nM	− 28.2 ± 1.8	−33.7 ± 8.8
UK14304	100 nM	− 56.6 ± 1.5	−40.2 ± 17.5
UK14304	1000 nM	− 66.8 ± 1.7	−56.9 ± 10.6
Rauwolscine	0.1 μM	+ 145.9 ± 16.3	+ 56.5 ± 11.6**
Rauwolscine	1 μM	+ 242.7 ± 30.0	+ 88.8 ± 29.7**
Rauwolscine	10 μM	+ 267.1 ± 15.1	+ 99.5 ± 31.4**
Adenosine	1 μM	− 16.3 ± 4.5	−34.0 ± 8.6
Adenosine	3.2 μM	− 25.3 ± 3.3	−45.5 ± 6.0**
Adenosine	10 μM	− 34.3 ± 2.4	−50.9 ± 5.6**
CCPA	32 nM	− 15.1 ± 1.6	−45.3 ± 6.4**
CCPA	100 nM	− 22.5 ± 2.8	−49.9 ± 4.0**
CCPA	320 nM	− 33.5 ± 2.6	−68.2 ± 9.3**

Experiments were carried out on guinea-pig vas deferens preincubated with [^{3}H]noradrenaline as in Fig. 2, with four periods of electrical stimulation (210 pulses, 7 Hz; S_1 to S_4). In contrast to Fig. 2, the superfusion medium contained prazosin (0.3 μM) and suramin (300 μM). UK14304 is 5-bromo-6-(2-imidazolin-2-ylamino)-quinoxaline; CCPA is 2-chloro-N^6-cyclopentyladenosine. From data of Driessen et al (1993, 1994) and Driessen & Starke (1994). Significant differences from % change in evoked tritium overflow: $^{**}P < 0.01$.

ATP more than the release of noradrenaline (Table 1). In all experiments, contractions were totally blocked.

Taken together, these observations lead to the conclusion that activation of prejunctional receptors either does not change the composition of the co-transmitter mixture released (EP_3 receptors), or changes it in favour of ATP (α_2-adrenoceptors), or changes it in favour of noradrenaline (A_1 purinoceptors). So far we have not observed opposite changes of the release of noradrenaline and ATP. Previous studies have shown that simple changes of the frequency or chain length of stimulation can change the composition of the released co-transmitter mixture (von Kügelgen & Starke 1991b, 1994). Neither the mechanism nor the biological significance of the differential release of noradrenaline and ATP is known. It seems clear, however, that noradrenaline and ATP are not stored in the same proportions in all vesicles in a homogeneous population of axons. As discussed by Burnstock (1990), there must be especially noradrenaline-rich and especially ATP-rich vesicles, and what is more, there must be mechanisms by which release can be switched from one population to another.

P2 autoreceptors

Table 1 demonstrates the familiar operation of prejunctional α_2 autoreceptors and adenosine A_1 receptors. Given the transmitter nature of ATP, do postganglionic sympathetic axons also possess receptors for ATP? Two somewhat different positive answers have been given. One was that the axons possess a new type of purinoceptor, called by the authors P3, through which both adenosine and nucleotides such as ATP would inhibit transmitter release (Shinozuka et al 1988, Forsyth et al 1991). The second was that the axons possess two kinds of release-inhibiting purinoceptors, the well-known A_1 receptor for adenosine and related compounds and a separate P2 purinoceptor selectively activated by nucleotides (von Kügelgen et al 1989; see also Fujioka & Cheung 1987).

After previous investigations on the mouse and rat vas deferens (Kurz et al 1993, von Kügelgen et al 1994a) we have recently carried out a study in rat atria preincubated with [^{3}H]noradrenaline in order to distinguish between these two possibilities. The adenosine A_1 receptor agonist CPA (cyclopentyladenosine) as well as the nucleotides ATP and ATPγS reduced the electrically evoked release of [^{3}H]noradrenaline (Fig. 3). The P2X selective nucleotide β,γ-methylene-L-ATP (30 μM) caused no change. The A_1-selective antagonist DPCPX (1,3-dipropyl-8-cyclopentylxanthine; 3 nM) shifted the concentration–response curves of CPA as well as of ATP and ATPγS to the right, and the extent of the shift was similar for the three agonists (apparent pK_B values between 9.3 and 9.7; not shown). It follows that all three agonists reduced the release of [^{3}H]noradrenaline by activation of prejunctional A_1 purinoceptors. A differential change was obtained, however, with the P2 antagonist Cibacron blue. As shown in Fig. 3, Cibacron blue (30 μM) caused little if any change of the concentration–response curve of CPA, but clearly shifted the curves of ATP and ATPγS to the right, with apparent pK_B values of 5.0 and 5.1, respectively. Therefore, whereas CPA acts exclusively through the A_1 prejunctional receptor, the two nucleotides act through an additional, Cibacron blue-sensitive site, in all likelihood a P2 purinoceptor. A combination of DPCPX (3 nM) and Cibacron blue (30 μM) was also tested. The mixture shifted the concentration–response curve of ATP to a greater extent than did DPCPX alone (not shown) or Cibacron blue alone (Fig. 3). From the additional shift caused by the mixture, beyond the shift caused by DPCPX alone, an apparent pK_B for Cibacron blue of 4.7 was calculated, similar to the pK_B values from experiments with Cibacron blue alone, in accord with the existence of two separate receptors (von Kügelgen et al 1995a).

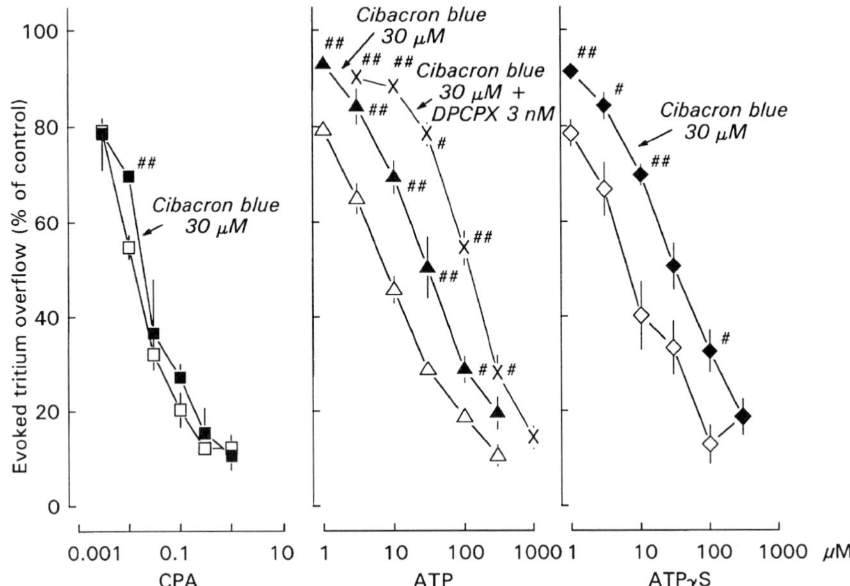

FIG. 3. Interaction of purinoceptor agonists with Cibacron blue 3GA (Cibacron blue) or 1,3-dipropyl-8-cyclopentylxanthine (DPCPX) on evoked overflow of tritium from rat atria preincubated with [^{3}H]noradrenaline. There were four periods of electrical stimulation (30 pulses, 1 Hz; S_1 to S_4). Cyclopentyladenosine (CPA), ATP and ATPγS were added at increasing concentrations (abscissae) 6 min before S_2, S_3 and S_4. Open symbols, experiments with agonists alone; solid symbols, experiments in which the medium contained 30 μM Cibacron blue throughout superfusion; X, experiments in which the medium contained both DPCPX (3 nM) and Cibacron blue (30 μM) throughout superfusion. Significant differences from agonists given alone: # $P<0.05$; ##$P<0.01$. (Modified from von Kügelgen et al 1995b.)

Are the prejunctional P2 receptors autoreceptors *sensu stricto*, i.e. sites of inhibition by endogenous ATP? Figure 4 suggests this is so: given alone, Cibacron blue increased the release of [^{3}H]noradrenaline. The increase averaged 91%, whereas under the same conditions the α_2-selective antagonist yohimbine (1 μM) caused an increase of 226% (Fig. 4). DPCPX (3 nM) failed to cause any increase, an observation that again supports the view that prejunctional P2 and A_1 purinoceptors are distinct entities.

Similar evidence for separate prejunctional A_1 and P2 purinoceptors has been obtained in mouse and rat vas deferens (Kurz et al 1993, von Kügelgen et al 1994a), the rat iris (Fuder & Muth 1993), mouse atria (von Kügelgen et al 1995a) and, beyond the sympathetic nervous system, rat brain cortex (von Kügelgen et al 1994b). In several but not all instances, the P2 receptors were shown to be autoreceptors by experiments such as that of Fig. 4.

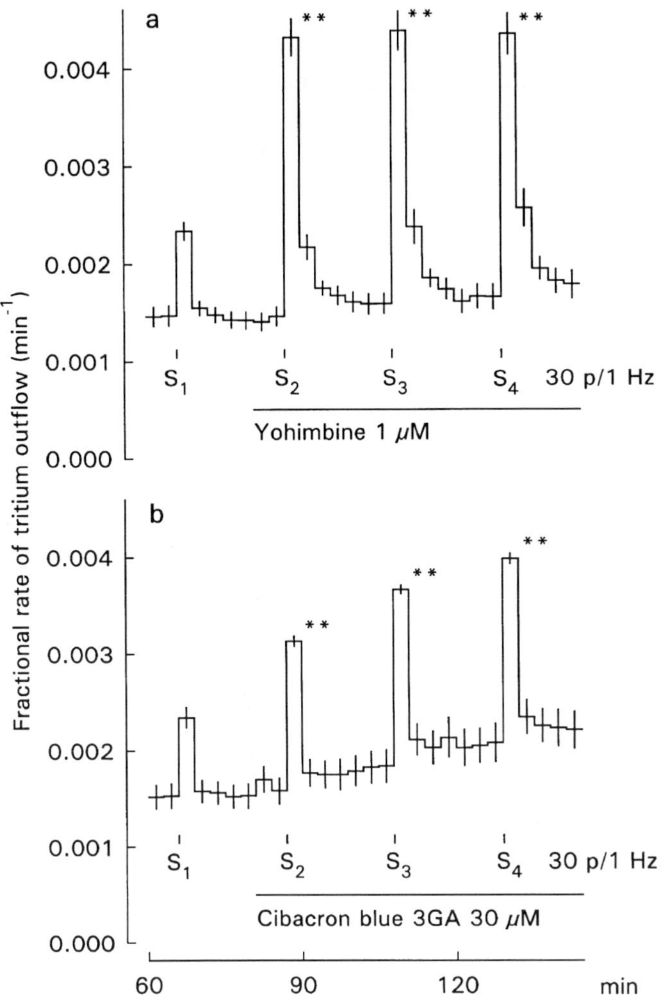

FIG. 4. Effect of yohimbine and Cibacron blue 3GA on tritium outflow and the response to electrical stimulation in rat atria preincubated with [^{3}H]noradrenaline. There were four periods of electrical stimulation (30 pulses, 1 Hz; S_1 to S_4). Yohimbine and Cibacron blue 3GA were added as indicated. Significant differences of S_n/S_1 overflow values from corresponding controls (solvent instead of drugs): $^{**}P<0.01$. (From data of von Kügelgen et al 1995b.)

α_2 Autoreceptors and P2 autoreceptors may operate in parallel to depress transmitter release. Their conditions of operation also seem to be similar. As in the case of the α_2 autoreceptors, inhibition through P2 autoreceptors requires preceding transmitter release and sets in after a certain latency; transmitter release elicited by very brief (some 10 ms) pulse trains is subject to little, if any, autoinhibition (von Kügelgen et al 1993, 1995b). One obviously interesting piece of information is still lacking: do prejunctional P2 autoreceptors depress the release of ATP in addition to the release of noradrenaline, and if so, is there a differential modulation like that shown in Table 1?

Differential co-transmitter release and the existence of post- and prejunctional receptors for noradrenaline as well as ATP perhaps confer flexibility, but certainly confer complexity, on postganglionic sympathetic neuroeffector transmission.

References

Bültmann R, Starke K 1993 Evans blue blocks P_{2X} purinoceptors in rat vas deferens. Naunyn-Schmiedeberg's Arch Pharmacol 348:684–687

Bültmann R, Driessen B, Goncalves J, Starke K 1995 Functional consequences of inhibition of nucleotide breakdown in rat vas deferens: a study with Evans blue. Naunyn-Schmiedeberg's Arch Pharmacol 351:555–560

Burnstock G 1990 Co-transmission. Arch Int Pharmacodyn Ther 304:7–33

Driessen B, Starke K 1994 Modulation of neural noradrenaline and ATP release by angiotensin II and prostaglandin E_2 in guinea-pig vas deferens. Naunyn-Schmiedeberg's Arch Pharmacol 350:618–625

Driessen B, von Kügelgen I, Starke K 1993 Neural ATP release and its α_2-adrenoceptor-mediated modulation in guinea-pig vas deferens. Naunyn-Schmiedeberg's Arch Pharmacol 348:358–366

Driessen B, von Kügelgen I, Starke K 1994 P_1-purinoceptor-mediated modulation of neural noradrenaline and ATP release in guinea-pig vas deferens. Naunyn-Schmiedeberg's Arch Pharmacol 350:42–48

Ellis JL, Burnstock G 1990 Modulation by prostaglandin E_2 of ATP and noradrenaline co-transmission in the guinea-pig vas deferens. J Auton Pharmacol 10:363–372

Exner HJ, Schlicker E 1995 Prostanoid receptors of the EP_3 subtype mediate the inhibitory effect of prostaglandin E_2 on noradrenaline release in the mouse brain cortex. Naunyn-Schmiedeberg's Arch Pharmacol 351:46–52

Forsyth KM, Bjur RA, Westfall DP 1991 Nucleotide modulation of norepinephrine release from sympathetic nerves in the rat vas deferens. J Pharmacol Exp Ther 256:821–826

Fredholm BB, Fried G, Hedqvist P 1982 Origin of adenosine released from rat vas deferens by nerve stimulation. Eur J Pharmacol 79:233–243

Fuder H, Muth U 1993 ATP and endogenous agonists inhibit evoked [^{3}H]-noradrenaline release in rat iris via A_1 and P_{2Y}-like purinoceptors. Naunyn-Schmiedeberg's Arch Pharmacol 348:352–357

Fujioka M, Cheung DW 1987 Autoregulation of neuromuscular transmission in the guinea-pig saphenous artery. Eur J Pharmacol 139:147–153

Kurz K, von Kügelgen I, Starke K 1993 Prejunctional modulation of noradrenaline release in mouse and rat vas deferens: contribution of P_1- and P_2-purinoceptors. Br J Pharmacol 110:1465–1472

Lew MJ, White TD 1987 Release of endogenous ATP during sympathetic nerve stimulation. Br J Pharmacol 92:349–355

Sedaa KO, Bjur RA, Shinozuka K, Westfall DP 1990 Nerve and drug-induced release of adenine nucleosides and nucleotides from rabbit aorta. J Pharmacol Exp Ther 252:1060–1067

Shinozuka K, Bjur RA, Westfall DP 1988 Characterization of prejunctional purinoceptors on adrenergic nerves of the rat caudal artery. Naunyn-Schmiedeberg's Arch Pharmacol 338:221–227

von Kügelgen I, Starke K 1991a Noradrenaline–ATP co-transmission in the sympathetic nervous system. Trends Pharmacol Sci 12:319–324

von Kügelgen I, Starke K 1991b Release of noradrenaline and ATP by electrical stimulation and nicotine in guinea-pig vas deferens. Naunyn-Schmiedeberg's Arch Pharmacol 344:419–429

von Kügelgen I, Starke K 1994 Corelease of noradrenaline and ATP by brief pulse trains in guinea-pig vas deferens. Naunyn-Schmiedeberg's Arch Pharmacol 350: 123–129

von Kügelgen I, Schöffel E, Starke K 1989 Inhibition by nucleotides acting at presynaptic P_2-receptors of sympathetic neuro-effector transmission in the mouse isolated vas deferens. Naunyn-Schmiedeberg's Arch Pharmacol 340:522–532

von Kügelgen I, Kurz K, Starke K 1993 Axon terminal P_2-purinoceptors in feedback control of sympathetic transmitter release. Neuroscience 56:263–267

von Kügelgen I, Kurz K, Starke K 1994a P_2-purinoceptor-mediated autoinhibition of sympathetic transmitter release in mouse and rat vas deferens. Naunyn-Schmiedeberg's Arch Pharmakol 349:125–132

von Kügelgen I, Späth L, Starke K 1994b Evidence for P_2-purinoceptor-mediated inhibition of noradrenaline release in rat brain cortex. Br J Pharmacol 113:815–822

von Kügelgen I, Stoffel D, Starke K 1995a P_2-purinoceptors at sympathetic axons in mouse atria. Naunyn-Schmiedeberg's Arch Pharmacol 351:138R(abstr)

von Kügelgen I, Stoffel D, Starke K 1995b P_2-purinoceptor-mediated inhibition of noradrenaline release in rat atria. Br J Pharmacol 115:247–254

Westfall DP, Stitzel RE, Rowe JN 1978 The postjunctional effects and neural release of purine compounds in the guinea-pig vas deferens. Eur J Pharmacol 50:27–38

DISCUSSION

Burnstock: I would like to raise a point that worries me a bit. Jim Ellis and I have published three or four papers showing the same kind of thing: angiotensin, prostaglandin, and calcitonin gene-related peptide produced different prejunctional effects on the release of noradrenaline and ATP (Ellis & Burnstock 1989a,b, 1990a,b). What bothers me in retrospect is that we looked at endogenous release of ATP using the firefly luciferin–luciferase technique, but we used tritium label to look at release of noradrenaline. It is well known that freshly taken up noradrenaline isn't necessarily incorporated into vesicles right away. For instance, tyramine can release non-vesicular pools of noradrenaline. Consequently, we should be comparing the endogenous release of both ATP and noradrenaline. The reason one hesitates to do this is that it's not easy to get enough noradrenaline out to measure with the sensitivity of the assay systems currently available; that's why we both used tritium labelling. But it worries me to conclude (as we did in all those papers) that there's an incongruity of release, and therefore there may be storage of ATP and noradrenaline in separate vesicles and/or separate nerves, if in fact we aren't comparing like with like. We were comparing freshly taken up noradrenaline and its release with endogenous ATP. What do you think?

Starke: Apart from the practical reasons you mentioned, Sol Langer and I came to the conclusion a long time ago that it is highly advantageous to look at total tritium release, because it includes measurement of all the metabolites. About 80% of noradrenaline release is not appearing as noradrenaline but as metabolites, and all that is measured when we measure tritium. When we measure endogenous noradrenaline we run into another difficulty, namely that we never know whether the fate of the released noradrenaline has changed. In short, measurement of endogenous noradrenaline has its advantages, but so has tritium. I may add that there are no reported deviations of the release of recently taken up [^{3}H]noradrenaline and endogenous noradrenaline in terms of prejunctional modulation. They are always modulated in similar ways.

Burnstock: It is well known that one end of the vas deferens has largely adrenergic responses and the other end is largely purinergic. How do you explain that if there aren't different populations of nerves, one dominated by ATP the other by noradrenaline?

North: The density of P2X receptors observed immunohistochemically declines dramatically as you go down the vas deferens. There are fewer receptors at the epididymal end.

Burnstock: There is a remarkable plasticity of expression of both transmitters and receptors. The actual proportions and combinations of co-transmitters can vary in one organ in different regions, and this varies with age. So the machinery for varying the proportions of transmitters and receptors is present.

Jacobson: We know that there is a constant tonic activation of adenosine A_1 receptors. For example, if you administer DPCPX *in vivo*, there are dramatic effects on memory, neuroprotection and so forth. What if you had at the synapse a constant source of adenosine that was not related to the ATP released? Wouldn't this explain the DPCPX shift of the ATP-release curve? This would not necessitate postulating action of ATP on an A_1 receptor directly.

Starke: I agree that might be the case in the brain, where there is endogenous adenosine at about 100 nM in the extracellular space, but not in the periphery. In vas deferens there is not enough endogenous adenosine to cause inhibition, so DPCPX has no effect, whereas in the brain slice DPCPX causes a doubling of transmitter release.

Jacobson: If you look in binding assays, there's no evidence that ATP has any sort of respectable affinity at A_1 receptors—even AMP is only marginally active.

Starke: Bailey & Hourani (1990) showed that at least functionally you do get activation of A_1 receptors by nucleotides. We are pretty sure that under our conditions the ATP is not acting by way of adenosine, because when we block adenosine formation by inhibiting 5′ nucleotidase, or when we treat the tissue with adenosine deaminase, ATP remains active.

Hourani: Probably the best evidence for this comes from the rat duodenum. Here, there are relaxant A_1 and A_2 receptors, and adenosine seems to act on the A_2 receptors because its effects are inhibited by the non-selective adenosine antagonist 8-SPT, but not by A_1-selective (nanomolar) concentrations of DPCPX. ATP acts on a P2 receptor rather than a P1 because its effects are not blocked by 8-SPT or by DPCPX. However, the effects of β,γ-methylene ATP are blocked by 8-SPT and by nanomolar concentrations of DPCPX (Hourani et al 1991, Nicholls et al 1992). My conclusions at the time were that β,γ-methylene ATP was acting directly on an A_1 receptor. However, I could not rule out the possibility that there might be a nucleotide diphosphohydrolase sitting right next to the A_1 receptor. But the fact that β,γ-methylene ATP does activate the A_1 receptor while ATP doesn't means that the effect of β,γ-methylene ATP can't easily be explained by the normal ectonucleotidase effect. Whether this means that β,γ-methylene ATP really binds to an A_1 receptor is unclear.

Edwards: I had the impression that suramin also blocked the breakdown of ATP, so I was surprised that you didn't get an increase in overflow from guinea-pig vas deferens in the presence of suramin. Secondly, the differential effect of suramin could also be explained by a block of the enzyme.

Starke: Suramin is certainly having a dual effect. It reduces postjunctional ATP release and blocks its degradation. Depending on the conditions, we sometimes get an increase in ATP overflow by suramin. For example, when we give it alone in guinea-pig vas deferens, we get an increase. Or when we stimulate the tissue by noradrenaline and release ATP, we can increase that ATP release by suramin. But under conditions where there is very little release due to prazosin, as I showed, then apparently the blockade of postjunctional release prevails and overflow goes down.

Kennedy: You showed that suramin and Cibacron blue increased noradrenaline release. At those concentrations of suramin and Cibacron blue you would have a substantial inhibition of ectoATPase. Could it be that the increase in noradrenaline release was because you are decreasing adenosine production and depressing presynaptic feedback?

Starke: We have repeated these experiments in the presence of DPCPX with the same results.

Westfall: Work in my laboratory is generally consistent with the work of Klaus Starke and his colleagues. We differ in some ways, but in terms of the general principles we are in agreement. This is especially so with the idea that the release of the co-transmitters ATP and noradrenaline may be differentially modulated.

The technique we are using has evolved over the years with the latest refinements being made by my colleague Latchezar Todorov. We superfuse vas deferens in a small-volume bath (the fluid volume with tissues absent is approximately 200 μl) at a rate of about 2 ml/min. The superfused tissues are subjected to electrical field stimulation (EFS) for 1 min at various frequencies and samples of superfusate are collected every 10 s. These are analysed for adenine nucleotides and adenosine by HPLC–fluorescence detection and for catecholamines by HPLC–electrochemical detection methods. We can also prepare one vas deferens for measuring the mechanical response of the tissue. Figure 1 (*Westfall*) illustrates the results from such an experiment. In panel A are shown representative contractions of a vas deferens response to EFS for 1 min at 2, 4 or 8 Hz. In panel B are summarized responses from a number of experiments which were obtained by integrating the area under the contractile curve. Panel C depicts the overflow of ATP at 10 s intervals from tissues stimulated at 2, 4 or 8 Hz for 1 min. Panel D shows the same for noradrenaline. For both transmitters, the overflow was increased as the frequency of stimulation increased. It is apparent, however, that the time-course of overflow differed markedly for the two transmitters. The release of noradrenaline increased with increasing number of pulses when stimulated at 2 or 4 Hz. When stimulated at 8 Hz the release of noradrenaline reached a peak at 30 s and remained constant for the remainder of the 1 min stimulation period. At the completion of the stimulation, the overflow of noradrenaline gradually returned towards prestimulation levels. The release of ATP, on the other hand, reached a peak much more quickly, by about 20 s, and then declined dramatically, even though the stimulation continued for 1 min. Under the conditions of our experiments, unlike those of Klaus Starke, we do not see a prazosin-sensitive component of the release of ATP. Thus we believe we are looking at the neuronal release of the nucleotide. Also, the phasic nature of the overflow of ATP is not due to depletion or to metabolism.

The differing pattern of release of noradrenaline and ATP is not compatible with the idea that they are released from the same exocytotic vesicles but rather suggests that the release occurs from two populations of vesicle, one of which releases predominantly ATP and the other predominantly noradrenaline.

Consistent with our idea of separate sites of release are our findings with Ca^{2+} channel antagonists. Shown in Fig. 2 (*Westfall*) are the effects of ω-conotoxin GVIA (10 nM), a putative N-type channel antagonist, and ω-agatoxin IVA (10 nM), a putative P-type channel antagonist, on release of the co-transmitters. Conotoxin substantially decreased the release of noradrenaline evoked by EFS at 8 Hz, but had considerably less, although some, influence on the release of ATP. On the other hand, agatoxin significantly reduced the release of ATP and to a lesser degree the release of noradrenaline. Our tentative hypothesis is that the release of the sympathetic nerve co-transmitters may be coupled to different Ca^{2+} channels—the release of noradrenaline being

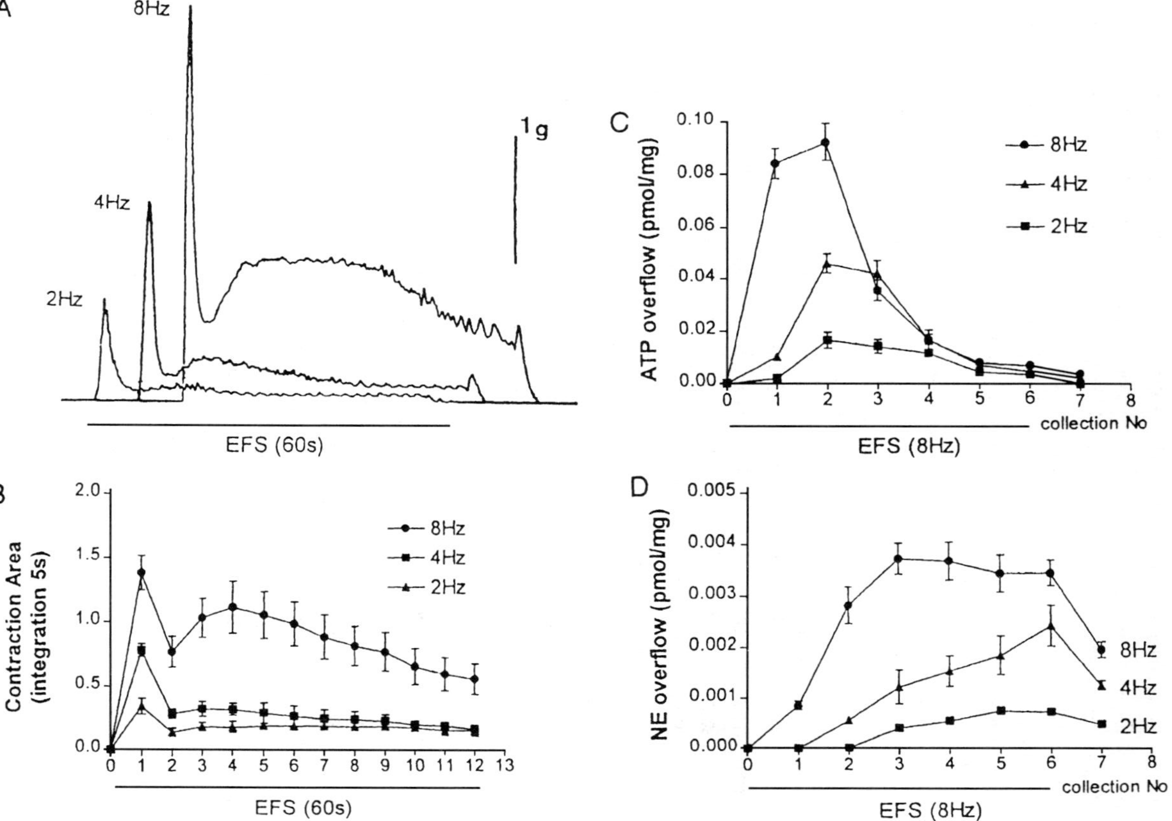

FIG. 1. (*Westfall*) The effects of electrical field stimulation at 2, 4 and 8 Hz on the contraction of vas deferens (A), mean data of the contraction obtained by integrating the area under the contractile record (B), on the release of ATP (C) and noradreneline (NE; D).

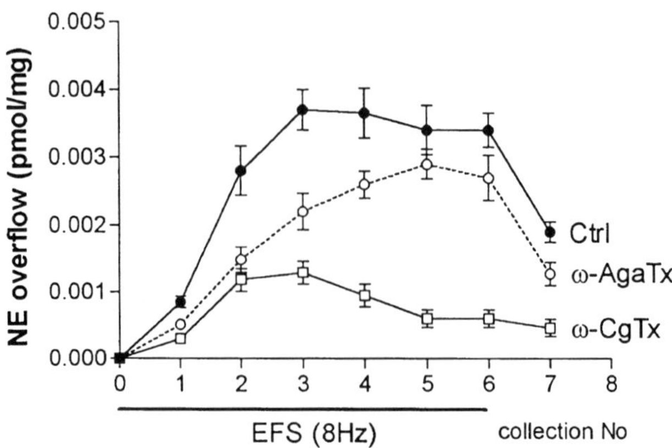

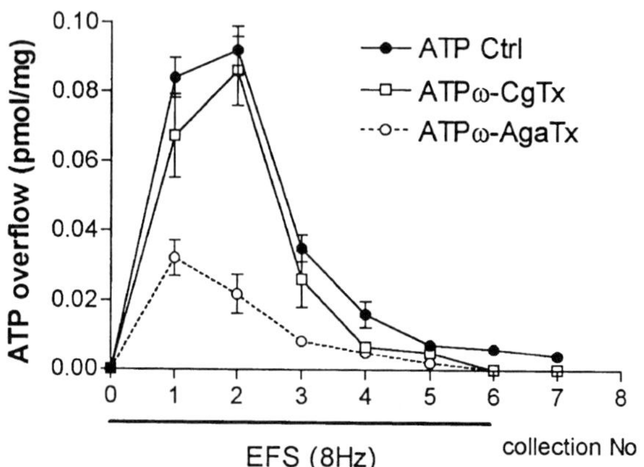

FIG. 2. (*Westfall*) The effects of ω-agatoxin IVA and ω-conotoxin GVIA on the electrical field stimulation-induced release of noradrenaline (NE; upper panel) and ATP (lower panel) from guinea-pig vas deferens. The tissues were exposed to the toxins at 10 nM for 30 min prior to and during the stimulation.

more dependent on Ca^{2+} entry through N-type channels, whereas the release of ATP may be coupled to non-N-type, possibly P-type Ca^{2+} channels.

Figure 3 (*Westfall*) deals with the effects of idazoxan (0.1 nM), an α_2 antagonist, on the simultaneous release of noradrenaline and ATP. Idazoxan produces a substantial increase in the release of noradrenaline and, although the α_2 antagonist also enhances the release of ATP, this enhancement is much less than that of noradrenaline. Thus, our results are consistent with those of Klaus Starke which indicate that there can be differential modulation of the release of co-transmitters. Our view is that this is possible because there is release from two populations of vesicles.

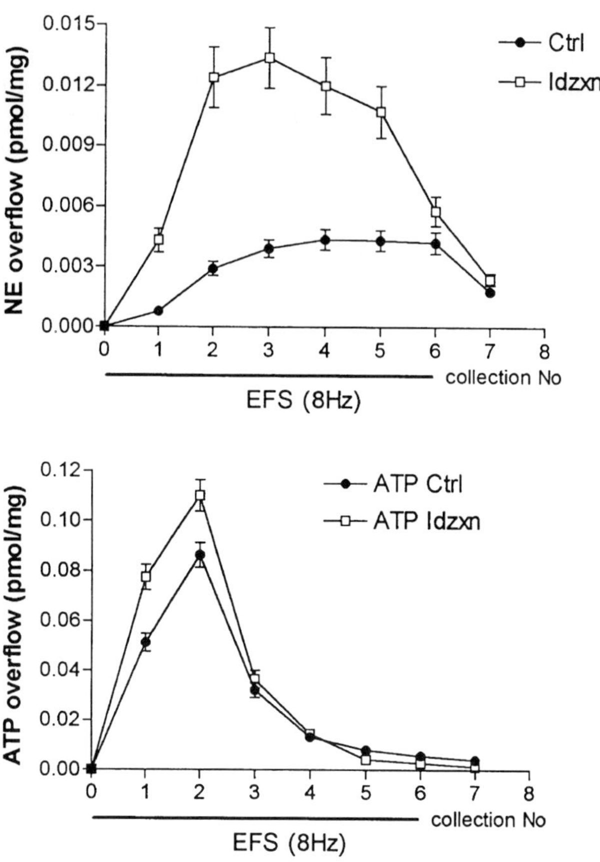

FIG. 3. (*Westfall*) The effect of idazoxan (Idxn; 0.1 μM) on the electrical field stimulation-induced release of noradrenaline (NE; upper panel) and ATP (lower panel) from guinea-pig vas deferens.

Burnstock: I'm relieved to see that confirms more or less what we said.

Starke: It was very reassuring to see the change of the co-transmitter mixture in the course of a pulse train. We saw the same. We didn't do it with endogenous noradrenaline, but we carried out a study in the guinea-pig vas deferens looking at pulse trains of the length of 10–540 pulses. We measured release of [^{3}H]noradrenaline and endogenous ATP. The pattern was exactly the same as that in David Westfall's experiments, which means that the average noradrenaline release per pulse stays approximately the same in the course of a pulse train, whereas the release of ATP decreases. When we calculated an ATP : noradrenaline release ratio, it had a characteristic sharp decline with the number of pulses (von Kügelgen & Starke 1991).

Burnstock: Have you ever thought about the physiological role of ATP compared with noradrenaline, since they have such different characteristics? Low frequency discharge favours the effects of ATP and it gives a fast, short-acting contraction, whereas the noradrenaline is more dominant with high frequency discharge and produces slower, longer-lasting contractions.

Starke: If we include neuropeptide Y (NPY), then it may be that purinergic, adrenergic and NPY-ergic contraction phases develop one after the other in that sequence. In principle, however, it is not clear why there is a need for three transmitters.

Burnstock: They do different things with different frequencies of discharge. For instance, there is some evidence that during stress there are short bursts of low frequency discharge in sympathetic nerves that would favour the ATP component of sympathetic co-transmission, whereas during gentle exercise there are long trains of impulses which would favour the noradrenaline component. Even if they are stored in the same vesicles you would get this differential effect, because of the different nature of the receptors they act on.

Westfall: It was my belief, before we were able to examine the temporal aspects of co-transmitter release, that the phasic, short-lasting response of the vas deferens, which is mediated by ATP, was due to rapid metabolism of ATP coupled with rapid desensitization of the P2X receptor. Although these events undoubtedly continue, the phasic nature of the response to nerve stimulation appears to be related to the phasic, short-lasting release of ATP.

In terms of neurotransmission in general, this could have important implications. The system is binomial. If you give a train of stimuli, at one point early in the train you have one transmitter being released. In the middle of the train you have two transmitters being released, and at the end of the train you have only the second transmitter being released. The nerve can make a choice about which mixture of transmitters it wants to release: early on ATP, in the middle both and at the end noradrenaline.

Burnstock: I must say I am not convinced that NPY is a true co-transmitter in many sympathetically innervated tissues. It is a very potent prejunctional modulator of the release of both noradrenaline and ATP and it is also a potent postjunctional potentiator of the actions of noradrenaline and ATP, but it does little itself in many of these organs (the exceptions include heart, brain and spleen, where non-sympathetic nerves containing NPY are present too).

IJzerman: What effect do adenosine transport blockers have in your system?

Westfall: Adenosine transport inhibitors will enhance the presynaptic actions of adenosine by increasing the extracellular concentration of adenosine, but they do not enhance the nucleotide-induced effects on transmitter release.

Burnstock: If you look at early development in a system like this, the progression of appearance of receptor types is going to give you some useful clues as to what they do separately. I am a great believer in looking at development.

Westfall: I would now like to make some comments about the identity of the presynaptic receptor that can modulate neurotransmitter release. We and others have found that ATP, acting without breakdown to adenosine, can decrease nerve stimulation-evoked release of transmitter; this action can be antagonized by DPCPX and other xanthine antagonists of adenosine receptors. The question is whether this is an effect on a P1 receptor, a P2 receptor or, in our view, an unusual hybrid receptor which we call P3.

In Fig. 4 (*Westfall*) are shown the results of ligand binding studies conducted by Andrew Smith in my laboratory. [^{3}H]DPCPX is bound to membranes from rat vas deferens and competed for with 2-chloroadenosine and the nucleotides ATP, α,β-methylene ATP and β,γ-methylene ATP. The competition curve for 2-chloroadenosine is bimodal, exhibiting a very high affinity site and a second site of lesser affinity. The nucleotides also competed for binding with [^{3}H]DPCPX in a monophasic fashion with an affinity somewhat less than that for 2-chloroadenosine at the second site. Although these experiments are quite preliminary, they suggest that there may be two 'receptors' for nucleosides, perhaps a P1 receptor (probably of the A_1 type because DPCPX is supposedly

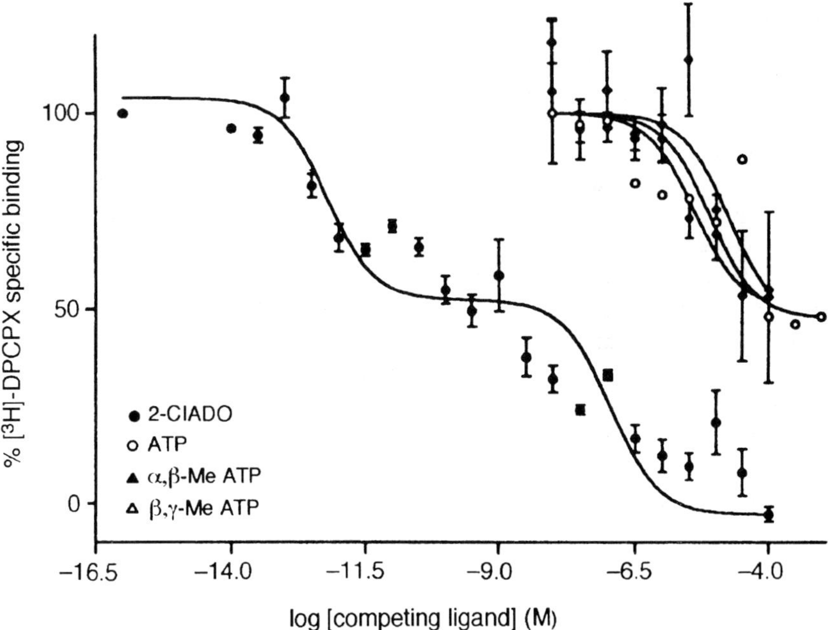

FIG. 4. (*Westfall*) The inhibition of specific [^{3}H]DPCPX binding by competing ligands in rat vasa deferentia. The curve for 2-chloroadenosine (2-ClADO) fits better to a two site binding model.

A_1 specific) and a second site that also binds nucleotides such as ATP that may be an atypical A_1 receptor or the so-called ‘P3’ receptor.

Williams: The high affinity, which is around 10^{-12} M, is very reminiscent of some data published by Bruce Morton some 15 years ago on A_1 and A_2 binding in spermatozoa (Morton et al 1982). These were very interesting results that people were sceptical of because of the very high potency of binding.

Hourani: We've recently published a paper showing binding of tritiated DPCPX to a number of rat peripheral tissues, including vas deferens. We find single site binding with a K_d of around 1 nM (Peachey et al 1994).

In the rat vas deferens we have evidence for inhibitory prejunctional A_1 receptors (as you would expect), postjunctional A_2 receptors inhibiting contraction, and also postjunctional A_1 receptors enhancing contraction to ATP (Hourani et al 1993, Hourani & Jones 1994). We've recently done an ontogenic study looking at the development of these separate receptors. The receptors develop at different times but the binding parallels the postjunctional A_1 receptor. So in our hands the DPCPX binding study picks up only the postjunctional A_1 receptors (Peachey et al 1995).

Westfall: Was it competed for by nucleotides?

Hourani: We didn't look.

Westfall: That's the problem we've always had. These cross-over studies haven't been done.

References

Bailey SJ, Hourani SMO 1990 A study of the purinoceptors mediating contraction in the rat colon. Br J Pharmacol 100:753–756

Ellis JL, Burnstock G 1989a Angiotensin neuromodulation of adrenergic and purinergic co-transmission in the guinea-pig vas deferens. Br J Pharmacol 97:1157–1164

Ellis JL, Burnstock G 1989b Modulation of neurotransmission in the guinea-pig vas deferens by capsaicin: involvement of calcitonin gene-related peptide and substance P. Br J Pharmacol 98:707–713

Ellis JL, Burnstock G 1990a Neuropeptide Y neuromodulation of sympathetic co-transmission in the guinea-pig vas deferens. Br J Pharmacol 100:457–462

Ellis JL, Burnstock G 1990b Modulation by prostaglandin E_2 of ATP and noradrenaline co-transmission in the guinea pig vas deferens. J Auton Pharmacol 10:363–372

Hourani SMO, Jones DAD 1994 Postjunctional excitatory adenosine A_1 receptors in the rat vas deferens. Gen Pharmacol 25:417–420

Hourani SMO, Bailey SJ, Nicholls J, Kitchen I 1991 Direct effects of adenylyl 5′-(α,β-methylene)diphosphonate, a stable ATP analogue, on relaxant P_1-purinoceptors in smooth muscle. Br J Pharmacol 104:685–690

Hourani SMO, Nicholls J, Lee SBS, Halfhide EJ, Kitchen I 1993 Characterisation and ontogeny of P1-purinoceptors in rat vas deferens. Br J Pharmacol 108:754–758

Morton BE, Thenawidjaja M, Dimsdale S 1982 Evidence for functional A1 and A2 adenosine receptors on hamster and human spermatozoa. Soc Neurosci Abstr 8:697

Nicholls J, Hourani SMO, Kitchen I 1992 Characterisation of P_1-purinoceptors on rat duodenum and urinary bladder. Br J Pharmacol 105:639–642

Peachey JA, Hourani SMO, Kitchen I 1994 The binding of 1,3-[^{3}H]-dipropyl-8-cyclopentylxanthine to adenosine A_1 receptors in rat smooth muscle preparations. Br J Pharmacol 113:1249–1256

Peachey JA, Brownhill VR, Hourani SMO, Kitchen I 1995 The ontogeny of adenosine receptor subtypes in the rat vas deferens. Br J Pharmacol 115:143P(abstr)

von Kügelgen I, Starke K 1991 Release of noradrenaline and ATP by electrical stimulation and nicotine in guinea-pig vas deferens. Naunyn-Schmiedeberg's Arch Pharmacol 344:419–429

General discussion II

Distribution of purinoceptors

Burnstock: I would like to raise an interesting topic which hasn't yet been discussed. There are a few systems in which it is beginning to look as though individual cells have a differential distribution of purinoceptors. For instance, in the epithelial cells in lung tissue there are many receptors at one end but few at the other. The same is true in the inner hair cells of the cochlea. There is even a suggestion that these hair cells have adenosine receptors at one end and ATP receptors at the other, which is amazing. I think we are going to see more of this kind of subcellular distribution of receptors, with different cells having different responses at different ends. What are the implications of this and how is it controlled?

Boucher: In the polarized epithelium there are a couple of interesting observations. There are presumably different numbers of receptors on one membrane versus the other but those relationships are complex. When the apical membrane is stimulated by ATP, it elicits smaller Ca^{2+} transients but larger Cl^- currents than ATP administration to the basolateral membrane. Paradiso et al (1995) refer to a more general biological issue: it appears that when you stimulate one membrane, the receptors on the contralateral membrane don't know it. Consequently, you can selectively activate P2 receptors on the apical membrane, the receptors on the contralateral membrane are not desensitized and, perhaps more interestingly, the pools of Ca^{2+} that are tapped by the contralateral receptor have not been accessed by the activated apical membrane receptors. With respect to capacitative Ca^{2+} entry, the corollary also appears to be true, i.e. the influx path on the membrane ipsilateral to P2 receptor stimulation but not the contralateral membrane is activated. It seems that within an epithelium you can distribute the receptors on one membrane versus the other membrane and receptors in the apical membrane can respond to the outside world with respect to ATP or P2U mediated response, without perturbing the housekeeping membrane of the cell, which is the basolateral or blood-facing membrane. To us the most interesting thing is that you can actually separate the responses mediated by a common receptor when it faces the outside versus the inside world.

Burnstock: When we have the cloned receptors, we will have a better chance of understanding the precise cellular distribution and hopefully the physiological significance.

North: There's some preliminary evidence the vas deferens antibody picks up staining within the spinal cord on primary afferents which disappears after rhizotomy (R. P. Elde, unpublished results). But we don't see anything much in the cell bodies; it is possible that the receptors have been shipped off in large numbers down the processes.

Burnstock: We've started using hibernation as a model of plasticity. In hibernating animals, the heart practically stops beating, the gut is an empty tube and the bladder is not doing anything for months. We are interested in finding out what happens to responses to ATP during hibernation in these organs and in blood vessels. It would be great if we could identify the genes that control the expression of purinoceptors in this way if their expression declines and is turned on again as the animal warms up and revives.

ATP as a neurotransmitter

Barnard: Since we are approaching the end of this meeting and we can begin to relax, I would like to raise a philosophical question. If I were designing a nervous system, I would never use ATP simply as a transmitter. It is in a category different from any other molecule that is used as a transmitter, because the cell has to put so much free energy into using it: it has a high energy bond. This is not true of any other transmitter and the energy that comes from breaking it in the termination of the transmission will be wasted as heat, if the transfer potential is not used. Is there an additional role for ATP related to the functioning of ectokinases, in which its presence extracellularly is related to its free energy potential? I have seen in the neurobiology literature a recent interest in ectokinases: I would like to know what the afficionados of that phenomenon think about the status of it in this context. It seems to me that, for example, there could be some functions that are mediated during development of neural circuitry by phosphorylation of extracellular domains of proteins. If that were to be the case, then I would feel much better about ATP, because it would have a dual role.

Burnstock: This was the kind of argument that people threw at me 25 years ago. I have talked to physical chemists about this[1]. They said that the level of ATP within cells is enormous compared with the small amount that comes out during effective nerve transmission. Cells are being broken and damaged all the time, with enormous wastage of ATP so, by comparison, the amount of ATP broken down during neurotransmission was trivial.

---

[1]*Burnstock:* Soon after the meeting, I contacted Barbara Banks, who is an expert in this field (see Banks & Vernon 1970). To quote from her letter to me: 'The most important fact about ATP is that it is involved in so many reactions and cellular processes that it must be supplied continuously. Except in brown fat which is needed as a source of heat, ATP is formed as fast as it is needed. Release of ATP from cells will cause the rate of its synthesis (from ADP and phosphate) to increase. If adenosine is taken up again by cells

The second thing they told me was that when acetylcholine and neuropeptides break down, the energy expended is comparable to that expended during the breakdown of ATP. In terms of heat production, the enthalpy (ΔH) of the myosin-catalysed hydrolysis of ATP (to ADP) is 20.5 kJ/mol (Gajewski et al 1986), whereas, for comparison, the ΔH for acetylcholine hydrolysis by erythrocyte acetylcholinesterase is 38.6 kJ/mol (Das et al 1985).

Barnard: I don't think large numbers of cells are breaking down in the brain. It's more likely that there's an additional role for ATP there in which the transfer potential is used[2].

---

there is not even an extra need for ATP to take part in the resynthesis of the purine base. Cellular events are regulated by protein phosphorylation and dephosphorylation which may switch enzymes on or off and has nothing to do with "energy" at all. G proteins may have either excitatory or inhibitory effects on cell chemistry but again their activation has nothing to do with "energy".

'When it is suggested that ATP is associated with "energy", the word is used incorrectly. It is an abbreviation for "the change in Gibb's function" or "the change in Gibb's free energy". Unlike "energy" proper, Gibb's function is not conserved, but abbreviating "the change in Gibb's energy" to "the energy" suggests that it is just like heat and work. It is not. Gibb's free energy is a Second Law concept where the equality signs of the First Law (heat + work = a constant) are replaced by the inequalities, stating for instance that the entropy of the Universe is increasing. Gibb's free energy decreases for any spontaneous change, just as the entropy (of the universe) increases.'

[2]*Barnard:* [This comment was added to answer the points raised in the first footnote.] You are quoting correctly some of the elementary thermodynamics involved, but this does not negate what I have said. What is important in the bioenergetics *is* the Gibbs free energy change, ΔG, which is what I had referred to. Since it seems I now have to go into more detail about this, Lubert Stryer, universally accepted as a standard modern authority, writes in his textbook 'ATP is often called a high-energy phosphate compound and its phospho-anhydride bonds are referred to as high-energy bonds. *They are high-energy bonds in the sense that much free energy is released when they are hydrolysed.... This ATP–ADP cycle is the fundamental mode of energy exchange in biological systems'* [his italics]. Stryer cites the ΔG for the conversion of ATP to ADP (calculated for physiological conditions) as −32 kJ/mol, and contrasts this exceptionally high level with ΔG for the corresponding and relevant reaction of glucose-6-phosphate to glucose, which is only −13 kJ/mol.

If *any* fairly complex molecule is broken down after transmission, considerable energy will be dissipated, representing the cost of formation of all the parts not re-used as such. That loss can be a necessary consequence of using it functionally, e.g. as a released and destroyed transmitter. It is irrelevant, therefore, to compare here the value of ΔH, for example, for the degradation of a neuropeptide. What is different in the special case of ATP is that the cell is losing the *high group-transfer potential* of each molecule in it, which arises from its resonance structures as a phospho-anhydride: the fission of one bond in ATP can be coupled to any one of a variety of useful anabolic reactions, so that the free energy is harnessed by the cell. That is why ATP and not other phosphates can function in the kinase-mediated phosphorylations which Professor Banks cites correctly as controlling cellular functions. Acetylcholine does not have this group transfer potential and its hydrolysis will always be a simple loss to the system, and hence the

Burnstock: Adenosine is sucked up rapidly by cells, converted to ATP and re-used. You can stimulate the nerves supplying the vas deferens for 48 h without the neuron cell body present, without diminution of the contractile responses; the retrieval system is brilliant.

Barnard: But that isn't related to the energy considerations I have noted.

Burnstock: But the amount of energy loss involved in the breakdown of such a few molecules is nothing compared with what is happening in energy terms in cells.

Williams: I would argue that neuropeptides represent one of the more excessive uses of energy in terms of providing molecules to function as neurotransmitters. If one is looking *only* at considerations related to entropy, peptides make no sense whatsoever as communication signals. They've not been especially resource effective or much of a success in terms of drug discovery either.

Barnard: Again, you don't necessarily know what functions are required in cell signalling that only a peptide could maintain.

Williams: ATP is a very dynamic molecule. While we've not discussed adenosine today, there is the potential for a purinergic cascade with ATP eventually forming adenosine, a neuromodulator in its own right (Williams 1995). ATP and adenosine acting via P2 and P1 receptors, respectively, can then have subtle and potentially complex influences on the physiological actions of one another. There is also an ectokinase component that was reported by Ehrlich et al (1988) that reflected ATP effects on long-term potentiation. I don't see peptide neurotransmitters having the same sophistication in terms of modulating cellular interactions as ATP, although one may argue that prepropeptides may subserve a similar function at the translation level.

---

comparison with it in this particular context is misleading. ATP is a uniquely valuable small molecule to the cell.

These considerations should never have been used, however, to claim that ATP could not act as a transmitter and I agree with Geoff Burnstock on that. But they do recognize a feature of it found in no other transmitter—its group transfer potential.

With regard to the stated unimportance to the cell economy of the fraction of the ATP pool lost by release, I would argue that it is difficult to be dogmatic about this when one has to consider it at the level not of the ATP content of the tissue, but of the available ATP pool in a single releasing neuron. Very large numbers of transmitter molecules can be released in the impulses, which may be sustained and are at a frequency in the millisecond range, in P2X neurotransmission. The small neurons which are often involved may, in fact, have to divert a significant fraction of their ATP steady-state pool to this storage and release: we do not have a sufficient knowledge of the presynaptic parameters for this system to assert that the contrary must be true.

Barnard: There may be functions that you could only use a peptide for—there is scope in the enormous diversity of the peptides for all kinds of interactions with proteins.

Burnstock: There's another way of looking at it: in evolutionary terms. ATP is a primitive molecule that was utilized for intracellular purposes as a high energy store early on. But it was also used very early on as an extracellular messenger. There are primitive plants in which ATP leaks out of cells and opens channels in neighbouring cells. You can easily imagine this as a very useful form of communication that was developed further during evolution. The ectoenzymes have been there from very early on, and ATP as a messenger arose very early on.

Barnard: That isn't an objection to what I'm saying, it is a support for it.

Burnstock: Not at all. The body uses what makes sense for it to use, and it only retains what makes sense in terms of survival. If the body uses a little bit of ATP for communication the energy question is totally irrelevant.

Barnard: I would think that it would still be an overall economy if ATP is used also for its phosphorylation ability. I am not convinced by the idea that ATP could only be a metabolite which, when it happens to be liberated on cell damage or stress, can serve as an intercellular communication molecule. Glucose and a hundred other metabolites can equally get out but aren't used in this way.

Burnstock: But they weren't used because they didn't turn out to be so useful.

Barnard: That would seem to be a very safe conclusion!

Jacobson: It has been suggested that the phosphorylation aspect is utilized in establishing neural contacts.

Barnard: That's what I'm saying: the ectokinases may be more important than we have been allowing.

North: Let's say that normal membrane recycling is going on all the time and the cell would actually have no way of stopping ATP getting into the vesicles and being popped out of the cell, because there's so much of it in the cytoplasm.

Illes: I'd like to extend the idea raised by Alan North. Concerning the co-transmitter concept, the function of ATP may be simply to keep the vesicular uptake mechanism running. During the exocytotic release of transmitters, ATP is reaching the extracellular space because the vesicle pours out its entire content into the synaptic gap. Then, in the course of ontogenesis, sooner or later receptors which recognize ATP will rise on the postsynaptic site.

Burnstock: It's a point of view. I take the point of view that ATP was a primitive transmitter which has been retained as a co-transmitter in variable proportions in different nerves. Some ATP is present in every nerve that has been studied so far.

Di Virgilio: I came into this field relatively late. Previously I worked on both the bioenergetics of mitochondria and Ca^{2+} homeostasis. There are connections between these different fields. Once cells learned to use Ca^{2+} as a signalling molecule, they exploited this cation in all possible processes (secretion, proliferation, motility, etc.). Similarly, when, thanks to the decrease in the intracellular Ca^{2+} concentration, energy metabolism based on phosphate bonds became possible, the cells might have taken the decision to exploit extensively the new phosphorylated metabolites not only as energy intermediates, but also as signalling molecules. There is such an excess production of these metabolites that any cell can very easily afford to waste at least part of it.

Some friends of mine working in reproductive physiology reckon that in the sperm fluid as much as 30% of total ATP is extracellular, and it is actively released into this fluid. We know that sperm desperately need ATP for moving. Why should they waste so much ATP by releasing it to the external medium if there was not an overproduction of ATP?

Barnard: Waste is an anthropomorphic term: I don't know that you can say that it's wasted. You don't know what the overall economy is.

References

Banks BE, Vernon CA 1970 Reassessment of the role of ATP in vivo. J Theor Biol 29:301–326

Das YT, Brown HD, Chattopadhyay SK 1985 Enthalpy of acetylcholine hydrolysis by acetylcholinesterase. Biophys Chem 23:105–114

Ehrlich YH, Snyder RM, Kornecki E, Garfield MG, Lenox RH 1988 Modulation of neuronal signal transduction systems by extracellular ATP. J Neurochem 50:295–301

Gajewski E, Steckler DK, Goldberg RN 1986 Thermodynamics of the hydrolysis of adenosine 5′-triphosphate to adenosine 5′-diphosphate. J Biol Chem 261:12733–12737

Paradiso AM, Mason SJ, Lazarowski ER, Boucher RC 1995 Membrane-restricted regulation of Ca^{2+} release and influx in polarized epithelia. Nature 377:643–646

Williams M 1995 Purinoceptors in central nervous system function. Targets for therapeutic intervention. In: Bloom FE, Kupfer DJ (eds) Psychopharmacology, the fourth generation of progress. Raven, New York, p 643–655

Involvement of distinct receptors in the actions of extracellular uridine nucleotides

Jean-Marie Boeynaems*, Didier Communi, Sabine Pirotton, Serge Mottet† and Marc Parmentier

*Institute of Interdisciplinary Research, School of Medicine, and Departments of *Medical Chemistry and †Vascular Pathology, Université Libre de Bruxelles, Campus Hôpital Erasme, Bâtiment C, Route de Lennik 808, B-1070 Bruxelles, Belgium*

Abstract. The P2 purinoceptors were initially defined as a family of receptors responsive to extracellular adenine nucleotides. In the late 1980s, it became clear that extracellular uridine nucleotides are also able to modulate cell function. The existence of a nucleotide receptor, common to ATP and UTP, was suggested by indirect pharmacological arguments (for instance the lack of additivity and the cross-desensitization of the responses to the two nucleotides) and later demonstrated by the cloning of a P2U receptor equally responsive to ATP and UTP. Vascular endothelial cells are a paradigm of cells on which both ATP and UTP exert physiologically relevant effects (stimulation of prostacyclin and nitric oxide release). Their response to nucleotides is mediated by two distinct receptors, both coupled to phospholipase C: a specific purinoceptor responsive to ATP and ADP (P2Y) and a nucleotide receptor responsive to ATP and UTP (P2U). We have recently cloned from the human genome a new subtype of receptor (tentatively called $P2Y_4$), which is structurally related to the P2U receptor. Functional expression revealed its coupling to phospholipase C and its selective responsiveness to UTP and UDP. According to the new nomenclature, the P2 receptors that are coupled to G proteins belong to the P2Y family. It now appears that this family encompasses specific purinoceptors ($P2Y_1$, formerly called P2Y), nucleotide receptors common to ATP and UTP ($P2Y_2$, previously P2U) and selective pyrimidinoceptors ($P2Y_4$). The existence of these pyrimidinoceptors suggests that uridine nucleotides may play a role as intercellular mediators, independently from adenine nucleotides.

1996 P2 purinoceptors: localization, function and transduction mechanisms. Wiley, Chichester (Ciba Foundation Symposium 198) p 266–277

In the late 1980s, it became apparent that extracellular uridine nucleotides exert effects on many tissues and cells, in a similar way to adenine nucleotides. At the time it was proposed that these actions were mediated by specific pyrimidinoceptors distinct from the purinoceptors (Seifert & Schultz 1989). The existence of nucleotide or P2U receptors common to ATP and UTP constituted an alternative possibility, in favour of which experimental evidence began to accumulate—for instance, the lack of additivity and the cross-desensitization of the responses to the two nucleotides (Brown et al 1991, O'Connor et al 1991, Motte et al 1993). The final demonstration of the non-selective nucleotide receptor concept came with the cloning and functional expression of a P2U receptor at which ATP and UTP were equally active and equipotent (Lustig et al 1993, Parr et al 1994). However, the pyrimidinoceptor hypothesis reemerged with the observation that UTP and UDP stimulate the formation of inositol phosphate in C6-2B glioma cells, whereas ATP and ADP are completely inactive (Lazarowski & Harden 1994). We have contributed to this area of research in two distinct ways: first by showing that the responsiveness of vascular endothelial cells to adenine nucleotides involves both P2U and P2Y receptors and, more recently, by cloning a human receptor specifically responsive to uridine nucleotides.

P2U receptors in endothelial cells

Whereas the physiological significance of adenine nucleotide receptors expressed in a wide range of cell types remains unclear, the rationale for their presence on vascular endothelial cells seems rather obvious. Indeed, dense bodies of platelets contain huge amounts of ADP and ATP which are released in the vicinity of endothelial cells following platelet activation. In addition, during hypoxia (Hopwood et al 1989) or in response to increased shear stress (Bodin et al 1991), endothelial cells themselves release ATP which might behave as an autocrine mediator. ATP and ADP exert two major effects on the vascular endothelium: they stimulate the release of prostacyclin (PGI_2) and nitric oxide (NO) (Boeynaems & Pearson 1990). PGI_2 is a potent inhibitor of platelet aggregation and NO is a mediator of endothelium-dependent relaxation: in addition, NO also inhibits platelet aggregation and its action is synergistic with that of PGI_2. Platelet-derived ATP and ADP are thus the mediators of a negative feedback mechanism which can limit the extent of intravascular platelet aggregation and help to localize platelet thrombi to areas of endothelial damage (Fig. 1).

Signalling by adenine nucleotides in endothelial cells involves mainly phospholipase C (PLC) activation, with a rapid and transient accumulation of inositol-1,4,5-trisphosphate ($InsP_3$) (Pirotton et al 1987, Forsberg et al 1987) and a biphasic increase in cytosolic Ca^{2+} concentration, starting with a transient mobilization of intracellular Ca^{2+} and followed by an influx of extracellular Ca^{2+} (Hallam & Pearson 1986). The stimulation of PGI_2 production results from the mobilization of free arachidonic acid by

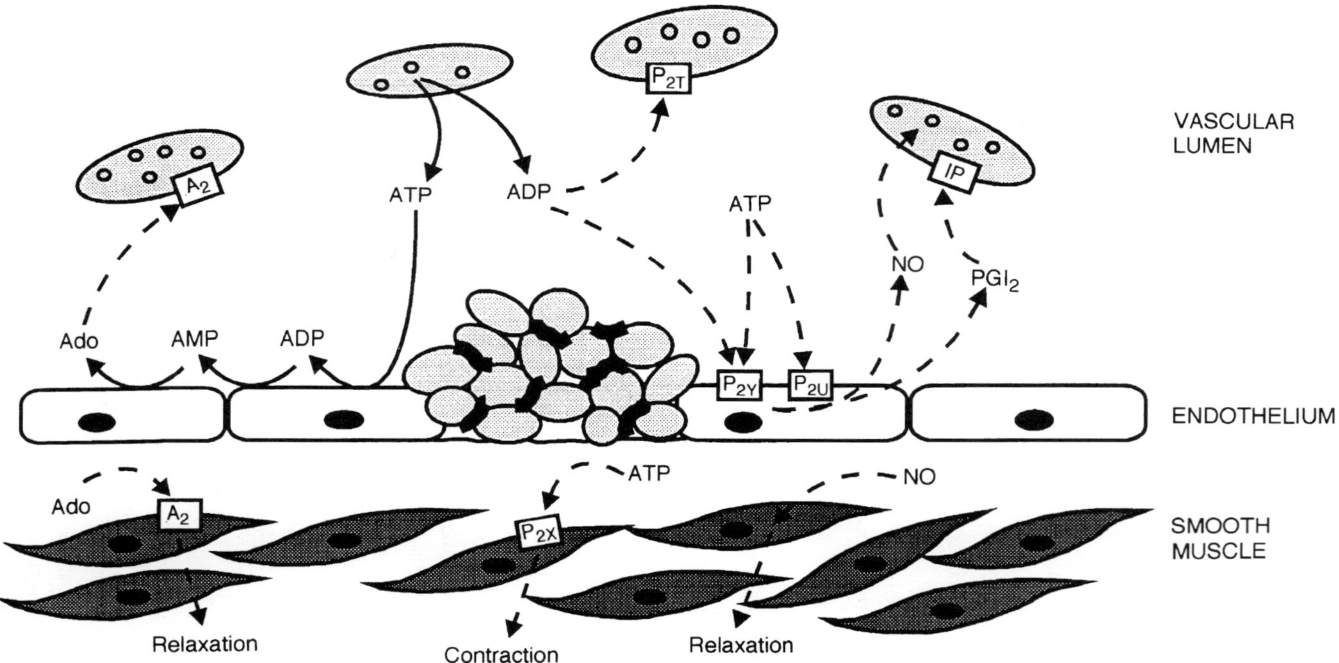

FIG. 1. Multiple receptors and feedback loops are involved in the local regulation of vascular tone and haemostasis by adenine nucleotides. Large quantities of ATP and ADP are locally released from platelets in response to aggregating agents. ADP recruits additional platelets via P2T receptor activation. On the other hand, ATP and ADP released from platelets, or from endothelial cells themselves, activate the P2 receptors (P2Y and/or P2U) expressed on endothelial cells. This results in the release of two vasodilators and potent inhibitors of platelet aggregation: prostacyclin (PGI_2), which acts via a specific receptor (IP), and nitric oxide (NO). NO is responsible for the endothelium-dependent relaxation by ATP of the underlying smooth muscle, which can be directly contracted by ATP via P2X receptors. Finally, ATP is rapidly degraded by endothelial ectonucleotidases into adenosine, which inhibits platelet aggregation and induces smooth muscle relaxation via A_2 receptors.

phospholipase A_2, directly activated by Ca^{2+} (Fig. 2). The ATP-induced production of NO results from the activation of NO synthase by the Ca^{2+}–calmodulin complex: it differs from the PGI_2 release by its more sustained time-course and its dependence on extracellular Ca^{2+} inflow (Fig. 2).

It was believed for a long time that the actions of ATP on the vascular endothelium involved only P2Y receptors (Burnstock & Kennedy 1985). However, several facts did not fit with this concept. In some cases, 2-methylthioATP (2-MeSATP), a P2Y-specific agonist, though more potent than ATP, produced a lower maximal effect (Needham et al 1987). On the other hand, some endothelial responses to ATP were mimicked by UTP, which is not a P2Y agonist (O'Connor et al 1991). This led to the proposal that endothelial

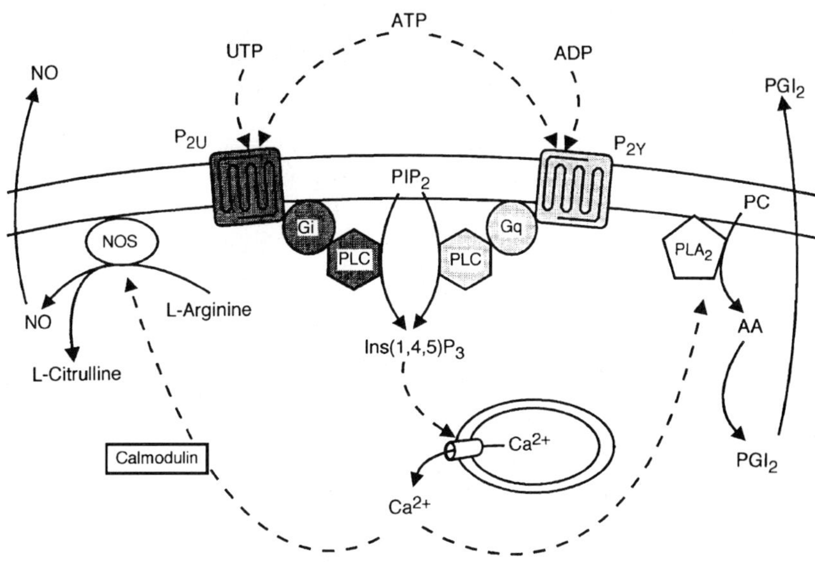

FIG. 2. Adenine nucleotides interact with two distinct G protein-coupled receptors, P2Y and P2U, which are coexpressed on bovine endothelial cells. Both classes of receptors appear to be coupled to a phospholipase C (PLC) via different G proteins: G_q/G_{11} would be involved in the response to P2Y receptor activation, whereas a member of the G_i family would be involved in the stimulation by P2U receptors. PLC activation results in the hydrolysis of phosphatidylinositol bisphosphate (PIP_2) into inositol-1,4,5-trisphosphate (Ins(1,4,5)P_3) and diacylglycerol (DAG). Inositol-1,4,5-trisphosphate mobilizes Ca^{2+} from the endoplasmic reticulum. The increase in cytoplasmic free [Ca^{2+}] induces the release of free arachidonic acid by phospholipase A_2 (PLA_2), resulting in an enhanced synthesis and release of prostacyclin (PGI_2). The activation of nitric oxide synthase (NOS) by the Ca^{2+}–calmodulin complex results in the production of NO from arginine.

cells can express both P2Y and P2U receptors (O'Connor et al 1991). The relative contribution of P2Y and P2U receptors to the endothelial responses to adenine nucleotides is variable, depending on the type of vessel. In microvascular endothelial cells from bovine adrenal medulla (Purkiss et al 1993), rabbit myocardium (Mannix et al 1993) and rat brain (Frelin et al 1993), the $InsP_3$ response to ATP is mediated exclusively by P2U receptors. In contrast, the bovine aortic endothelial cell line AG4762 expresses only P2Y receptors (Allsup & Boarder 1990). During the first passages, the two subtypes are co-expressed on cultured bovine aortic endothelial cells (Motte et al 1993, Wilkinson et al 1993). This coexistence is demonstrated by three lines of evidence. The effects of 2-MeSATP and UTP on inositol phosphates are additive, whereas the effects of ATP and either UTP or 2-MeSATP are not. ATP desensitizes the responses to both UTP and 2-MeSATP, whereas there is only a minimal cross-desensitization between 2-MeSATP and UTP (Motte et al 1993). Finally, the responses to 2-MeSATP and UTP are differentially modulated by pertussis toxin and 12-*O*-tetradecanoylphorbol 13-acetate (TPA, also known as phorbol myristate acetate): the response to UTP was strongly inhibited by pertussis toxin but only slightly sensitive to TPA inhibition; in contrast, the 2-MeSATP effect was reduced by TPA and unaffected by pertussis toxin (Motte et al 1993, Communi et al 1995). These latter results can be explained by the involvement of distinct G proteins in the coupling of the two receptors to PLC (Fig. 2). The stimulation of PLC by the P2Y receptor would be mediated by a member of the G_q/G_{11} family, whereas a G_i protein would be involved in the control by the P2U receptors.

Beyond this initial step in the transduction mechanism, we failed to detect any differences between the signalling pathways activated by the P2Y and P2U receptors (Communi et al 1995). In particular, the time course of $InsP_3$ accumulation and the relative proportion of inositol trisphosphate isomers and inositol tetrakisphosphate were similar. Measurements of $[Ca^{2+}]_i$ in single cells showed that almost all bovine aortic endothelial cells were responsive to both 2-MeSATP and UTP, indicating that P2Y and P2U receptors are not segregated on distinct cell subpopulations (Communi et al 1995). A response to UTP was also observed in freshly isolated aortic endothelial cells, showing that the P2U receptors are already expressed in the endothelium *in situ* and do not appear as a consequence of cell culture (Communi et al 1995). The reason for the expression of two distinct ATP receptors coupled to the same effector mechanism on the same cells remains unclear. One hypothesis would be that in addition to their coupling to a common effector—PLC—the P2Y and P2U receptors are also coupled to other transduction mechanisms specific to each of them. In any case, this receptor coexistence is not unique to aortic endothelial cells: for instance, P2Y and P2U receptors are also co-expressed on osteoblasts (Reimer & Dixon 1992).

Cloning of a human uridine nucleotide receptor

In order to isolate new subtypes of P2 receptors, we synthesized sets of degenerate oligonucleotide primers on the basis of the best conserved segments in the published sequences of the chick brain P2Y (Webb et al 1993) and murine neuroblastoma P2U (Lustig et al 1993) receptors. Their use in a PCR on human genomic DNA amplified *inter alia* a 669 bp sequence that was used as a probe to screen a human genomic library. Several clones were isolated and the sequence analysis of one of them revealed a 1095 bp open reading frame. The deduced amino acid sequence was consistent with that of a G protein-coupled receptor (Fig. 3). It exhibited 58% identity with the sequence of the human P2U receptor and 41% with the chick P2Y receptor. According to the new nomenclature (Fredholm et al 1994), the P2 receptors are subdivided in

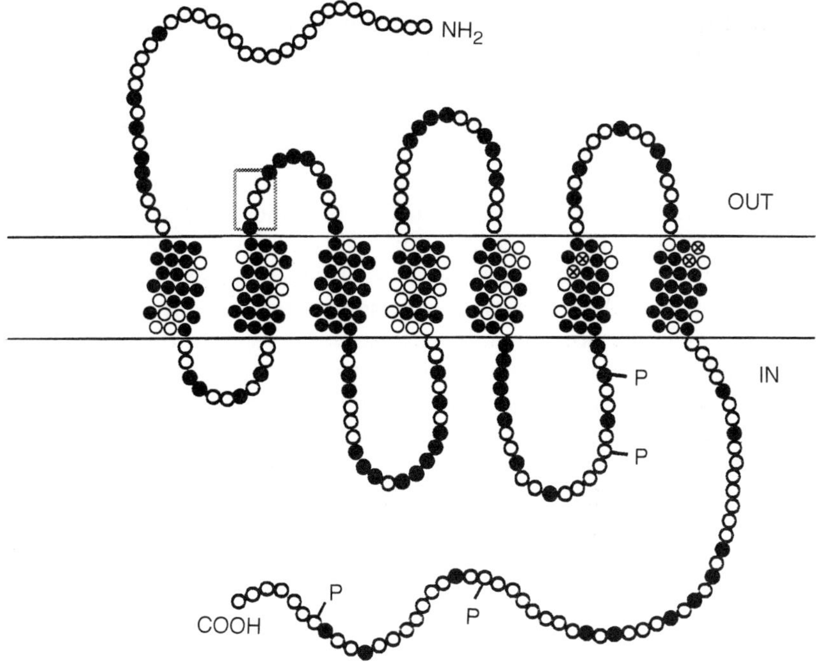

FIG. 3. Schematic diagram of the sequence of the P2Y$_4$ receptor showing its differences from the P2Y$_2$ receptor. Open circles identify the residues which differ between the two sequences and black circles correspond to identical residues. The crossed circles denote the four residues which play a crucial role in nucleotide binding to the P2Y$_2$ receptor and are conserved in the P2Y$_4$ sequence. The box corresponds to the RGD motif of the P2Y$_2$ receptor, which is not conserved in the P2Y$_4$ subtype. The letter 'P' identifies putative phosphorylation sites.

two classes: the P2Y receptors coupled to G proteins and the P2X receptors, which are ligand-gated ion channels. The P2Y receptor thus becomes the $P2Y_1$ subtype and the P2U receptor becomes $P2Y_2$. Since another subtype, called $P2Y_3$, has been cloned from the chick brain (Barnard et al 1994), the new receptor subtype depicted on Fig. 3 could be designated as $P2Y_4$. Its sequence exhibits some interesting features. Site-directed mutagenesis studies have demonstrated the crucial influence of charged amino acids in transmembrane helices 6 and 7 of the $P2Y_2$ receptor on agonist potency and specificity (Erb et al 1995): His^{262}, Arg^{265} and Arg^{292} seem to be directly involved in the binding of the negatively charged phosphate groups, whereas a mutation of Lys^{289} into Arg decreased the affinity for ATP/UTP and increased that for ADP/UDP. These four residues are conserved in the $P2Y_4$ subtype, whereas Arg^{265} and Lys^{289} are substituted respectively by Lys and Gln in the $P2Y_1$ receptor. On the other hand, despite its structural relatedness to the $P2Y_2$ receptor, the new subtype exhibits a conspicuous difference in the first putative extracellular loop: the RGD motif, an integrin binding consensus sequence present in the $P2Y_2$ sequence, is not conserved in the $P2Y_4$ receptor. Northern blot analysis revealed a strong signal at 1.8 kb in the human placenta: except for lung, other human and rat organs were negative, as were several human cell lines (e.g. K562, HL-60). The coding sequence was inserted into pcDNA3 in order to transfect 1321N1 human astrocytoma cells. In control cells, or in cells transfected with the vector, the accumulation of inositol phosphates was not modified by ATP, UTP or UDP. In cells stably expressing the $P2Y_4$ receptor, the formation of inositol phosphates was stimulated by UTP and UDP, whereas an effect of ATP was barely detectable. Therefore the $P2Y_4$ receptor is a selective pyrimidinoceptor, although structurally it belongs to the P2Y family. This family thus encompasses selective purinoceptors ($P2Y_1$), nucleotide receptors responsive to both adenine and uridine nucleotides ($P2Y_2$), and selective pyrimidinoceptors ($P2Y_4$). Although the occurrence of uridine nucleotides in extracellular fluids is poorly documented, the existence of such pyrimidinoceptors suggests that UTP and UDP may play a role as intercellular mediators, independently from adenine nucleotides.

Acknowledgements

This work was supported by an Action de Recherche Concertée of the Communauté Française de Belgique, by the Belgian Programme on Interuniversity Poles of Attraction initiated by the Belgian State, Prime Minister's Office, Federal Service for Science, Technology and Culture, and by a grant of the Fonds de la Recherche Scientifique Médicale. D. C. is a fellow of the Institut pour l'Encouragement de la Recherche Scientifique dans l'Industrie et l'Agriculture. S. P. is a Chargé de Recherches of the Fonds National de la Recherche Scientifique.

References

Allsup DJ, Boarder MR 1990 Comparison of P_2-purinergic receptors of aortic endothelial cells with those of adrenal medulla: evidence for heterogeneity of receptor subtype and of inositol phosphate response. Mol Pharmacol 38:84–91

Barnard EA, Burnstock G, Webb TE 1994 G protein-coupled receptors for ATP and other nucleotides: a new receptor family. Trends Pharmacol Sci 15:67–70

Bodin P, Bailey D, Burnstock G 1991 Increased flow-induced ATP release from isolated vascular endothelial cells but not smooth muscle cells. Br J Pharmacol 103:1203–1205

Boeynaems J-M, Pearson JD 1990 P_2 purinoceptors on vascular endothelial cells: physiological significance and transduction mechanisms. Trends Pharmacol Sci 11:34–37

Brown HA, Lazarowski ER, Boucher RC, Harden TK 1991 Evidence that UTP and ATP regulate phospholipase C through a common extracellular 5′-nucleotide receptor in human airway epithelial cells. Mol Pharmacol 40:648–655

Burnstock G, Kennedy C 1985 Is there a basis for distinguishing two types of P_2-purinoceptor? Gen Pharmacol 16:433–440

Communi D, Raspe E, Pirotton S, Boeynaems J-M 1995 Coexpression of P_{2Y} and P_{2U} receptors on aortic endothelial cells: comparison of cell localization and signaling pathways. Circ Res 76:191–198

Erb L, Garrad R, Wang Y, Quinn T, Turner JT, Weisman GA 1995 Site-directed mutagenesis of P_{2U} purinoceptors—positively charged amino-acids in transmembrane helix-6 and helix-7 affect agonist potency and specificity. J Biol Chem 270:4185–4188

Forsberg EJ, Feuerstein G, Shoami E, Pollard HB 1987 Adenosine triphosphate stimulates inositol phospholipid metabolism and prostacyclin formation in adrenal medullary endothelial cells by means of P_2 purinergic receptors. Proc Natl Acad Sci USA 84:5630–5634

Fredholm BB, Abbracchio MP, Burnstock G et al 1994 Nomenclature and classification of purinoceptors. Pharmacol Rev 46:143–156

Frelin C, Breittmayer JP, Vigne P 1993 ADP induces inositol phosphate-independent intracellular Ca^{2+} mobilization in brain capillary endothelial cells. J Biol Chem 268:8787–8792

Hallam TJ, Pearson JD 1986 Exogenous ATP raises cytoplasmic free calcium in fura-2 loaded piglet aortic endothelial cells. FEBS Lett 207:95–99

Hopwood AM, Lincoln J, Kirkpatrick KA, Burnstock G 1989 Adenosine 5′-triphosphate, adenosine and endothelium-derived relaxing factor in hypoxic vasodilatation of the heart. Eur J Pharmacol 165:323–326

Lazarowski ER, Harden TK 1994 Identification of a uridine nucleotide-selective G-protein-linked receptor that activates phospholipase C. J Biol Chem 269:11830–11836

Lustig KD, Shiau AK, Brake AJ, Julius D 1993 Expression cloning of an ATP receptor from mouse neuroblastoma cells. Proc Natl Acad Sci USA 90:5113–5117

Mannix RJ, Moatter T, Kelley KA, Gerritsen ME 1993 Cellular signaling responses mediated by a novel nucleotide receptor in rabbit microvessel endothelial cells. Am J Physiol 265:11675–11680

Motte S, Pirotton S, Boeynaems J-M 1993 Heterogeneity of ATP receptors in aortic endothelial cells. Involvement of P_{2y} and P_{2u} receptors in inositol phosphate response. Circ Res 72:504–510

Needham L, Cusack NJ, Pearson JD, Gordon JL 1987 Characteristics of the P_2 purinoceptor that mediates prostacyclin production by pig aortic endothelial cells. Eur J Pharmacol 134:199–209

O'Connor SE, Dainty IA, Leff P 1991 Further subclassification of ATP receptors based on agonist studies. Trends Pharmacol Sci 12:137–141

Parr CE, Sullivan DM, Paradiso AM et al 1994 Cloning and expression of a human P_{2U} nucleotide receptor, a target for cystic fibrosis pharmacotherapy. Proc Natl Acad Sci USA 91:3275–3279

Pirotton S, Raspe E, Demolle D, Erneux C, Boeynaems J-M 1987 Involvement of inositol 1,4,5-trisphosphate and calcium in the action of adenine nucleotides on aortic endothelial cells. J Biol Chem 262:17461–17466

Purkiss JR, Wilkinson GF, Boarder MR 1993 Evidence for a nucleotide receptor on adrenal medullary endothelial cells linked to phospholipase C and phospholipase D. Br J Pharmacol 108:1031–1037

Reimer WJ, Dixon SJ 1992 Extracellular nucleotides elevate $[Ca^{2+}]_i$ in rat osteoblastic cells by interaction with two subtypes. Am J Physiol 262:1040C–1048C

Seifert R, Schultz G 1989 Involvement of pyrimidinoceptors in the regulation of cell functions by uridine and by uracil nucleotides. Trends Pharmacol Sci 10:365–369

Webb TE, Simon J, Krishek BJ et al 1993 Cloning and functional expression of a brain G-protein-coupled ATP receptor. FEBS Lett 324:219–225

Wilkinson GF, Purkiss JR, Boarder MR 1993 The regulation of aortic endothelial cells by purines and pyrimidines involves co-existing P_{2Y}-purinoceptors and nucleotide receptors linked to phospholipase C. Br J Pharmacol 108:689–693

DISCUSSION

Barnard: Firstly, I would like to congratulate you on finding an interesting new P2Y receptor subtype. You said that it was expressed in placenta: did you check a variety of other tissues on Northern blots?

Boeynaems: The problem with human tissue is one of availability. On a blot of human tissues from Clontech, which was of poor quality, we had a signal in placenta and a small signal in the lung: the rest was negative. The other thing we did was to take blots from rat organs, and do hybridization at low stringency. The results were entirely negative. We also got negative results with several human cell lines: HL-60, K562 and a few of neural origin.

Harden: How can we explain a PLC-linked receptor like the P2U receptor, whose ability to activate PLC is blocked by pertussis toxin yet the receptor in no way seems to interact with adenylate cyclase? This has bothered me for a long time.

Boeynaems: In endothelial cells the situation is even more complex, because we have demonstrated that ATP was able to increase cAMP by a mechanism which is inhibited by methylxanthines but apparently does not involve the conversion into adenosine. This adds further complexity and is somewhat similar to the P3 receptor of David Westfall which was also inhibited by xanthine. So in the case of endothelial cells, the interpretation is fairly complex because of this additional mechanism. Indeed, you would expect an inhibition, but we have not observed one.

Harden: You almost have to speak of some sort of colocalization of the G protein with PLC and say that population of G_i or G_o doesn't recognize or can't touch adenylate cyclase.

Surprenant: How common is it for the P2U-coupled pathways to be pertussis toxin sensitive? Does this differ for P2Y receptor-coupled pathways?

Harden: Pertussis toxin will block anywhere from 50–80% of the response in most cells where the P2U receptor is natively expressed. I don't know of any P2Y-linked receptor that activates PLC in which pertussis-toxin substantially blocks coupling.

Surprenant: So, if one were measuring ATP-evoked changes in intracellular Ca^{2+} (or Ca^{2+} flux), and pertussis significantly inhibited the response, would this be strong evidence that the response was mediated by a P2U receptor?

Harden: No. For example, the uridine nucleotide-selective receptor on C6 cells activates PLC and this activity is blocked very nicely by pertussis toxin, yet it doesn't regulate adenylate cyclase.

Lustig: The accepted dogma is that pertussis toxin ADP-ribosylates the α subunit of G_i. Perhaps in some systems it's actually acting on a different substrate.

Harden: That is true. Most experiments are simply correlative in that you can correlate the blockade of coupling to PLC with the occurrence of pertussis toxin-promoted ADP ribosylation of the 42 kDa protein.

Burnstock: Can I just change the subject and make a physiological point. The top people interested in the role of NO, such as John Vane and Salvador Moncada, make a big point about the blood vessels being controlled by a resting vasodilatory tone due to spontaneous release of NO, on the basis that when they introduce a NO synthase inhibitor into a whole animal they get a huge rise in blood pressure. There are whole meetings these days in the NO field about how shear stress can lead to release of NO directly from endothelial cells. I think one should not forget that ATP is also spontaneously released from endothelial cells. These cells are packed with ATP that has been known to be released during shear stress (and hypoxia); it then acts on P2U/P2Y receptors which release NO to produce vasodilation. My guess is when the time comes and we have an *in vivo* antagonist to the P2Y and/or P2U purinoceptors, this will lead to a huge increase in blood pressure, i.e. that the release of ATP precedes the release of NO from endothelial cells. In therapeutic terms using a P2Y antagonist may be just as effective as using NO synthase inhibitors to raise blood pressure. Platelets are another important source of ATP that can act on endothelial cells.

Boeynaems: Yes, but we have to take into account the amount of nucleotides present in platelets, where their concentration is really enormous. When platelets aggregate the local concentration of ATP can reach 10 μM or so. This is greater than the concentration that can be released from endothelial cells.

Burnstock: But there's a difference in that the ATP pours out of platelets under particular circumstances which are not so tightly controlled as ATP coming from endothelial cells: they're not geared to hypoxia and changes in flow, they're geared to aggregation.

Wiley: Could I ask about another potential modulator: does P2U or P2Y give any differential activation of phospholipase A and generation of arachidonic acid and prostanoids from the membrane phospholipids?

Boeynaems: I have no precise answer, but clearly both are able to stimulate prostacyclin release.

Neary: I have a question about your cross-desensitization experiments. If I understood your data, it looked like you did not get cross-desensitization when you measured the response with $InsP_3$, but when you measured changes in intracellular Ca^{2+} it seemed you were observing cross-desensitization. Is that correct?

Boeynaems: Homologous desensitization is the predominant form of desensitization in these endothelial cells but there is some degree of heterologous desensitization. Perhaps heterologous desensitization was particularly striking in the Ca^{2+} traces that I showed, but the major component of desensitization is homologous desensitization.

Burnstock: I would like to raise the issue of purinoceptor subtype numbering. We seem to be coming round to the view that there are P2X and P2Y families. As subtypes are cloned and published we give them numbers, and I think that's the simplest way to go.

Leff: Since we've got a large enough contingent of the nomenclature committee here, could we adopt a policy (which has been accepted by some of the subcommittees in IUPHAR) that we give a lower case assignation to an entity which has been cloned but not characterized, but for which there's sufficient interest to create further study? If we draw the line where a decent pharmacological characterization hasn't been done, that might help to define which are and which are not in a full state of characterization.

We use the word pyrimidino, and we're suggesting that these newer UTP-recognizing G protein-linked receptors should be numbered in the P2Y sequence, but there is obviously still a level of debate going on about that. It might be sensible to anticipate the need for a designation of UTP receptors which are not related to P2Y purinoceptors. I'm not particularly concerned about nomenclature *per se*, but I am keen for us to have a consistent way of communicating with each other.

Burnstock: At the meeting in Atlanta members of the IUPHAR Purinoceptor Subcommittee wanted to leave this issue open. If it turns out (as it looks at present) that this receptor is rather similar in its structure to the P2Y family of receptors, it would seem inappropriate to start a whole new different nomenclature for pyrimidines.

Harden: The problem comes when someone who is not in communication with any of this group clones something and decides to name it whatever they want to name it, which they have the right to do. How is that to be dealt with?

Burnstock: The nomenclature committee would be meeting every so often and it will look at all these issues and try to do the sensible thing.

Leff: So we stick with Xs and Ys as the main differentiation and if we discover something like Charles Kennedy was describing yesterday, an ion channel-linked, UTP-recognizing receptor, will we provisionally stick that in amongst the Xs?

Burnstock: Indeed, we may need to change it altogether if it builds up into a whole new thing.

Features of P2X receptor-mediated synapses in the rat brain: why doesn't ATP kill the postsynaptic cell?

Frances A. Edwards

Department of Pharmacology, University of Sydney DO6, NSW 2006, Australia

Abstract. In addition to the widespread excitatory transmission mediated by glutamate receptors, P2X receptors also mediate fast excitatory synaptic transmission in the brain. The receptors in the brain have some features which are different from the more extensively characterized peripheral P2X receptors, possibly suggesting a difference in the subunits making up the protein. Perhaps the most notable feature of the central receptors is a higher Ca^{2+} permeability than seen in other areas, with a linear current–voltage relation. The potential danger to the postsynaptic cell of the high Ca^{2+} permeability of neuronal P2X receptors is discussed and various forms of inbuilt features of the synapse and receptors are outlined which would combine to protect the neurons from excessive Ca^{2+} influx and consequent danger of cell death. These features include a rapid breakdown of transmitter in the cleft, which not only removes the ATP from the cleft quickly, but also results in the production of adenosine. Evidence for the interaction of ATP-mediated transmission and stimulation of purinoceptors by this adenosine is presented from experiments using patch-clamp recording in brain slices. This demonstrates an elegant negative feedback mechanism by which a fast excitatory transmitter can be inactivated by breakdown to a product which acts as a slow inhibitory modulator, controlling release of the original transmitter.

1996 P2 purinoceptors: localization, function and transduction mechanisms. Wiley, Chichester (Ciba Foundation Symposium 198) p 278–289

Although P2X receptors had been shown to mediate synaptic transmission in a variety of peripheral synapses, it was only fairly recently that we reported direct evidence that they also mediate synaptic transmission in the brain (Edwards et al 1992). Earlier studies had shown effects of P2X receptor agonists and antagonists in various brain regions (Bean 1992, Edwards & Gibb 1993) but direct measurement of synaptic currents had not been reported. Many questions have yet to be answered concerning the role of ATP-mediated transmission in the brain. So far, in the CNS such synapses have only been directly demonstrated in the medial habenula nucleus. Although P2X receptors are widespread in the brain (Bo & Burnstock 1994, Balcar et al 1995), it is not yet clear whether synapses occur in many different regions or whether they are restricted to the medial habenula and perhaps a few other specialized regions. Moreover, the function of the medial habenula nucleus is far from clear. What we know so far about P2X receptor-mediated synaptic transmission in the brain and some interesting features of this type of synaptic transmission are outlined below.

P2X receptors in the medial habenula nucleus compared with peripheral P2X receptors

Slight variations between P2X receptors in the medial habenula and those in the periphery suggest that different subtypes of the receptor may be present in the brain. It will be interesting to see how many subunit types of the P2X receptor appear as information from the recent clones is used to extend the search throughout different tissues (Brake et al 1994, Valera et al 1994 and elsewhere in this book). The difficulties of stimulating substantial whole-cell currents in habenula neurons in brain slices have so far prevented us from thoroughly investigating the pharmacology of the habenula receptors. The small responses are presumably partly due to ATP being effectively broken down in the slice to adenosine. The introduction of the new Astra nucleotidase inhibitor (ARL67156, 6-*N*,*N*-diethyl-D-β,γ-dibromomethylene ATP; formerly known as FPL67156) should improve this situation with more information forthcoming in the near future. However, from our original work it was clear that, as in the periphery, the P2X receptors which underlie synaptic currents in the habenula are blocked by suramin (12 & 30 μM) and are stimulated (but rapidly desensitized) by α,β-methylene ATP (α,β-meATP; 30 μM) (Edwards et al 1992).

Another feature distinguishing subsynaptic P2X receptors in the brain from peripheral receptors is their Ca^{2+} permeability. In the majority of neurons the channels underlying the synaptic currents may be up to 10 times more permeable to Ca^{2+} than to Na^+ (F. A. Edwards & A. J. Gibb, unpublished results). The highest $Ca^{2+}:Na^+$ permeability ratio for the P2X receptor seen in any other cell is 3 : 1 for the rabbit ear artery (Benham & Tsien 1987). In their very high Ca^{2+} permeability these channels are not unlike the *N*-methyl-D-aspartate (NMDA) receptor channel. Unlike the NMDA channel (Fig. 1B), however, the current–voltage relationship (I–V) is linear (Fig. 1C,D), implying that the greatest influx of Ca^{2+} would be likely to occur under conditions when the cell was hyperpolarized and, in terms of other transmitters, at its least excitable. The linear I–V is another feature of the habenula P2X receptor channel which is different to peripheral channels. Peripheral channels generally show a strong inward rectification (Fig. 1A; see e.g. Krishtal et al 1988, Benham & Tsien 1987). Influx of Ca^{2+} through the NMDA receptor under conditions of ischaemia is thought to be one of the main causes of cell death

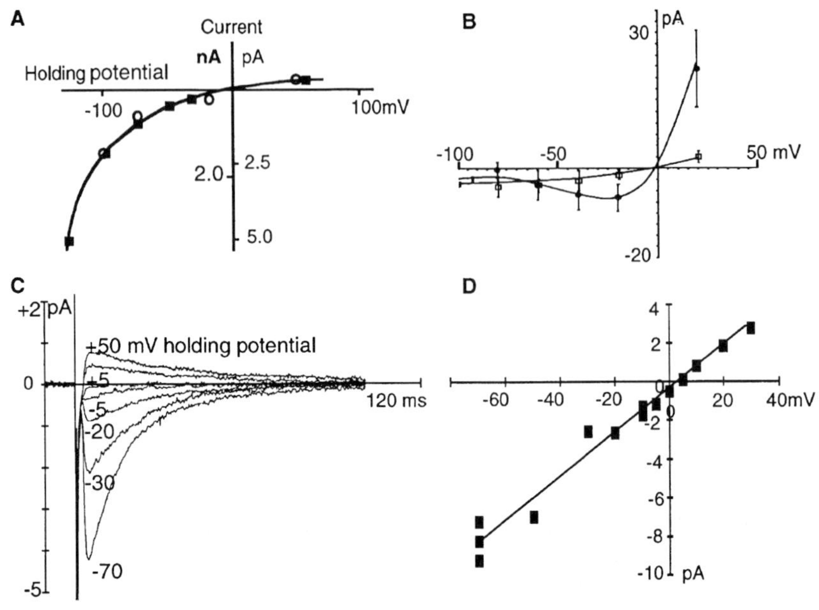

FIG. 1. Comparison of the current–voltage relation (I–V curve) of different receptor types. (A) Voltage dependence of P2X receptor channels in rat sensory neurons showing both single-channel current amplitudes (○) and macroscopic whole-cell current amplitude (■) (data taken from Krishtal et al 1988). Note the inward rectification. (B) Voltage dependence of the *N*-methyl-D-aspartate (NMDA) receptor which underlies the slow component of the synaptic response in the rat visual cortex. (Reproduced from Stern et al 1992.) The two curves show the amplitude of the late component of synaptic responses (●) and the same amplitude measurements in the presence of an NMDA receptor antagonist (□). Note the classic bell shape of the negative voltage range of the curve in the absence of antagonist. This is due to blockage of the channel by Mg^{2+} at negative potentials and means that very little Ca^{2+} flows through the channel around resting membrane potentials. (C,D) Voltage dependence of the P2X receptors which underlie synaptic currents in the rat medial habenula nucleus. (C) Averages of groups of 20 consecutive synaptic currents, recorded using the whole-cell patch-clamp configuration in slices of medial habenula, at different membrane holding potentials. The membrane voltages are as marked with one unmarked average of currents recorded at 30 mV falling between the 50 mV and 5 mV recordings. (D) The I–V curve for maximum amplitude of ATP-mediated synaptic currents from a different cell in the medial habenula. Note the lack of inward rectification, in comparison to the graph in Panel A. Ca^{2+} influx will be substantial through both peripheral and central P2X receptor channels if activated at resting membrane potentials, in the absence of other controlling mechanisms. For both C and D the recording electrode contained Cs^{+} as the primary cation. Intracellular Ca^{2+} was buffered with 10 mM EGTA and the extracellular solution contained 140 mM Na^{+} as the primary cation and 2 mM Ca^{2+}. Methods and conditions were as previously described in Edwards et al (1992).

(Wallis & Panizzon 1995). Thus, considering the millimolar concentrations of ATP present in the cytoplasm of all cells, it might be expected that the death of a particular cell in the brain and the consequent leakage of ATP into the extracellular space could be fatal for any nearby cell which had P2X receptors. Moreover, prolonged high-frequency stimulation of an ATP-releasing neuron could result in influx of dangerously high levels of Ca^{2+} into the postsynaptic cell. Clearly, protective mechanisms must exist.

Mechanisms to limit the Ca^{2+} influx caused by P2X receptors

The influx of Ca^{2+} is limited by a number of mechanisms built into the ATP-mediated synaptic transmission in the medial habenula.

The decay time of the synaptic current

Unlike the I–V curve of the peak of the synaptic current, the decay time is voltage sensitive (Fig. 2), so that currents decay faster at hyperpolarized potentials. As the Ca^{2+} influx will be proportional to the area under the curve of the current, this will tend to balance the increase in influx resulting from the greater driving force for Ca^{2+} at hyperpolarized potentials.

A low density of receptors on the synaptic membrane

At other fast synapses in the brain (glutamatergic and γ-aminobutyric acid [GABA]ergic), there are very few receptors in the active zone of the synapse, but a large number of extrasynaptic receptors. Thus although the quantal size and even the maximum amplitude of evoked currents is very small compared with peripheral tissues, application of glutamate agonists to the whole cell results in hundreds of picoamps or even nanoamps of current. Consequently, excess leakage of glutamate in an area of tissue, such as may occur during ischaemia or due to local tissue damage, will cause depolarization of the nearby cells followed by Ca^{2+} influx through activated NMDA receptors. This frequently results in cell death. In contrast, P2X receptors in medial habenula neurons seem to be very sparsely distributed both synaptically and extrasynaptically. The synaptic response to released ATP is small, seldom exceeding 500 pS, and local whole-cell application of agonist (α,β-meATP) results in an even smaller current response. Although this is undoubtedly in part due to desensitization of the receptors, the result supports the theory that the receptors are not very plentiful. This is further supported by the fact that single channels were found in only one of six outside-out patches pulled from these cells, suggesting that much of the membrane does not contain this type of channel. The sparsity of P2X receptors may be another protective mechanism against exposure to high concentrations of ATP which might occur due to death of a neighbouring neuron.

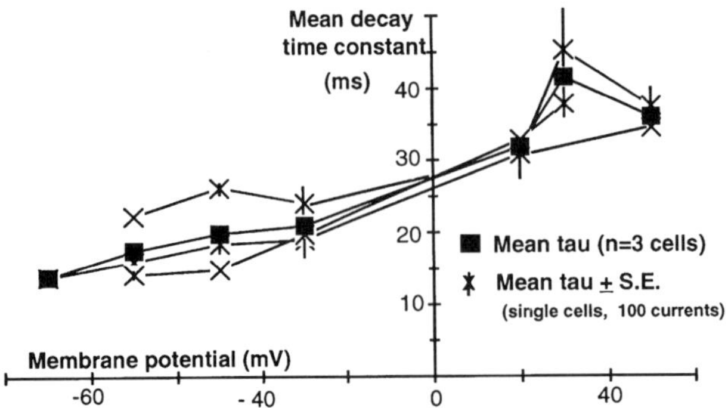

FIG. 2. Voltage dependence of the duration of P2X receptor-mediated synaptic currents in the rat medial habenula. Synaptic currents, such as those shown in the previous figure, were fitted with a single exponential and the average decay time constants ± SEM of 100 currents were plotted against the holding potential for each of three cells. The average of these means is also plotted (■). The voltage dependence is not very steep; nevertheless, the decay time constant at −70 mV (in the vicinity of resting potential) is about half that at positive potentials.

Small numbers of synaptic connections

Not only are there few receptors present on the cells, but there also seem to be very few synaptic connections mediated by P2X receptors. The evidence for this comes from the fact that it is generally quite difficult to find ATP-mediated synaptic connections. Nevertheless, after moving the stimulating electrode down tracks in many different regions of the slice, a connection may finally be located. In contrast, in the absence of antagonists of the glutamate or GABA receptors, almost every track in which the electrode is placed results in release from a presynaptic neuron. It is possible that the apparent sparsity of synaptic connections is due to the angle at which the slice has been cut, but even stimulation of the whole incoming nerve trunk results only in stimulation of glutamatergic currents. Thus, while influences of the slicing procedure cannot be ruled out, it seems more likely that each cell receives not more than two or three inputs from purinergic cells. The infrequency of miniature synaptic currents is further indirect support for a sparsity of connections.

Fast breakdown of released transmitter

As mentioned above, like acetylcholine, but unlike other fast transmitters of the brain, extracellular ATP is broken down by nucleotidases as soon as it reaches the extracellular space. It was only recently that an ATP-nucleotidase inhibitor became available; this will be invaluable in assessing the affect of ATP breakdown on the synaptic current and the potency of ATP in the preparation. It will also allow assessment of the true number of receptors in the slice and the extent to which ATP itself desensitizes the receptors. Desensitization could be yet another factor limiting the Ca^{2+} influx through P2X receptor channels.

Increased failures at high frequency due to local adenosine production?

At the neuromuscular junction it has been suggested that neuromuscular depression, associated with high-frequency stimulation, is largely due to adenosine which is produced by breakdown of ATP coreleased with acetylcholine (e.g. Redman & Silinsky 1994). A similar mechanism may occur in the CNS. This would provide a novel feedback system in the brain: a fast neurotransmitter being broken down to a slow inhibitory neuromodulator which in turn inhibits the release of the original transmitter. Evidence for such a mechanism can be observed by stimulating transmission at ATP-receptor gated synapses at different frequencies. As the frequency of stimulation is increased from 0.5, through 1, 2 and 5 Hz up to 10 or even 100 Hz, the failure rate increases dramatically, which could be due to the adenosine produced from ATP breakdown building up in the cleft. Evidence that this increased failure rate is due to adenosine comes from the inclusion of an A_1- receptor antagonist in the bath solution. Bath application of 8-cyclopentyl-1,3-dimethylxanthine (8CPT, 1–10 μM) dose-dependently prevents the increased failure rate seen at high stimulation rates (Fig. 3). The high concentrations of antagonist needed to overcome the depression of release suggest that, like other fast neurotransmitters, ATP and its breakdown product (adenosine) are present in high concentrations in the synaptic cleft. Interestingly, adenosine antagonists have effects even at very low frequencies, suggesting a fairly high level of adenosine in the slice under basal conditions. Moreover, the release probability at low frequencies in the presence of an adenosine antagonist is very high.

Such a failure to release ATP at high stimulation frequencies would act as yet another protective mechanism against excessive Ca^{2+} influx. This means that if high-frequency stimulation occurs *in vivo*, the frequency of ATP release would be virtually the same as at lower stimulation frequency levels. In all other systems investigated so far, ATP has been a co-transmitter and this may also be the case in the CNS. It may be that one of the roles of ATP transmission is to act as a negative modulator via its breakdown product adenosine on other neurotransmitter systems.

Evidence that ATP is in fact the transmitter at P2X receptor-mediated synapses in the brain

One of the requirements for a substance to be defined as a neurotransmitter is evidence that the substance is released at the appropriate times from the

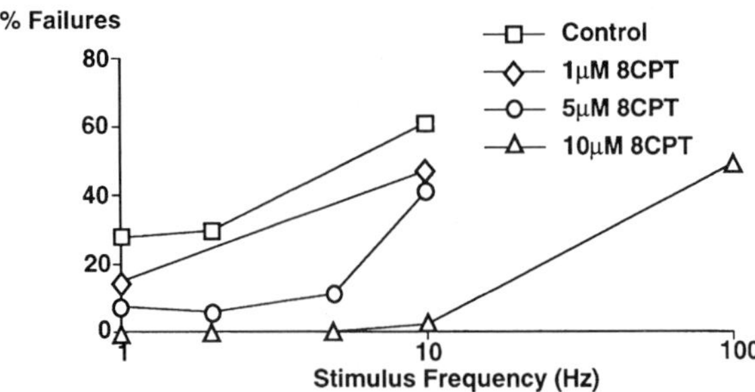

FIG. 3. Dependence of the failure rate of synaptic P2X receptor-mediated synaptic currents on stimulus frequency. The effect of adenosine receptor blockade. Synaptic currents were measured in medial habenula neurons using the whole-cell patch-clamp configuration as for the above figures. Under control conditions the failure rate became greater with increased stimulation rates. The A_1 receptor antagonist 8CPT decreases the failure rate at low frequencies (0.5 & 1 Hz) so that at high concentrations of 8CPT (10 μM) almost no failures are recorded. As the frequency is raised, higher concentrations of the antagonist are needed to overcome the frequency-dependent increase in failure rate. The observed competition between frequency of stimulation and the adenosine receptor antagonist suggest that increasing the frequency of stimulation results in accumulation of adenosine in the cleft (presumably as a result of breakdown of released ATP) and that this adenosine results in the increased failure rate seen. At 100 Hz stimulation, in the absence of 8CPT (control), most synapses almost completely fail to release (not shown). Even 10 μM 8CPT fails to overcome this effect completely. This suggests that very high concentrations of adenosine accumulate in the cleft under conditions of high-frequency stimulation or that another factor also comes into play, such as depletion of vesicles at docking sites.

presynaptic terminal. If it can be shown that the source of the adenosine which inhibits release of transmitter is extracellular ATP, this will be indirect evidence that ATP is in fact released in a frequency-dependent manner. The recent development by Astra of a compound (ARL67156) which blocks breakdown of ATP to adenosine in the extracellular space, provides a tool for the experiment which will be performed on return from this conference. If the new Astra compound prevents frequency-dependent failure of transmission at these synapses, it will be evidence that it is ATP that is released rather than other possible agonists of the P2X receptor. If the frequency-dependent failure rate is not affected by inhibition of ATP breakdown, this will suggest that adenosine itself is also released, rather than its presence being dependent on ATP breakdown or that the Astra product is not effective on this particular type of ATPase.

Summary and conclusions

P2X receptor-mediated synaptic transmission has a number of features which distinguish it from transmission via the conventional fast transmitters glutamate and GABA. Perhaps the most significant of these is that P2X receptors allow Ca^{2+} influx even at resting membrane potentials. In the brain, this otherwise only occurs via a specific subtype of AMPA receptor present in interneurons (Jonas et al 1994). This may imply a wide range of functions for this very specialized transmission as Ca^{2+} acts as a second messenger activating various intracellular processes. It could, however, also be very dangerous for the postsynaptic cell as excess Ca^{2+} influx results in cell death. A wide range of features of the synapse and the postsynaptic cells may serve as protective mechanisms against excess Ca^{2+} influx, resulting in a pathway for small low-frequency pulses of Ca^{2+} influx. These protective mechanisms include: voltage–dependent decay of synaptic currents causing decreased ion flux at hyperpolarized potentials; a low density of receptors in the membrane, possibly all situated in the synapse; a small number of synapses per postsynaptic neuron; fast breakdown of transmitter in the synaptic cleft; and an increasing failure rate at higher frequencies so that release of transmitter is always between 0 and 0.5 Hz, even at stimulation rates of 10 to 100 Hz.

These combined protective devices would serve to ensure the survival of the postsynaptic cell both during high stimulation frequencies and also in the event of the death of a neighbouring cell when leakage of concentrated ATP would be likely to occur.

Acknowledgements

Work on Ca^{2+} permeability of P2X receptors in medial habenula has been done in collaboration with Alasdair Gibb, who is also an ongoing source of interesting discussion of all the areas included. I would also like to acknowledge valuable editorial and technical assistance from Jenean Spencer and Anna Cunningham. F. A. E. is supported for both salary and research by the Australian Research Council.

References

Balcar VJ, Li Y, Killinger S, Bennett MR 1995 Autoradiography of P_{2X} ATP receptors in the rat brain. Br J Pharmacol 115:302–306

Bo X, Burnstock G 1994 Distribution of $[^3H]\alpha,\beta$-methylene ATP binding sites in rat brain and spinal cord. NeuroReport 5:1601–1604

Bean BP 1992 Pharmacology and electrophysiology of ATP-activated ion channels. Trends Pharmacol Sci 13:87–90

Benham CD, Tsien RW 1987 A novel receptor-operated Ca^{2+}-permeable channel activated by ATP in smooth muscle. Nature 328:275–278

Brake AJ, Wagenbach MJ, Julius D 1994 New structural motif for ligand-gated ion channels defined by an ionotropic ATP receptor. Nature 371:519–523

Edwards FA, Gibb AJ 1993 ATP: a fast neurotransmitter. FEBS Lett 325:86–89

Edwards FA, Gibb AJ, Colquhoun D 1992 ATP receptor-mediated synaptic currents in the central nervous system. Nature 359:144–147

Jonas P, Racca C, Sakmann B, Seeburg PH, Monyer H 1994 Differences in Ca^{2+} permeability of AMPA-type glutamate receptor channels in neocortical neurons caused by differential GluR-B subunit expression. Neuron 12:1281–1289

Krishtal OA, Marchenko SM, Obukhov AG 1988 Cationic channels activated by extracellular ATP in rat sensory neurons. Neuroscience 27:995–1000
Redman RS, Silinsky EM 1994 ATP released together with acetylcholine as the mediator of neuromuscular depression at frog motor nerve endings. J Physiol 477:117–127
Stern P, Edwards FA, Sakmann B 1992 Fast and slow components of unitary EPSCs on stellate cells elicited by focal stimulation in slices of rat visual cortex. J Physiol 449:247–278
Valera S, Hussy N, Evans RJ et al 1994 A new class of ligand-gated ion channel defined by P_{2X} receptor for extracellular ATP. Nature 371:516–519
Wallis RA, Panizzon KL 1995 Delayed neuronal injury in the hippocampal slice. In: Schurr A, Rigor BM (eds) Brain slices in basic and clinical research. CRC Press, Boca Raton, FL

DISCUSSION

Illes: May I ask a provocative question: you have shown that this ATP response is rather small. If you move to the peripheral vegetative nervous system (and many things that happen in the CNS also happen in a similar way in the peripheral nervous system), many places in the periphery, such as the dorsal root ganglion, demonstrate large responses to purinoceptor agonists. Is it just possible that you are not in the right place? The medial habenula is probably just not the optimal place to study ATP responses.

Edwards: It's quite possible that there are many other ATP responses in the brain, but the fact that they have not yet been detected suggests they're also small. It's only quite recently we've had patch-clamping in brain slices which gives us this sort of resolution, and which enabled me to find these currents at all. Consequently, I think it's highly likely that the responses in other places will also be small. This is probably related to the way the brain works. The channels are highly Ca^{2+} permeable and cells are packed tightly together. It may be an essential requirement that the responses be small to avoid cell death from Ca^{2+} influx during high levels of activity or when nearby cells die, releasing ATP. Small synaptic currents are also a general feature of fast synaptic transmission in the brain.

Illes: If you were to measure the endplate potentials instead of endplate currents, would you see anything?

Edwards: Yes, I think we would. These cells have nearly a giga ohm of input resistance.

Burnstock: People have been aware of purines in the brain for 50 years or so, and there were reports of ATP effects in the olfactory lobe. However these were short-lived, because ATP rapidly broke down to adenosine and the dominant response was inhibition. If you weren't lucky enough to get your microelectrode in to a neuron quickly after applying the ATP you only saw the inhibitory action of adenosine.

Starke: You ascribed a large physiological role to adenosine inhibiting release by subsequent pulses, but on your slide you plotted the frequency and not pulse number on the X axis. What happens in the course of a pulse chain at constant frequency? Will you see initially a low rate of failures and then later on in the pulse train a larger number of failures?

Edwards: That is a good question; I haven't analysed that yet. I haven't even looked to make sure that I get the first pulse, which is of course essential, if ATP must first be released and broken down before adenosine will be present.

Starke: This frequency release relationship, or frequency failure relationship (depending on how you express it) is unusual if one considers what happens in the periphery when one records excitatory junction potentials: facilitation becomes greater as the frequency is increased, so the number of failures seems to decrease when the frequency is increased. The medial habenula seems to behave quite differently from peripheral tissue.

Edwards: It's also different in that the probability of release is clearly very high in the absence of adenosine, which is different from peripheral synapses and also different from most central synapses, though they haven't been done in the presence of adenosine antagonists. It may be that there's a background level of adenosine we hadn't realized.

Surprenant: You are saying that there are very few release sites and a high probability of release, so your effects are what would be expected. Whereas in the periphery it's exactly the opposite: there are numerous release sites with a very low probability of each one releasing.

Westfall: Where do the nerves come from that input to the habenula, and where are the outputs going? What is the function of this nucleus?

Edwards: The function of the nucleus is really unclear. It's referred to as a relay nucleus, which means that no one knows what it does. The connections are reasonably well described: there's a lot of cholinergic input from the striatum and then cholinergic output to the interpeduncular nucleus. There's glutamate input from quite a few higher areas. Basically it's a relay nucleus between higher and lower brain areas. These ATP inputs seem to be a local phenomenon, because if I stimulate the whole of the incoming pathway everything can be blocked by CNQX. This doesn't actually guarantee that it doesn't come in with the input first—it's very likely that it has to go over a glutamate synapse before it stimulates the purinergic receptor from the input. It seems to be some sort of local interneuron effect.

Westfall: What is the pathological consequence of lesioning the habenula?

Edwards: Very little work has been done on this rather obscure nucleus.

Surprenant: You're keen to put ATP on it so that it doesn't break down: that makes sense. But what about the opposite situation? One could imagine that lots of ATP is always coming out, let's say from glutamate release, and that this acts eventually to desensitize the receptors. So for synaptic release it's possible that if you actually broke down endogenously released ATP by applying apyrase and then immediately washing it off prior to stimulation, your excitatory postsynaptic currents would be bigger.

Edwards: If there is a background level of ATP which is desensitizing receptors, then that's possible. There are lots of unknowns. It might be worth a try but it is more likely that ATP is broken down by ectonucleotidases in the slice, not allowing build up.

Wiley: On the single-channel recording that you showed, did you calculate the unitary conductance?

Edwards: Yes, it's around 20 pS.

Burnstock: If co-transmission is taking place with glutamate and ATP, which wouldn't surprise me, then you always get synergism as we saw yesterday with noradrenaline and ATP from sympathetic nerves. It might be worth seeing whether a low dose of glutamate will actually potentiate the ATP response. I would expect it, if they come out together, and that would also provide indirect evidence for co-transmission.

Edwards: You mean it would synergize by increasing release? It is hard to increase it much more.

Surprenant: A very interesting paper recently showed that subsequent to ATP activation of P2X receptors in spinal cord neurons, the glutamate response is vastly potentiated, and this potentiation appeared to be direct rather than G protein coupled (Li & Perl 1995).

Edwards: It's also very interesting to know whether or not there are P2Y receptors here and what they're doing. I've never looked at that at all. It's quite possible that there is a subsidiary slow effect as well.

IJzerman: About the breakdown that you mentioned: the time frame of your experiments is so short that I wouldn't expect that an ectoATPase is capable of degrading ATP. I'm inclined to say this is probably not the case.

Edwards: It is certainly the case with acetylcholine, because if you block the cholinesterase you get the response. I need to use ARL67085 to check this. Because it can happen with cholinesterase I think it is realistic to suggest that it can happen with ATPase. Are you suggesting that ATPases are much slower?

IJzerman: In our hands, in entirely different conditions, they are.

Barnard: Acetylcholinesterase is famous for having by far the highest catalytic constant. ATPases are much slower.

I had also thought the same thing as Ad IJzerman: it is remarkable that in a millisecond or two you get the large effect of the ATPase.

Edwards: It's not a millisecond or two. With the release of ATP into the synapse, that's under a millisecond and there you get the response. My application is more in the second range.

Barnard: Then it is perfectly understandable.

Edwards: And it also has to go through the slice and it may well be that they're exactly positioned at the edges of the synapse so that applied ATP must pass the enzymes before reaching the receptors.

Barnard: I presume you can never get your electrode right into the subsynaptic area?

Edwards: I now have a fast piezo applicator, and I'm hoping to develop a system where I can actually lift these cells up and apply it to them, or otherwise dissociate the cells. Then I'll be able to get really fast application, but at the moment I don't have that.

Barnard: I do not think the enzyme could compete with that.

Kennedy: All your recordings have been voltage-clamp. Have you ever performed voltage recordings and seen purinergic excitatory postsynaptic potentials?

Edwards: No, I haven't. I don't know exactly what I'd gain by that.

Illes: With respect to whole-cell recording, where you showed partial sensitivity to suramin, I wanted to ask whether it was just a factor of the concentration of suramin or whether you think that there is a suramin-resistant component to the EPCs?

Edwards: In those applications I've just got the suramin and the ATP together in the pipette. So I think it's just binding rate constants and so forth.

Miras-Portugal: Concerning the co-transmission with ATP and glutamate: there is a recent review of vesicular transporters by Schuldiner et al (1995). The vesicles transporting glutamate do not appear to transport ATP. Nevertheless, ATP is necessary for vesicular ATPase functioning and gradient formation.

Kennedy: In your studies with α,β-meATP, you seemed to imply that part of the reason the currents were small was because you thought α,β-meATP was being broken down.

Edwards: No, I think it was small because the response would have been largely desensitized.

Surprenant: It's desensitizing over a several seconds, it's not desensitizing with a time constant of a few hundred milliseconds.

Edwards: I have never applied it fast enough to know, and that's really important here. It may be that it is already largely desensitized. It's already partially desensitized, no question, because the total whole cell response is less than total synaptic response. We might be seeing just the tail of the response.

References

Li J, Perl ER 1995 ATP modulation of synaptic transmission in the spinal substantia gelatinosa. J Neurosci 15:3357–3365

Schuldiner S, Shirvan A, Linial M 1995 Vesicular neurotransmitter transporters: from bacteria to humans. Physiol Rev 95:369–392

P2 purinoceptors in the immune system

Francesco Di Virgilio*, Davide Ferrari*, Simonetta Falzoni*, Paola Chiozzi*, Maddalena Munerati*, Thomas H. Steinberg† and Olavio R. Baricordi‡

Institutes of General Pathology and Medical Genetics‡, University of Ferrara, Ferrara, Via L Borsari 46, I-44100 Ferrara, Italy and †Division of Infectious Diseases, Department of Medicine, Washington University, St. Louis, MO 63110, USA*

Abstract. Immune cells express plasma membrane receptors for extracellular nucleotides. Both G protein-linked metabotropic and channel-forming ionotropic receptors have been described, although no P2 receptor subtype has been cloned from the immune system thus far. Metabotropic receptors have been described in human B but not T lymphocytes; they have not been found in mouse B and T cells. Ionotropic receptors seem to be ubiquitously expressed in the immune system; however, their functional properties, if not their pharmacology, appear to be different in different immune cells. Human T normal and B leukaemic lymphocytes, human macrophages, mouse B and T lymphocytes, mouse microglial and macrophage cells, and rat mast cells express ionotropic receptors that recognize ATP^{4-} as the preferred ligand, are activated by 3′-*O*-(4-benzoyl)benzoyl ATP and inhibited by oxidized ATP. The pharmacological profile of ionotropic receptors expressed by different immune cells is similar, but their permeability properties may be different: the pore formed by receptors expressed by macrophages, microglial cells and mast cells is typically permeable to charged molecules of molecular mass up to 900 Da; on the contrary, that expressed by lymphocytes has a molecular cut-off of 200–300 Da. The ionotropic receptor of immune cells is modulated by inflammatory cytokines (e.g. interleukin [IL]-2 and γ-interferon) and is also modulated during monocyte to macrophage differentiation. Transient stimulation of the ionotropic receptor of macrophages and microglial cells elicits IL-1β release. Sustained activation leads to cell death, either by necrosis or apoptosis, depending on the given cell type.

1996 P2 purinoceptors: localization, function and transduction mechanisms. Wiley, Chichester (Ciba Foundation Symposium 198) p 290–305

Purine nucleotides and nucleosides were first recognized as potent extracellular mediators as early as 1929 in a pivotal study by Drury & Szent-Györgyi (1929). A voluminous literature followed, reporting actions of extracellular nucleotides in almost any cell and tissue. Effects of adenosine in the immune system have been thoroughly investigated and it is being evaluated as a potential powerful anti-inflammatory agent. As this report focuses on the action of extracellular nucleotides, the reader is referred to other excellent recent reviews for an evaluation of the role of nucleosides (and nucleoside receptors) in immune cells (e.g. Cronstein 1994).

Early studies on the possible role of extracellular nucleotides in immune cell responses date from the late 1960s. Cohn & Parks (1967) reported that extracellular ATP (ATP_e) increased the number of cytoplasmic vacuoles in mouse macrophages, and Sugiyama (1971) and Dahlquist & Diamant (1974) showed that ATP_e caused histamine secretion from mast cells.

However, despite these early indications of a possible role for ATP_e in immune-mediated reactions, ATP_e receptors (purinergic P2) were granted little interest until recently. Scattered observations hinted at a possible role of ATP_e in lymphocyte proliferation (Gregory & Kern 1978), natural killer cell-dependent cytotoxicity (Schmidt et al 1984) or macrophage phagocytosis (Sung et al 1985), but no attempt was made rigorously to characterize plasma membrane receptors responsible for these effects and the intracellular mechanisms involved.

In the mid 1980s, Silverstein's and Dubyak's groups began a thorough investigation on ATP_e-mediated responses in leukocytes (see Dubyak & Fedan 1990 for review). These studies rekindled interest in the possible role of purinergic P2Z receptors in immune and inflammatory cells.

P2 receptor subtypes in immune cells

Monocytes and macrophages, rat mast cells, neutrophils and tonsillar B lymphocytes express receptors of the P2Y/P2U subtype (metabotropic receptors) that can be activated by ATP_e, extracellular UTP (UTP_e) and other nucleotides (ADP, CTP, GTP, etc.) (see Dubyak & El-Moatassim 1993 for review). Though these receptors have not yet been cloned from immune cells, they do not seem to differ pharmacologically and functionally from the well-known P2Y/P2U receptors characterized and cloned from other cell types.

Besides P2Y/P2U receptors, immune cells also express pore/channel-forming ATP_e receptors (ionotropic receptors). Nomenclature of ionotropic receptors in immune cells is confusing, as some of them, e.g. lymphocytes, express ATP_e receptors showing a lower molecular mass cut-off with respect to the permeabilizing receptors of mast cells and macrophages. In the past, these atypical receptors have been sometimes referred to as 'P2X-like' receptors. However, the pharmacological profile of the various ionotropic receptors characterized in different immune cells is the same and typically of the P2Z type (see Table 1). In this chapter we will refer to the ionotropic receptor of immune cells as the P2Z receptor. Whether or not all of these ATP-induced pores truly belong to the same family of receptors will not be determined until the molecules responsible for these activities have been cloned.

The P2Z receptor is defined on the basis of the ability to mediate plasma membrane permeabilization to low molecular weight hydrophilic solutes and selective activation by 3′-*O*-(4-benzoyl)benzoyl ATP (BzATP).

TABLE 1 Properties of ionotropic purinergic receptors expressed by immune cells

Cell type	*Agonist*	*Antagonist*	*Pore size*	*Effects*	*References*
Human macrophages	BzATP≫ATP	oxo-ATP	~900 Da	IL-1β release MGC formation Cell death	Falzoni et al 1995, S. Falzoni & F. Di Virgilio, unpublished
Mouse macrophages	BzATP≫ATP	oxo-ATP	~900 Da	Cell death	Nuttle & Dubyak 1994, T. H. Steinberg & F. Di Virgilio, unpublished
Human T lymphocytes	BzATP≫ATP	oxo-ATP	?	Cell proliferation	O. R. Baricordi & D. Ferrari, unpublished
Human B lymphocytes (leukaemic, dystrophic)	BzATP≫ATP	oxo-ATP	~300 Da	?	Wiley et al 1994, Ferrari et al 1994
Mouse B lymphocytes	ATP	?	?	Cell death	Di Virgilio et al 1989
Mouse T lymphocytes	ATP	?	200–300 Da	Cell death	Pizzo et al 1991, El-Moatassim et al 1989
Rat mast cells	ATP	?	~900 Da	Histamine release Cell death	Cockroft & Gomperts 1979
Human Langerhans' cells	ATP	?	>400 Da	?	Girolomoni et al 1993
Mouse microglial cells	BzATP≫ATP	oxo-ATP	~900 Da	IL-1β release Cell death	D. Ferrari & F. Di Virgilio, unpublished

BzATP, 3′-*O*-(4-benzoyl)benzoyl ATP; oxo-ATP, oxidized ATP; MGC, multinucleated giant cell; IL-1β, interleukin 1β.

As early as 1979 it was suggested that ATP-dependent permeability changes in mast cells were mediated by specific receptors (Cockcroft & Gomperts 1979), but until very recently there were some doubts that membrane permeabilization was due to a receptor-mediated mechanism, as a non-specific perturbation of the plasma membrane could not be formally excluded. Reports from several laboratories during the last few years have conclusively dismissed these reservations (see Di Virgilio 1995). Evidence that ATP_e-dependent plasma membrane permeabilization is receptor mediated is provided by the following observations:

(1) It is restricted to a few cell types.
(2) Stable ATP_e-resistant mutants can be selected from susceptible cell types.
(3) Cytotoxic lymphocytes are characterized by high spontaneous ATP_e resistance.
(4) Chronic treatment of mouse splenocytes with IL-2 causes ATP_e resistance.
(5) Pretreatment with the selective P2Z receptor blocker oxidized ATP fully antagonizes the ATP_e permeabilizing effect.

P2Y/P2U receptors

Stimulation with ATP_e causes inositol-1,4,5-trisphosphate formation and $[Ca^{2+}]_i$ increases in several inflammatory cell types (e.g. neutrophils, monocytes and macrophages) and myeloid progenitor cells (Dubyak & El Moatassim 1993). These changes depend on activation of G protein-linked P2Y/P2U receptors. Stimulation of these receptors has also been suggested to be linked to activation of phospholipase A_2 and D. Therefore, metabotropic receptors could be involved in the generation of mediators of inflammation (through activation of lysosomal enzyme release or activation of the arachidonic acid cascade) or might synergize with different inflammatory stimuli to elicit other Ca^{2+}-mediated responses, e.g. formation of reactive oxygen intermediates (Ward et al 1988). In addition, P2Y receptors might also facilitate migration of inflammatory cells by up-regulating expression of adhesion molecules on monocyte and granulocyte plasma membrane (Freyer et al 1988).

The P2Z receptor

In contrast to P2Y/P2U and P2X receptors that have been recently cloned (this volume: Barnard et al 1996, Lustig et al 1996, Surprenant 1996), the P2Z receptor has not been cloned, thus its molecular structure is unknown. Functional studies suggest that it is a membrane pore gated by ATP_e. In the absence of this extracellular nucleotide, the pore is closed; in its presence, the pore opens and stays open as long as ATP_e is present. At variance with other ligand-activated channels, the P2Z receptor does not inactivate; this means that closure can only occur as a consequence of removal of the ligand. ATP_e removal may be due to hydrolysis by plasma membrane ATPases or nucleotidases, bulk phase equilibration of ATP_e or complexation of the fully ionized fraction (ATP^{4-}) by extracellular Ca^{2+} and Mg^{2+}. The mechanism by which ATP_e binding triggers pore opening is unknown. The strong temperature dependency (complete inhibition of extracellular low molecular mass solute uptake below 20 °C) suggests that the pore may be formed by ATP_e-dependent assembly of multimeric complexes (Nuttle & Dubyak 1994). A cooperative process in pore opening is suggested by Hill coefficients of approximately 2 calculated for mast cells and thymocytes (Pizzo et al 1991, Tatham & Lindau 1990), that indicate that binding of at least two ATP_e molecules is necessary for activation (Fig. 1).

Cell types that have been shown to express the P2Z receptor include rat mast cells, mouse and human macrophages, human epidermal Langerhans' cells, mouse microglial cell lines and a number of tumour cell lines of different origin. Expression of the P2Z receptor by lymphocytes is controversial. Mouse splenocytes and thymocytes, as well as human peripheral lymphocytes from patients with B cell chronic lymphocytic leukaemia, express an ATP_e-gated channel permeable to Na^+, Ca^{2+}, choline (M_r 100), methylglucamine (M_r 190) and ethidium bromide (M_r 314), while higher molecular mass solutes are excluded (e.g. propidium iodide, M_r 414) (Wiley & Dubyak 1989, Pizzo et al 1991, El-Moatassim et al 1989, Wiley et al 1993). Normal, quiescent, peripheral, human T and B lymphocytes appear to express this permeabilizing ATP_e

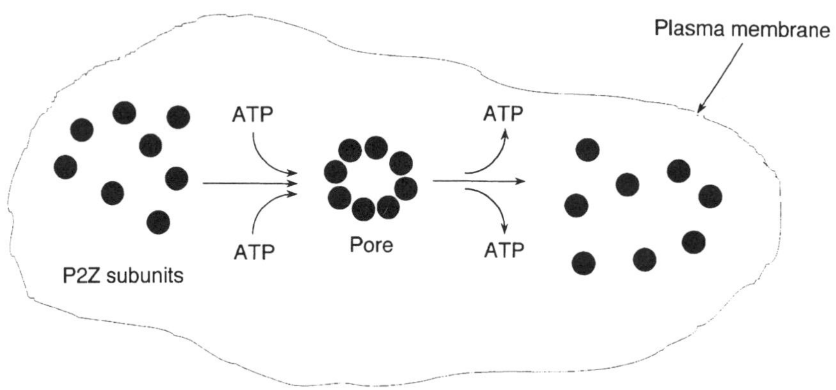

FIG. 1. Hypothetical mechanism of activation of the P2Z receptor. We speculate that the P2Z receptor is a multisubunit structure formed by monomers that in unstimulated cells are present in the disaggregated form. Binding of at least two ATP molecules drives aggregation of the monomers that join together to form membrane pores of different size. Removal of ATP causes disaggregation of the monomers and closure of the pore.

receptor at a low level. Nothing is known about activated lymphocytes. Human polymorphonuclear neutrophil granulocytes do not express the P2Z receptor.

A most interesting observation is that expression of P2Z receptor by immune cells is modulated during differentiation and by inflammatory cytokines. Early experiments showed that stimulation with interleukin 2 (IL-2) of ATP_e-sensitive mouse spleen lymphocytes led to the emergence of cytotoxic cells with unrestricted activity (lymphokine-activated killer [LAK] cells) that were exceedingly resistant to ATP_e (Di Virgilio et al 1989, Zanovello et al 1990). Later, Blanchard et al (1991) reported that γ-interferon (IFN-γ) enhanced the susceptibility of human macrophages to the lytic effect of ATP_e, although the mechanism involved was not identified. We recently showed that the increased susceptibility of IFN-γ-treated macrophages to ATP_e-mediated cytotoxicity depends on an increased expression of P2Z receptors (Falzoni et al 1995). Furthermore, expression of P2Z receptor by human monocyte-derived macrophages has been shown to be developmentally regulated (Hickman et al 1994, Falzoni et al 1995). Human peripheral blood monocytes express very little receptor activity (less than 15% become permeable to Lucifer yellow or YO-PRO-1 upon ATP_e treatment). After a few days (7–14) of *in vitro* culture, this percentage increases to 40–60%. Thus, although a systematic examination of the effect of inflammatory cytokines on P2Z expression in different immune cells has yet to be carried out, these observations clearly hint at the possibility that inflammatory mediators may also modulate the ATP_e-gated pore *in vivo*.

The P2Z receptor in the immune response

In spite of circumstantial evidence pointing to a potentially important role, the basic question of the function of P2Z purinergic receptors in the immune and inflammatory reactions is still unanswered. There are few doubts that P2Z receptors can mediate cell death. That ATP_e could be cytotoxic to different cell targets had been known for some years to those working in this field. But it wasn't until 1989 that a thorough re-examination was inaugurated with the aim of unveiling the molecular basis of this intriguing effect (Di Virgilio et al 1989, Filippini et al 1990). These studies clearly demonstrated that ATP_e-dependent cytotoxicity was a receptor-mediated effect, and not simply the result of a non-specific perturbation of the plasma membrane. Furthermore, a cytotoxic role for the P2Y/P2U receptor could be conclusively ruled out as stable ATP_e-resistant clones were selected from highly ATP_e-sensitive cell lines. These clones lacked any detectable P2Z activity but normally expressed P2Y/P2U receptors (Greenberg et al 1988, Murgia et al 1992).

The chain of events initiated by opening of the ATP_e pore that culminate in cell death include:

(1) rapid Ca^{2+} and Na^+ influx;
(2) rapid K^+ efflux;
(3) plasma membrane depolarization;
(4) efflux of phosphorylated low-molecular weight metabolites;
(5) cell rounding and swelling;
(6) inhibition of pinocytosis and phagocytosis;
(7) disaggregation of the microtubule network;
(8) release of exocytotic granule content; and
(9) cell death.

Na^+ influx appears to be the crucial triggering event as cytotoxicity (at least within the initial 8–10 h following ATP_e stimulation) can be prevented by incubation in Na^+-free, sucrose-containing medium (Murgia et al 1992). Under these conditions, macrophages permeabilize, swell and take up Lucifer yellow, but short-term release of cytoplasmic markers is negligible. Presence of Na^+ in the extracellular medium is both necessary and sufficient, as ATP_e is fully cytotoxic in the absence of extracellular Ca^{2+}. These features would point to colloido-osmotic lysis as the mechanism responsible for ATP_e-dependent cell death. However, it is well documented that ATP_e can also cause cell death by apoptosis. It was initially shown by Zanovello et al (1990) that ATP_e caused DNA fragmentation in the mouse mastocytoma cell line P815. These observations were later confirmed by a number of different laboratories in many cell types (Zheng et al 1991, Pizzo et al 1992, Molloy et al 1994). It is not yet clear whether necrosis and apoptosis are mediated by the same or two different P2 ionotropic receptors. Our observations in P815 cells, fibroblasts and lymphocytes initially suggested that apoptosis was dependent on activation of either a P2X-type receptor expressed by non-excitable cells, or an atypical P2Z receptor exhibiting a molecular mass cut-off smaller than that of the typical macrophage/mast cell P2Z receptor (Zanovello et al 1990, Pizzo et al 1991, 1992). However, more recent experiments show that a typical P2Z receptor can mediate apoptosis in macrophage and microglial cells (Molloy et al 1994, D. Ferrari & S. Falzoni, unpublished observations). Thus, it would seem that both atypical P2Z receptors, previously tentatively identified as P2X, and *bona fide* P2Z receptors can mediate apoptosis. However, it must be stressed that conditions under which both types of receptors mediate necrotic lysis are also well documented (Murgia et al 1992, Ferrari et al 1994).

P2 purinergic receptors and cell death

The links between P2 ionotropic receptors and cell death (whether by apoptosis or necrosis) are intriguing. Recent molecular cloning data have revealed a surprising sequence homology between the RP-2 protein, expressed by apoptotic thymocytes (Owen et al 1991), and P2X receptors cloned from rat PC12 cells and rat smooth muscle (Valera et al 1994, Brake et al 1994). Furthermore, the two cloned P2X proteins show structural analogy to a growing family of plasma membrane ion channels epitomized by amiloride-sensitive Na^+ channels and mechanosensitive channels from *Caenorhabditis elegans*. Sustained stimulation or mutations of these latter channels mediate neuronal cell death in *C. elegans*.

An additional link between P2 ionotropic receptors and cell death is provided by the observation that ATP_e acting at the P2Z receptor is a very powerful stimulus for activation of IL-1β converting enzyme (ICE), the enzyme responsible for generation of the mature form of IL-1β (Hogquist et al 1991, Laliberte et al 1994). ICE is a cysteine protease that appears to be the mammalian homologue of *C. elegans* Ced-3, which is required for developmental cell death in this nematode. ICE activation is increasingly recognized as a key step in the apoptotic process in mammalian cells. Activation of ICE has been shown to occur during apoptosis in epithelial cells deprived of extracellular matrix (Boudreau et al 1995), as well as in thymocytes and fibroblasts stimulated via the Fas/APO-1 pathway (Kuida et al 1995, Enari et al 1995, Los et al 1995). Although it is not known whether ATP_e-dependent apoptosis also involves activation of the ICE pathway, it is certainly intriguing that this nucleotide stimulates IL-1β maturation. Preliminary experiments suggest that ICE activation by ATP_e is linked to P2Z stimulation as we selected ATP_e-resistant, P2Z-lacking, P2Y-expressing microglial cell clones that showed near-normal ATP_e-induced Ca^{2+} release from stores, but no IL-1β secretion (Fig. 2). These mutant cell clones were, however, fully capable of releasing IL-1β when challenged with the K^+ ionophore nigericin, a potent stimulus for ICE activation that bypasses plasma membrane receptors (D. Ferrari & F. Di Virgilio, unpublished results).

Possible role of the P2Z receptor in the immune system

Early observations that ATP_e caused apoptotic cell death in a variety of cell targets, while cytotoxic cells were resistant, suggested to us and other investigators that this nucleotide could be involved in cell-mediated cytotoxicity (Di Virgilio et al 1990, Filippini et al 1990). However, this hypothesis, in its naive early formulation, has had to be revised as we and others clearly showed that ATP_e-resistant cell clones are fully susceptible to killing by cytotoxic T lymphocytes (Avery et al 1992, Zambon et al 1994). Why should mast cells, macrophages and microglial cells express such a dangerous receptor on the plasma membrane? One possibility is that the P2Z receptor mediates other kinds of cytotoxic responses more specifically aimed at elimination of cells involved in innate immunity, such as macrophages and microglial cells. Macrophages are involved in the defence against

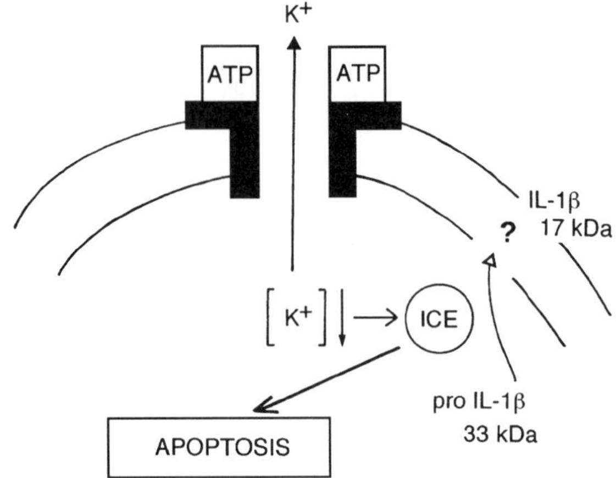

FIG. 2. Suggested pathway of activation of IL-1β converting enzyme (ICE) by ATP_e. In order to be released, IL-1β has to be processed by a cysteine protease (ICE) that cleaves the 33 kDa IL-1β precursor into the mature 17 kDa form. ICE is activated by a decrease in the cytoplasmic K^+ concentration. We postulate that efflux of K^+ through the P2Z pore is the main mechanism by which ATP_e causes ICE activation.

microorganisms that in some cases may survive phagocytosis and proliferate within the phagocyte (for example *Mycobacterium tuberculosis*). These macrophages can be activated to kill the intracellular parasite by IFN-γ stimulation, but the intracellular mechanism of the killing is unknown. Very recently, Molloy (1994) made the interesting observation that ATP_e stimulation of human macrophages that had ingested bacillus Calmette-Guérin, an attenuated form derived from *Mycobacterium bovis*, not only caused apoptosis of the macrophages but also killed the intracellular bacillus. Thus, it may be suggested that P2Z purinergic receptors represent true 'suicide receptors' exploited in immune-mediated reactions as an *ultima ratio* to kill macrophage cells infected by parasites that resist intracellular killing.

Another (but not alternative) hypothesis, is that the P2Z receptor originally evolved as a molecule engaged in cellular communication, and only accidentally acquired the capacity to signal cell death. The ATP_e-gated pore shares some functional properties of gap junctions (Beyer & Steinberg 1991), thus it may be postulated that it mediates transient or long lasting communication between immune cells that come in close contact during the inflammatory reaction. Such a role is suggested by recent observations by Falzoni et al reporting that inhibition of the P2Z receptor by the covalent blocker oxidized ATP prevented formation of multinucleated giant cells (MGCs) in human macrophage cultures stimulated with concanavalin A. This might suggest that regulated opening of the ATP_e-gated pore is somehow

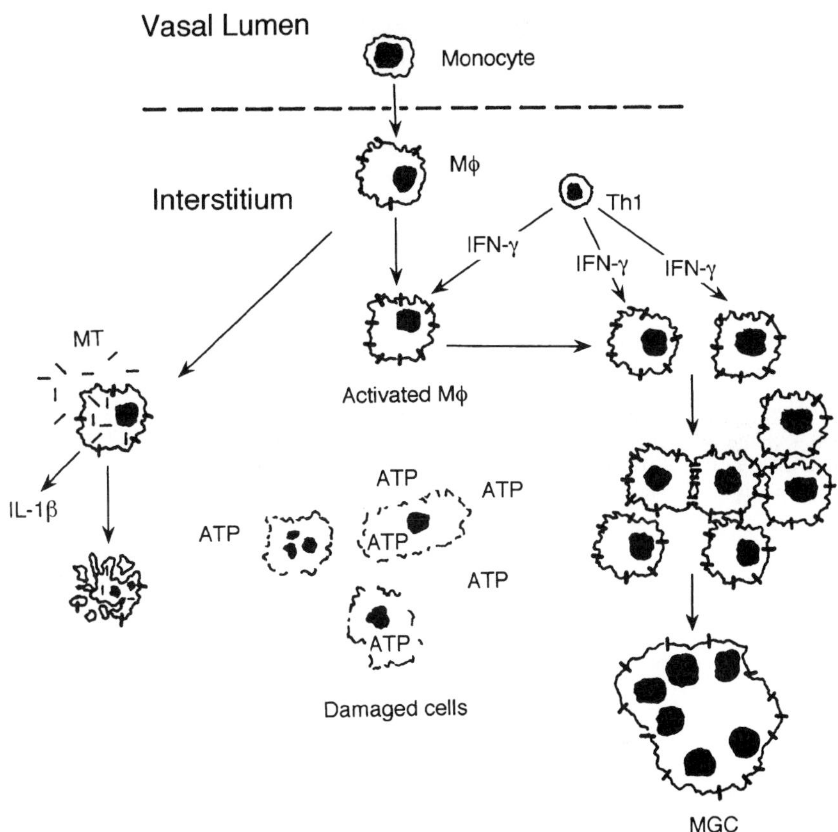

FIG. 3. Possible functions of P2Z receptor in the immune response. At inflammatory sites, monocytes leave the circulation and differentiate into macrophages (Mϕ). Mϕ differentiation increases expression of the P2Z receptor (small rectangles). During chronic inflammatory reactions, Mϕ are exposed to γ-interferon (IFN-γ) released by T helper cells (Th). IFN-γ stimulates macrophage activation, increases expression of P2Z receptor and drives formation of multinucleated giant cells (MGCs). The P2Z receptor could contribute to formation of early cytoplasmic bridges between adjacent macrophages in cooperation with plasma membrane adhesion molecules such as CD11a, CD18 and CD54. At the inflammatory site, Mϕ phagocytose invading microorganisms. Some of these, for example *Mycobacterium tuberculosis* (MT, rods), resist intracellular killing. The P2Z receptor could be used to mediate apoptotic elimination of infected Mϕ. Activation of the P2Z also causes maturation and release of IL-1β (17 kDa). ATP_e could be released by injured or dying cells during the inflammatory response. (Modified from Di Virgilio 1995.)

involved in establishing cytoplasmic bridges between adjacent cells thus allowing cell fusion.

Conclusions

Immune cells express purinergic receptors of the P2Y and P2Z subtype. The P2Y/P2U receptor in immune cells, as well as in many different cell types, is coupled to G protein activation, phosphoinositide hydrolysis, mobilization of intracellular Ca^{2+} and phospholipase stimulation. This will in turn lead to arachidonic acid release and generation of mediators of the inflammatory reaction.

The P2Z receptor appears to be uniquely expressed by immune cells and a few tumour cell types. Its stimulation causes transmembrane ion fluxes, plasma membrane depolarization and efflux of intracellular low molecular mass metabolites. Other as yet unknown transduction pathways might also be activated. Ligation of this receptor by ATP_e released at inflammatory sites might be needed for release of inflammatory cytokines (e.g. IL-1β), generation of MGCs or elimination of macrophages infected by intracellular pathogens.

Acknowledgements

This work was supported by grants from the National Research Council of Italy (Clinical Application of Cancer Research [ACRO] and special project Biology and Pathology of Calcium), the Ministry of Scientific Research (MURST), the Italian Association for Cancer Research (AIRC), the VII AIDS Project and Telethon of Italy.

References

Avery RK, Bleier KJ, Pasternack MS 1992 Differences between ATP-mediated cytotoxicity and cell-mediated cytotoxicity. J Immunol 149:1265–1270

Barnard EA, Webb TE, Simon J, Kunapuli SP 1996 The diverse series of recombinant P2Y purinoceptors. In: P2 purinoceptors: localization, function and transduction mechanisms. Wiley, Chichester (Ciba Found Symp 198) p 166–188

Beyer EC, Steinberg TH 1991 Evidence that the gap junction protein connexin-43 is the ATP-induced pore of mouse macrophages. J Biol Chem 266:7971–7974

Blanchard DK, McMillen S, Djeu JY 1991 IFN-γ enhances sensitivity of human macrophages to extracellular ATP-mediated lysis. J Immunol 147:2579–2585

Boudreau N, Sympson CJ, Werb Z, Bissel M 1995 Suppression of ICE and apoptosis in mammary epithelial cells by extracellular matrix. Science 267:891–893

Brake AJ, Wagenbach MJ, Julius D 1994 New structural motif for ligand-gated ion channels defined by an ionotropic ATP receptor. Nature 371:519–523

Cockcroft S, Gomperts BD 1979 ATP induces nucleotide permeability in rat mast cells. Nature 279:541–542

Cohn ZA, Parks E 1967 The regulation of pinocytosis in mouse macrophages. III. The induction of vesicle formation by nucleosides and nucleotides. J Exp Med 125:457–466

Cronstein BN 1994 Adenosine, an endogenous anti-inflammatory agent. J Appl Physiol 76:5–13

Dahlquist R, Diamant B 1974 Interaction of ATP and calcium on the rat mast cell: effect on histamine release. Acta Pharmacol Toxicol 34:368–384

Di Virgilio F 1995 The purinergic P2Z receptor: an intriguing role in immunity, inflammation and cell death. Immunol Today 16:524–528

Di Virgilio F, Bronte V, Collavo D, Zanovello P 1989 Responses of mouse lymphocytes to extracellular adenosine 5′-triphosphate (ATP). Lymphocytes with cytotoxic activity are resistant to the permeabilizing effects of ATP. J Immunol 143:1955–1960

Di Virgilio F, Pizzo P, Zanovello P, Bronte V, Collavo D 1990 Extracellular ATP as a possible mediator of cell-mediated cytotoxicity. Immunol Today 149:3372–3378

Drury AN, Szent-Györgyi A 1929 The physiological activity of adenine compounds with special reference to their action upon the mammalian heart. J Physiol (Lond) 68:213–237

Dubyak GR, El-Moatassim C 1993 Signal transduction via P_2-purinergic receptors for extracellular ATP and other nucleotides. Am J Physiol 265:577C–606C

Dubyak GR, Fedan JS 1990 Biological actions of extracellular ATP. Ann N Y Acad Sci 603:1–542

El-Moatassim C, Bernad N, Mani JC, Dornand J 1989 Extracellular ATP induces a nonspecific permeability of thymocyte plasma membrane. Biochem Cell Biol 67: 495–502

Enari M, Hug H, Nagata S 1995 Involvement of an ICE-like protease in Fas-mediated apoptosis. Nature 375:78–81

Falzoni S, Munerati M, Ferrari D, Spisani S, Moretti S, Di Virgilio F 1995 The purinergic P_{2Z} receptor of human macrophage cells. Characterization and possible physiological role. J Clin Invest 95:1207–1216

Ferrari D, Munerati M, Melchiorri L, Hanau S, Di Virgilio F, Baricordi OR 1994 Responses to extracellular ATP of lymphoblastoid cell lines from Duchenne muscular dystrophy patients. Am J Physiol 267:C886–C892

Filippini A, Taffs RE, Agui T, Sitkowsky MV 1990 EctoATPase activity in cytolytic T lymphocytes. Protection from the cytolytic effects of extracellular ATP. J Biol Chem 265:334–340

Freyer DR, Boxer LA, Axtell RA, Todd RF 1988 III. Stimulation of human neutrophil adhesive properties by adenine nucleotides. J Immunol 141:580–586

Greenberg S, Di Virgilio F, Steinberg TH, Silverstein SC 1988 Extracellular nucleotides mediate Ca^{2+} fluxes in J774 macrophages by two distinct mechanisms. J Biol Chem 263:10337–10343

Gregory SH, Kern M 1978 Adenosine and adenine nucleotides are mitogenic for mouse lymphocytes. Biochem Biophys Res Commun 83:1111–1116

Hickman SE, El Khoury J, Greenberg S, Schieren I, Silverstein SC 1994 P2Z adenosine triphosphate receptor activity in cultured human monocyte-derived macrophages. Blood 84:2452–2456

Kuida K, Lippke JA, Ku G et al 1995 Altered cytokine export and apoptosis in mice deficient in interleukin 1-β converting enzyme. Science 267:2000–2003

Los M, Van de Craen M, Penning LC et al K 1995 Requirement of an ICE/CED-3 protease for Fas/APO-1-mediated apoptosis. Nature 375:81–83

Lustig KD, Weisman GA, Turner JT, Garrad R, Shiau AK, Erb L 1996 P2U purinoceptors: cDNA cloning, signal transduction mechanisms and structure–function analysis. In: P2 purinoceptors: localization, function and transduction mechanisms. Wiley, Chichester (Ciba Found Symp 198) p 193–207

Molloy A, Laochumroonvorapong P, Kaplan G 1994 Apoptosis, but not necrosis, of infected monocytes is coupled with killing of intracellular bacillus Calmette–Guérin. J Exp Med 180:1499–1509

Murgia M, Pizzo P, Steinberg TH, Di Virgilio F 1992 Characterization of the cytotoxic effect of extracellular ATP in J774 mouse macrophages. Biochem J 288:897–901

Nuttle LC, Dubyak GR 1994 Differential activation of cation channels and non-selective pores by macrophages P_{2Z} purinergic receptors expressed in *Xenopus* oocytes. J Biol Chem 269:13988–13996

Owen GP, Hahn WE, Cohen JJ 1991 Identification of mRNAs associated with programmed cell death in immature thymocytes. Mol Cell Biol 11:4177–4188

Pizzo P, Zanovello P, Bronte V, Di Virgilio F 1991 Extracellular ATP causes lysis of mouse lymphocytes and activates a plasma membrane ion channel. Biochem J 274:139–144

Pizzo P, Murgia M, Zambon A et al 1992 Role of P_{2Z} purinergic receptors in ATP-mediated killing of tumor necrosis factor (TNF)-sensitive and TNF-resistant L929 fibroblasts. J Immunol 149:3372–3378

Schmidt A, Ortaldo JR, Herberman RB 1984 Inhibition of human natural killer cell reactivity by exogenous adenosine 5′-triphosphate. J Immunol 132:146–150

Sugiyama K 1971 Calcium-dependent histamine release with degranulation from isolated rat mast cells. Jpn J Pharmacol 21:209–226

Sung SS, Young JDE, Origlio AM, Heiple JM, Kaback HR, Silverstein SC 1985 Extracellular ATP perturbs transmembrane ion fluxes, elevates cytosolic Ca^{2+} and inhibits phagocytosis in mouse macrophages. J Biol Chem 260:13442–13449

Surprenant A 1996 Functional properties of native and cloned P2X purinoceptors. In: P2 purinoceptors: localization, function and transduction mechanisms. Wiley, Chichester (Ciba Found Symp 198) p 208–222

Tatham PER, Lindau M 1990 ATP-induced pore formation in the plasma membrane of rat peritoneal mast cells. J Gen Physiol 95:459–476

Valera S, Hussy N, Evans RJ et al 1994 A new class of ligand-gated ion channel defined by P_{2X} receptor for extracellular ATP. Nature 371:516–519

Ward PA, Cunningham TW, McCulloch KK, Phan SH, Powell J, Johnson J 1988 Platelet enhancement of O_2^- responses in stimulated human neutrophils: Identification of platelet factor as adenine nucleotides. Lab Invest 58:37–45

Wiley JS, Dubyak GR 1989 Extracellular adenosine triphosphate increases cation permeability of chronic lymphocytic leukemic lymphocytes. Blood 73:1316–1323

Wiley JS, Chen R, Jamieson GP 1993 The ATP^{4-} receptor-operated channel (P2Z class) of human lymphocytes allows Ba^{2+} and ethidium uptake: inhibition of fluxes by suramin. Arch Biochem Biophys 305:54–60

Zambon A, Bronte V, Di Virgilio F et al 1994 Role of extracellular ATP in cell-mediated cytotoxicity: a study with ATP-sensitive and ATP-resistant macrophages. Cell Immunol 156:458–467

Zanovello P, Bronte V, Rosato A, Pizzo P, Di Virgilio F 1990 Responses of mouse lymphocytes to extracellular ATP. II. Extracellular ATP causes cell type-dependent lysis and DNA fragmentation. J Immunol 145:1545–1550

Zheng LM, Zychlinsky A, Liu C-C, Ojcius DM, Young JD-E 1991 Extracellular ATP as a trigger for apoptosis or programmed cell death. J Cell Biol 112:279–288

DISCUSSION

Wiley: You've got data showing that extracellular ATP actually converts proIL-1 to the fully processed IL-1β. Microglia, like a lot of reticuloendothelial cells, probably have more than one purinoceptor. J774 cells have two, possibly three, purinoceptors. Do you have evidence linking this effect with the P2Z?

Di Virgilio: Yes, this is exactly why we turned to the microglial cells because it was very easy for us to select mutants either hyperexpressing or hypoexpressing P2Z. None the less, the P2Z-lacking clones express P2Y receptors normally: they have the G protein-linked receptor but they don't have the channel. This is what led us to claim that it is through the P2Z that IL is activated and released.

Abbracchio: Can you tell us more about the degree of necrosis and apoptosis induced by ATP in the same cell, and whether these are in any way correlated with expression of different receptor subtypes?

Di Virgilio: There is no straightforward answer to this. Initially, I thought apoptosis was mainly linked to a small P2X-type receptor. But now data are accumulating suggesting that you also get apoptosis when you open the large pore. Therefore, I think that it is mainly dependent on the cell type and experimental conditions, rather than on the particular receptor.

I attend quite a few apoptosis meetings—I'm interested in this field—and even with bona fide apoptotic receptors the picture is not all that clear. This situation is epitomized by the tumour necrosis factor receptor, but even with Fas/APO-1, the picture is not always so clear-cut.

Abbracchio: Is it possible that a single receptor, for example P2Z, can mediate both events depending upon some as yet unknown experimental condition?

Di Virgilio: This may well be the case. If you take human macrophages, pulse them with ATP for 15 min and then wash the ATP away, these cells will recover and behave perfectly normally for 24–48 h, and then die by apoptosis. If you take the same cells and instead challenge them for 2 h with ATP, they will die by necrosis. It might be that the length of exposure to ATP is critical in determining the fate of the cells.

Illes: The P2Z receptor is a pore. You seem to move from the pore to a kind of preformed channel that has a certain diameter where not only ions but also dyes cross the plasma membrane. Then you say that the P2Z receptor is activated by two molecules of ATP, so there's a certain cooperativity. Also, it recognizes mostly the tetrabasic form, but now we learn that conventional ATP antagonists such as suramin are antagonistic at this channel. Thus, the P2Z and P2X receptors move closer together.

Di Virgilio: I have to apologize if I used the words pore and channel interchangeably, because the two are quite separate terms to electrophysiologists. I think that at this point we still know too little about the functional properties of the P2Z receptor to be able to give it a precise name. The P2Z receptor of the macrophage—the 900 Da receptor—is no doubt a pore, because it has no ion selectivity and simply seems to work according to the molecular cut-off. The lymphocyte P2Z receptor seems to be more cation specific. Therefore, it looks more like a channel than a pore. But until we have more electrophysiological information on these structures I can't give a definitive answer.

Illes: The differentiation of monocytes into macrophages occurs over a number of distinct transitional steps that are now fairly well characterized. Do you have any idea at which of these intermediate stages the P2Z receptor is expressed?

Di Virgilio: In macrophage differentiation, we start seeing this receptor after two days in culture. After five days, about 60% of macrophages express the P2Z receptor, and by day 12, almost 80–90% of them express it. Under these conditions there is very little IFN-γ and macrophage activation is simply due to adhesion on the substrate. However, if you add IFN-γ at day 5, 100% of cells express the receptor.

Leff: When macrophages become foam cells in atheroma, do they retain the P2Z receptor?

Hickman: Yes, human foam cells do express P2Z receptors *in vitro*. We isolated human monocytes from peripheral blood and tested them for P2Z receptor activity by uptake of Lucifer yellow in response to ATP. Fresh monocytes do not have P2Z receptor activity. We then divided the monocytes into two groups. One group was cultured in Teflon beakers and the other group was fed acetylated low density lipoproteins in culture over a period of several days to convert them to foam cells. When we assessed P2Z receptor activity in both groups we found that foam cells do allow influx of Lucifer yellow in response to ATP, but so do the normal monocyte-derived macrophages. So in human monocytes the P2Z receptor may be differentially regulated.

Leff: So is there a role of ATP via the P2Z receptor in lipid uptake?

Hickman: I don't know. Interestingly, foam cells also express connexin 43.

Wiley: IFN-γ that causes maturation of the monocytes into macrophages is generally thought to come from activated T cells. Have you tried the experiment of adding T cells to your macrophage culture and activating them with mitogen and seeing if this causes expression?

Di Virgilio: Yes, it obviously works this way.

Miras-Portugal: Is it possible to relate the parts of the P2Z receptor that you have sequenced to the annexins or other proteins related to regulated exocytosis?

Hickman: We only have a partial nucleotide sequence of our clone (approximately 1200 of 1800 nucleotides) and initial comparisons of this with sequences of those in the databases do not reveal similarities to any cloned purinergic receptor or any of the connexins. So far, it appears to be unique. When we obtain and analyse the whole nucleotide and putative amino acid sequences we will know more. Also, once we figure out a possible protein structure, it may reveal structural similarities to other proteins that may not be obvious from just looking at the nucleic acid or amino acid sequences.

Leff: I'd like to ask a sort of drug industry-oriented question. I can understand that in the case of tuberculosis one would want a P2Z agonist, to promote the bacterial killing effect itself. For apoptosis, would you want an antagonist or an agonist? Which one should I be encouraging my chemists to make?

Di Virgilio: I wouldn't restrict myself to apoptosis because, although I'm interested in this field, apoptotic death is just one way for the cell to die. If you read the early literature by Richard Lockshin for instance, on apoptosis, there are at least two extremes, Type I and Type II, and many morphological intermediate states in which you have the typical morphology of apoptosis but you don't have DNA fragmentation or vice versa, and so on (Zakeri et al 1995). For instance, in *C. elegans*, which is the best studied system, many cells undergo programmed cell death but they don't die by apoptosis. There is no coincidence between programmed cell death and apoptosis. Therefore if I was in the drug industry I would look for something which would help prevent cell death in general by targeting this thing to specific receptors. Is the ATP receptor a good receptor? I would say that it is, because there is compelling evidence that in a number of cases activation of this receptor can cause cell death.

Barnard: I would like to raise the question of whether there are P2Y receptors active in cells in the immune system. We have noted in our chapter (Barnard et al 1996, this volume) information on a clone derived from an mRNA induced upon activation of T lymphocytes, which appears to encode a P2Y type of receptor, but with many significant sequence differences from the others known.

Di Virgilio: We do not see any G protein-linked ATP responses in lymphocytes unless they are mitogenically activated, although we have a paper in press showing that non-activated human T lymphocytes do have a channel or pore activated by ATP (Baricordi et al 1995).

Barnard: That's a P2Z type isn't it?

Di Virgilio: Yes, although I don't know whether we can retain this nomenclature at this point.

Barnard: Is it activated by BzATP?

Di Virgilio: Yes.

References

Baricordi OR, Ferrari D, Melchorri L et al 1995 An ATP activated channel is involved in mitogenic stimulation of human lymphocytes. Blood, in press

Barnard EA, Webb TE, Simon J, Kunapuli SP 1996 The diverse series of recombinant P2Y purinoceptors. In: P2 purinoceptors, localization, function and transduction mechanisms. Wiley, Chichester (Ciba Found Symp 198) p 166–188

Zakeri Z, Bursch W, Tenniswood M, Lockshin RA 1995 Cell death: programmed, apoptosis, necrosis or other? Cell Death & Differ 2:87–96

General discussion III

Purinoceptors and the pancreas

Petit: I am going to present a short summary of our data concerning purinoceptors and the endocrine pancreas, particularly concerning insulin secretion.

I have selected a few points highlighting the physiological role of purinoceptors on the pancreatic B cells, and also the potential clinical and therapeutic application of P2Y purinoceptor agonists in non insulin-dependent diabetes.

Administration of ATP elicits a clear and biphasic stimulation of insulin secretion from the isolated, perfused rat pancreas. This typical response is observed in the presence of a slightly stimulating glucose concentration (8.3 mM). When the glucose concentration is low (2.8 mM), the response to ATP is only weak and transient. ATP does not stimulate insulin secretion in the absence of glucose, which would be clinically relevant as regards the risk of hypoglycaemia.

In our laboratory, we have performed conventional pharmacological characterization of these receptors using concentration–response determinations, structure–activity relationships and relative potencies of natural nucleotides and nucleosides as well as non-hydrolysable structural analogues. This characterization led us to conclude that there is a P2Y purinoceptor on the B cell, 2-methylthioATP being about 50 times more potent than ATP in stimulating insulin secretion. The response to ATP could also be anatagonized by 2′,2′-pyridylisatogen (PIT). On the other hand, the pancreatic B cell is also provided with an A_1 receptor which inhibits insulin secretion. Thus, in this system there are both P2-stimulating receptors and A_1-inhibiting receptors.

The mechanisms of action of P2 purinoceptors on the B cell have been investigated in various preparations using different agonists and may involve different pathways. It is now generally agreed that the response to ATP is accompanied by an increase in the cytoplasmic Ca^{2+} concentration. This Ca^{2+} signal has been measured in single cells.

The physiological role of these P2Y receptors in insulin secretion can be summarized as follows. ATP is present in high concentrations in the secretory granules of the B cells and is released during glucose-induced insulin secretion. Furthermore, ATP is stored and co-released with acetylcholine from peripheral cholinergic nerve terminals. Thus, ATP released together with insulin or acetylcholine may activate the P2Y receptor and amplify the response to either glucose, which is the major nutrient stimulus of the B cell, or acetylcholine, which is a well-known and important modulator of the secretion. Indeed, we reported a potentiating synergism between P2 purinoceptor activation and a physiological glucose increment in isolated perfused pancreas. A potentiating interaction was also demonstrated between ATP or ADP and acetylcholine in stimulating insulin secretion.

I will now discuss some data obtained *in vivo* in rats and dogs. The effects of ADPβS in anaesthetized rats are dependent upon the nutritional state of the animals. In fasted animals, ADPβS after i.v. injection elicits only a transient insulin response without affecting significantly the glycaemia. By contrast, in fed animals, the same dose of ADPβS induced a sustained insulin response and significantly reduced the glycaemia. However, the basal glycaemia was higher in fed animals than in fasted animals, and in that case, the drug decreased glycaemia to a similar baseline as in fasted animals. Thus, the lack of ADPβS effect under fasted conditions may limit the risk of hypoglycaemia.

During an i.v. glucose tolerance test, when glucose is administered in anaesthetised rats, the simultaneous administration of ADPβS strongly increases insulin secretion, which results in a more rapid decrease in glucose concentration towards basal values.

In conscious dogs, the oral administration of ADPβS elicits an increase in plasma insulin with a slight and transient reduction in glycaemia.

During an oral glucose tolerance test in conscious dogs, ADPβS markedly enhances insulin secretion and consequently reduces the hyperglycaemia. Taken together, these results highlight the potential usefulness of P2Y purinoceptors as a target for the development of new therapeutic drugs in non-insulin-dependent diabetes.

Jacobson: You showed acute administration of the ATP agonist. Have you looked at chronic administration?

Petit: Not yet. Such experiments would be worthwhile and are currently under consideration.

Jacobson: Would you expect much of a problem of tachyphylaxis of the response?

Petit: Tachyphylaxis has been reported to occur with α,β-methylene ATP acting on P2X purinoceptors. We have shown that the potentiating effect of ATP and ADP on insulin secretion is mediated by the P2Y purinoceptor subtype of the B cell.

Illes: You said that ATP from the extracellular side amplifies the response to glucose with respect to insulin release. We are talking here all the time about ATP as an extracellular signal molecule and we shouldn't forget that ATP is also a signal molecule from the inner side of the membrane. In addition, glucose, by being metabolized, leads to the build-up of ATP, and ATP closes ATP-dependent K^+ channels. So ATP from the inner and outer sides of the plasma membrane does the same thing; namely, it depolarizes pancreatic B cells and releases insulin.

Petit: Indeed, in addition to its well known intracellular role as a source of energy, ATP was more recently considered as an intracellular signal closing the ATP-dependent K^+ channels. The subsequent depolarization of the plasma membrane leads to the opening of voltage-gated Ca^{2+} channels. The influx of Ca^{2+} rapidly increases the cytosolic Ca^{2+} concentration and triggers the exocytosis of insulin granules. Thus, considering the B cell, ATP is both an intracellular signal linking glucose metabolism to the electrophysiological events leading to insulin secretion, and an extracellular signal acting on P2 purinoceptors and modulating the insulin response to glucose and other secretagogues.

Wiley: Are there any cardiovascular effects of ADPβS?

Petit: Yes, *in vivo* in the dog we noticed a transient decrease in arterial blood pressure after i.v. injection of ADPβS.

Starke: Is there anything known about glucagon release in response to purinoceptor activation?

Petit: Yes, there are also purinergic receptors on the pancreatic A cell but these are of the P1/A_2 subtype. Glucagon secretion is stimulated by adenosine. Hence, adenosine affects in an opposite way the secretions of insulin and glucagon which by themselves have opposite effects on blood glucose: hypo- and hyperglycaemic actions respectively.

Leff: Has anybody attempted to identify the subtype on the B cell?

Petit: From a molecular biological viewpoint, no.

Challenges in developing P2 purinoceptor-based therapeutics

Michael Williams

Neuroscience Discovery, Pharmaceutical Products Division, Abbott Laboratories, 100 Abbott Park Road, Abbott Park, IL 60064, USA

Abstract. Advances in the molecular cloning, expression and functional characterization of the P2 purinoceptor superfamily have provided a wealth of data to support a diverse functional role for ATP and related nucleotides in the regulation of tissue function. As with other receptor superfamilies, it is likely that distinct subtypes of each receptor will subserve discrete functions depending on tissue distribution and disease pathophysiology. At the present time, ATP is being evaluated as an anticancer agent and as an anaesthesia adjunct whereas UTP is studied as a novel treatment for cystic fibrosis. ARL67085 is a potent and selective P2T receptor antagonist that has potential as a novel antithrombotic agent. The key to exploiting the P2 purinoceptor area to enhance understanding of disease aetiology and concurrent therapeutic potential will be to focus efforts on the identification of novel pharmacophores that have potent and selective interactions with the various receptor subtypes as potential new leads. To this end, the use of high-throughput screening in conjunction with combinatorial chemical, conventional chemical and natural product library compound sources will be critical.

1996 P2 purinoceptors: localization, function and transduction mechanisms. Wiley, Chichester (Ciba Foundation Symposium 198) p309–321

A considerable body of data, derived over the past half century, has clearly delineated the role of the purine nucleotide, ATP, as a neurotransmitter/ neuromodulator as distinct from its role as an intracellular building block and energy source (Burnstock 1996, Harden et al 1995). Physiological, molecular biological and (to a lesser extent) pharmacological data indicate a discrete role for ATP, acting via a family of cell surface receptors, in nervous system, immune and systemic function (Burnstock 1996). On a more global scale, ATP also appears to play a pivotal role in cell differentiation and apoptosis (Abbracchio et al 1995, Di Virgilio 1995).

Receptors for ATP, the P2 purinoceptor superfamily, can be divided into two main groups (Abbracchio & Burnstock 1994): the P2X, a ligand-gated ion channel family (Surprenant et al 1995, Evans et al 1995) for which six cloned subtypes, $P2X_{1–6}$ currently exist; and the P2Y, a G protein-coupled receptor family with up to seven members, four of which are designated $P2Y_{1–4}$ (Boarder et al 1995, Barnard et al 1996). Two additional nucleotide receptors are the P2Z receptor, an ATP-gated ion pore that is involved in the cytotoxic actions of ATP (Di Virgilio 1995) and a new class of UTP-sensitive, ATP-sensitive pyrimidine receptor (Lazarowski & Harden 1994).

The functional characterization of P2 purinoceptors, both those originally identified by classical physiological techniques (Burnstock 1996) and those cloned and expressed (Boarder et al 1995, Surprenant et al 1995, Barnard et al 1996), has been severely limited by a paucity of potent and selective ligands for the various receptor subtypes (Jacobson et al 1995). Receptor characterization has thus been largely dependent on localization and *in situ* studies, a strategy that is becoming increasingly common when cloning and expression are the driving forces in receptor identification. This situation contrasts markedly with that for the serotonin (5-HT) receptor superfamily, where selective ligands were instrumental in driving receptor characterization (Hoyer et al 1994).

An analogous situation to the present state of P2 receptor characterization is seen with the nicotinic acetylcholine receptor (nAChR) superfamily (Arneric et al 1995). nAChRs are pentameric ligand-gated ion channels that are activated by nicotine. Although the nAChR found in the neuromuscular junction has been well characterized (Changeux et al 1992), much less is known about nAChRs in the CNS. Eight α-subunits and three β-subunits have been cloned and offer a multitude of distinct pentamer combinations that have the potential to confer different pharmacological properties to the resultant ligand-gated ion channel. However, it is not always clear which subunits in which combinations—and consequently which ligand-gated ion channel subtypes—are responsible for the various pharmacological properties of nicotine. Thus, unlike the well-characterized 5-HT_{1D}-like receptor that is involved in migraine and for which selective ligands such as sumatriptan exist (Humphrey et al 1991), the discrete nAChR subtypes that mediate the cognitive, addictive, anxiolytic, cytoprotective, emetic and gastrointestinal actions of nicotine remain to be unequivocally identified.

However, again by analogy with the 5-HT receptor superfamily, novel ligands can provide the basis for correlating receptor structure with function. The $\alpha_4\beta_2$ agonist, ABT 418 and the α7-selective ligand, GTS 21, both have cognition-enhancing activity in animal models similar to that seen with nicotine (Arneric et al 1995), but both compounds lack the cardiovascular liabilities of the latter agent.

In the age of molecular biology and rational drug design, the inability to relate a defined molecular target to a given therapeutic utility can limit the acceptance, resource prioritization and the ultimate pace of a drug discovery programme. The identification of specific ligands that can define the physiological/pharmacological function of their particular recognition site thus remains a powerful influence in the discovery of new therapeutic agents.

In the P2 area, it is only very recently that a systematic effort in defining selective ligands has occurred (Jacobson et al 1995), with the concurrent identification of the platelet P2T receptor antagonist, ARL67085 (formerly known as FPL67085; Humphries et al 1995). This compound, the stable 2-propylthio bioisostere of ATP, blocks the aggregatory actions of ADP at the platelet and has antithrombotic actions (Humphries et al 1995). The P2T receptor is unusual, however, in that the actions of ADP in promoting platelet aggregation are mediated by ATP, a compound that is an agonist at other P2 receptor subtypes. For the remainder of the P2 purinoceptor family, the absence of potent and selective receptor antagonists that are active *in vivo* has hindered understanding of receptor function.

ATP as a neurotransmitter

ATP is an unusual molecule to function as a neurotransmitter. In addition to interacting with the two major families of P2 receptor and the P2Z receptor, it can be broken down via ectonucleotidase action to adenosine, a neuromodulator in its own right (Williams 1995). This phenomenon appears to be independent of the intracellular bioenergetic state of the tissue (Headrick et al 1994). Thus, unlike other traditional neurotransmitters, ATP has the potential to be a multifunctonal effector agent acting in a cascade-like manner to produce tissue-dependent effects, some of which may be mutually antagonistic. For instance, while ATP is a fast excitatory neurotransmitter at P2X receptors (Surprenant et al 1995, Zimmerman 1994), adenosine is a potent inhibitor of excitatory neurotransmitter release via its activation of adenosine A_1 receptors (Williams 1995). The degree to which such mechanisms are physiologically relevant will be dependent on both the activity of the various ectonucleotidases in a given tissue (Kennedy & Leff 1995) and the array of purinergic receptors and cell types present in the immediate extracellular space. In traumatized tissue, ATP released from neuronal sources would not only interact with other neurons and glia, but also astrocytes, reactive microglia and neutrophils. In addition, the nucleotide side chain has the potential to complex Ca^{2+}. If ectonucleotidase activity is increased as a result of tissue trauma, this will also contribute to defining the relative physiological contributions of the various members of the purinergic cascade. ATP can also modulate synaptic transmission by functioning as an ectokinase substrate (Ehrlich et al 1988).

An additional facet that is emerging as results from different research fields become integrated is the role of ATP and other purine nucleotides as trophic factors (Neary & Norenberg 1992, Neary et al 1996) and in events related to cell death (Di Virgilio 1995). ATP and GTP, as well as adenosine and guanosine, can exert beneficial trophic effects in nervous tissue following brain trauma and/or ischaemia. The effects of ATP are similar to those of the polypeptide growth factor, fibroblast growth factor 2 (FGF-2; also known as basic fibroblast growth factor, bFGF) in promoting astroctyte maturation. ATP can also act synergistically with FGF-2 to enhance mitotic processes (Neary et al 1996) and can activate astrocytic mitogen-activated protein kinase leading to activation of the early-response gene (e.g. c-*jun*, c-*fos*) pathways involved in cell growth, differentiation and stress responses.

The association of ATP with cell death processes involves several types of P2 receptor. The cytotoxic effects related to P2Z receptor activation are well documented (Di Virgilio 1995), and the similarities between the $P2X_1$ receptor and the thymocyte apoptotic clone RP-2 have already been mentioned. The G protein-coupled receptor $P2Y_3$ is also lethal when transfected into cell lines (Barnard et al 1996).

ATP pathophysiology

There is a paucity of data related to the role of ATP in human disease states. It has, however, been reported that P2 receptor mechanisms are elevated in spontaneously hypertensive rats (Burnstock 1996) and that ATP responses are increased in models of interstitial cystitis (Palea et al 1993).

The drug discovery process

The drug discovery process is a complex, expensive and lengthy endeavour involving a high risk of clinical failure (Williams et al 1993, Weisbach & Moos 1995). Only one compound in 5000 makes it to the marketplace as a drug entity. Accordingly, present data indicate that it costs in excess of $359 million and takes from 8–12 years to bring a compound to the marketplace (Di Masi et al 1991). It has also been estimated that approximately 80% of applied research programmes targeted towards drug discovery fail to achieve success due to technical hurdles (Williams et al 1993).

While significant advances have been made in enabling technologies to make the process of drug discovery more rational, the process is still highly dependent on serendipity. The choice of an innovative approach to a disease state involving significant unmet medical need is frequently a process that occurs in the form of hypothesis testing, involving the identification of a discrete molecular target that is thought to be involved in the disease pathophysiology and selective agents interacting with this target. The data to test the hypothesis are frequently only available in the targeted patient population, in essence extending the hypothesis testing from the bench to the clinic. While this may seem an obvious course of events for a scientifically rational drug discovery programme, many mechanistic approaches that have seemed totally obvious *in vitro* have failed in the clinic to the chagrin and eventual financial detriment of the companies involved. On the other hand, the failure to take a compound to its ultimate proof of principle in a disease state is an even greater waste of precious resources and does little to advance knowledge related to human disease aetiology.

Once a molecular target has been identified that appears to be related to a proposed disease pathophysiology, a pharmaceutical company will proceed to identify compounds that selectively interact with the target, acting as agonists or antagonists at a receptor target and, typically, as inhibitors of an enzyme target. As lead molecules are refined, their selectivity, safety and pharmacokinetic profile is enhanced until sufficient data are available to advance agents to the development stage. The major hurdle at this point is the relevance of animal models, if such entities exist, to the human condition. For many diseases there are no satisfactory models. For instance, in the CNS area, there is no such thing as a depressed or schizophrenic rat, and the test paradigms used for programmes in these disease areas either approximate some aspect of the human situation or reflect the *in vivo* phenotype of currently used therapeutic agents. To elaborate, the majority of antipsychotic models are actually animal models of dopamine receptor blockade, the original class of compound shown to have antipsychotic activity.

Interestingly, the one animal model that has proven highly useful in the drug research process to study compounds for hypotensive activity, the spontaneously hypertensive rat, has yet to be linked to a molecular lesion underlying increased blood pressure. None the less, compounds active in reducing blood pressure in this model are, subject to adequate bioavailability and duration of action, predictively active in humans.

Drug discovery technologies

The process of drug discovery is dependent on two basic elements: the identification of an appropriate molecular target; and the characterization of compounds interacting with this target. The combination of these elements and the iteration between them represent the basic elements and activities of drug discovery.

Molecular targets

In the P2 receptor area, a number of discrete molecular targets have been cloned and expressed and include various members of the P2X and P2Y receptor families. The $P2X_1$ receptor has been cloned from the rat vas deferens (Valera et al 1994), the $P2X_2$ from PC12 neuroblastoma cells (Brake et al 1994) and the $P2X_3$ from the dorsal root ganglion (Chen et al 1995). Other known clones of the P2X family have yet to be published. The cloned receptors differ in their pharmacology and agonist activity profiles (Surprenant et al 1995). The

$P2X_1$ is sensitive to α,β-methylene ATP and undergoes rapid desensitization, whereas the $P2X_2$ receptor is insensitive to α,β-methylene ATP and is not readily desensitized. The $P2X_1$ receptor is identical to RP-2, a cDNA clone expressed in apoptotic cells.

The density of P2Y receptors in brain tissue, based on binding data, is unusually high (Simon et al 1995) with B_{max} values in the range of 38 pmol/mg protein. In comparison, the density of receptors for the excitatory amino acid glutamate is in the range of 7 pmol/mg protein (Murphy et al 1987) and for adenosine it is in the range of 0.1–1 pmol/mg protein (Williams & Jacobson 1990). These findings suggest that P2Y receptors in nervous tissue subserve an important functional role that has yet to be defined.

The $P2Y_2$ receptor has been cloned from a variety of tissues and is identical to the P2U receptor responsive to both ATP and UTP. This receptor is unique in that it is the only P2 receptor containing an RGD integrin binding sequence. The $P2Y_3$ receptor is a developmentally expressed receptor that is responsive to ADP. A number of other P2Y receptors have been described and include clones designated as an endothelial $P2Y_4$ (Boeynaems et al 1996), HP212, R5 and 6H1, the latter two being cloned from rat brain and activated lymphocytes respectively (Barnard et al 1996). P2T, P2Z and the diadenosine polyphosphate (Ap_nA) receptors had not been cloned at the time of writing. Another 12 P2 receptors are rumoured to exist in partial sequence in various human genomic databases.

It is a major step from cloning and expression to function and, as noted, much of what is known related to cloned P2 receptor function is either structurally inferred, as in the case of the homology between $P2X_1$ and the apoptotic gene RP-2 (Surprenant et al 1995), or based on location. Thus the presence of the $P2X_3$ receptor in the dorsal root ganglion and its localization to small nociceptive neurons via anti-peripherin staining (Chen et al 1995) suggests that this receptor is involved in some aspect of ATP-related pain perception.

The assignment of function on the basis of structural homology and/or localization must, however, be substantiated by more physiological data. The latter can be derived by the development of transgenic knockouts or by genomic analysis of diseased cDNA libraries. However, even when these elegant molecular biological approaches have proven successful, it still requires the identification of selective ligands to fruitfully exploit the area for drug discovery.

To this end, given that all agonist ligands for the P2 receptor family are variations on the purine nucleotide pharmacophore (Cusack 1993, Jacobson et al 1995, Williams & Jacobson 1995), antagonist ligands like suramin and PPADS (pyridoxalphosphate-6-azophenyl-2′,4′-disulfonic acid) have selectivity and efficacy that remain sufficiently questionable to preclude their use *in vivo* to characterize P2 receptor function and it will be necessary to identify novel structural entities selective for the various P2 receptors to effectively advance the field.

Compound sources

The identification of novel pharmacophores can be accomplished in one of two ways: by rational drug design using computer-assisted molecular design, or by high-throughput screening, the latter involving compound library and combinatorial chemistry approaches. Both approaches have significant value and are typically used in an iterative mode.

Computer-assisted molecular design is an intellectually satisfying, rational approach to drug design. Using available information on target structure and the structure–activity relationship for ligands known to interact with the target, 3D models can be constructed to model the ligand recognition site. When combined with molecular point mutations of this target site (the replacement of individual amino acids using molecular biological techniques), valuable information can be derived on the mode of binding of known pharmacophores. This has been done for the P2Y purinoceptor (van Rhee et al 1995) and can provide a unifying hypothesis for the interaction of a variety of dissimilar structures to the active site of the protein target and, in some instances, derive a template for the design of new pharmacophores. This has been especially useful in designing inhibitors of the enzymes thymidylate kinase and HIV protease (Greer et al 1994). In general, however, computer-assisted molecular design is a retrospective rather than predictive technique that represents an approximation of the dynamics of the receptor–enzyme interaction with the ligand/substrate. Thus the minimal energy conformation of the target is modelled on an approximation of the water content and typically represents the target and their ligands as somewhat static entities. This approach can be enhanced by knowledge of the X-ray crystallographic structure of the target or, more appropriately, by deriving data related to the interaction of the ligand with its target using 2 or 3D nuclear magnetic resonance techniques (Fesik 1991).

While computer-assisted molecular design is an extremely valuable, emerging technology, its ultimate usefulness lies in conjunction with more traditional approaches to pharmacophore design rather than as a 'stand alone' computer-based drug design tool. The fact that the technique has typically been oversold in this latter context has seriously impeded its use as an enabling technology integrated with conventional medicinal chemistry approaches.

High-throughput screening is a 'numbers game' approach to the identification of novel pharmacophores. While this approach is not noted for its intellectual impact (Bowman & Harvey 1995), it has had unprecedented success in the past decade (Williams & Gordon 1996). In the area of neuropeptide receptor research, specifically in the area of the neurokinins, traditional medicinal chemistry approaches yielded little success in over a decade in terms of novel, stable pharmacophores that could be used as lead compounds for drug discovery efforts. However, in the space of 18 months, using high-throughput binding assays, three major pharmaceutical companies identified structurally dissimilar antagonists of substance P, derivatives of which are currently in clinical trials (Rees 1993).

In the P2 area, similar high-throughput screening efforts are being set up to identify non-nucleotide, non-purine pharmacophores to aid in the characterization of P2 receptor function. This approach is, however, limited by concerns related to the robust nature of the presently available agonist radioligand-based binding assays. One way of circumventing this problem is the use of whole-cell reporter assays (Williams et al 1995) using stably transfected P2 receptors linked to luciferin/luciferase or fluorescent green reporter systems. Newly identified pharmacophores with the appropriate selectivity and activity (preferably antagonists) identified in this manner can in turn be used to develop more robust binding assays.

Another compound source that is of considerable interest in today's drug discovery environment is that of combinatorial chemistry (Gordon et al 1994). This is a recently developed chemical technology that has the potential, through various approaches (Desai et al 1994) to allow the concurrent synthesis of many thousands to millions of compounds in the space of a few weeks. On a more modest scale, this technique can also be used to synthesize a targeted library numbering in the hundreds of compounds based on a common template using parallel reactions. Given that a single medicinal chemist can make 20–30 compounds a year using traditional medicinal chemistry techniques, the technique of combinatorial chemistry is to chemistry what the molecular biology revolution has been to traditional biochemical pharmacology approaches to studying cell and tissue function at the molecular level.

An additional facet of the search for novel ligands for P2 receptors is the potential existence of allosteric modulators. For ligand-gated ion channels, there is good precedent from the γ-aminobutyric acid $(GABA)_A$/benzodiazepine receptor, the *N*-methyl-D-aspartate receptor and the nAChR for physiologically relevant allosteric modulation of receptor function (Williams et al 1995). In this context, one of the seminal ATP antagonists, PIT (2′-2′-pyridylisatogen), can enhance P2Y receptor responses in PC12 cells (King et al 1994).

P2 receptor ligands under clinical evaluation

ATP is an effective and long-lasting inhibitor of human tumour cell growth, an effect that is postulated to occur via permeabilization of tumour cell membranes as well as the opening of plasma membrane Ca^{2+} and Na^{2+} channels (Rapaport 1993). These cytotoxic actions of ATP appear to involve the gap junction protein, connexin 43, and may thus be mechanistically related to the ATP-stimulated, P2Z receptor-mediated apoptosis described in macrophages (Di Virgilio 1995). The potential use of ATP as an antimetastatic agent is currently being investigated clinically by Medco.

A considerable body of data supports the use of ATP as an adjunct to inhalation anaesthetics (Fukunaga et al 1995). ATP can be used to reduce the doses of traditional anaesthetics offering an improved margin of safety, decreasing the risks associated with overdosage and toxicity and improving the time to recovery following surgery. Given i.v., ATP can be used to induce controlled hypotension to decrease the risks of bleeding during surgery (Muruyama et al 1979), to induce analgesia without causing respiratory depression (Fukunaga et al 1995) and to blunt autonomic responses. Upon infusion at 103 μg/kg per min, ATP can substitute completely for nitrous oxide and permit a reduction of 36% in the amount of enflurane required for anaesthesia, a net 61% reduction in the amount of conventional anaesthetic agents (Fukunaga et al 1995). Similar effects have also been reported for adenosine, the active metabolite of ATP (Segerdahl et al 1995), which is well characterized as an analgesic agent (Sawynok & Sweeney 1989).

Purine and pyrimidine nucleotides have been shown to be effective in promoting airway defence via their ability to stimulate mucin secretion, activate Cl^- channel function and increase ciliary beat frequency (Boucher et al 1995). This has led to studies evaluating UTP as a treatment for cystic fibrosis, UTP being preferred to ATP to avoid the bronchoconstrictor effects of the adenosine formed from ATP. The receptor mediating the effects of both ATP and UTP is of the P2U subtype (Parr et al 1994). Early clinical trials conducted by the biopharmaceutical company, Inspire, have shown that UTP alone, or in combination with amiloride, restored normal function to the peripheral airways of the lung in cystic fibrosis patients (Boucher et al 1995).

ARL67085 functions as a selective ADP (P2T receptor) antagonist (Humphries et al 1995). ARL67085 shows greater than 30 000-fold selectivity for the P2T receptor versus activity at P2X and P2Y receptors. In a canine model of thrombosis, the compound was able to abolish thrombosis and block ADP-induced platelet aggregation *ex vivo* at a mean infusion dose of 91 ng/kg per min i.v. These actions occurred in the absence of changes in blood pressure and heart rate with a good separation (28-fold) between those doses inhibiting thrombosis and those increasing bleeding time. In fact, preclinical data showed that ARL67085 had a superior margin of safety to fibrinogen receptor antagonists which showed only a twofold window of safety, and was superior in efficacy to aspirin. Phase I studies conducted by Fisons before their acquisition by Astra substantiated the preclinical profile and the compound is currently in Phase II trials.

Future directions

The challenge for the future of biomedical research in the P2 receptor area, specifically in the context of therapeutic agents, will involve the application of the significant advances in basic knowledge to focus on defined therapeutic opportunities. While information generation, integration and hypothesis testing is the *raison d'être* of scientific research, drug discovery-related research is ultimately more applied with a considerable focus on predefined endpoints. While these endpoints may not always be reached, the Nobel laureate Sir James Black has often been quoted as having said 'there's no point in looking for a needle in a haystack [an analogy for the drug discovery process] if you don't have any idea what the needle looks like'. Accordingly, drug discovery is typically initiated with a high degree of conceptual focus on the anticipated end product, with attendant milestones and testing procedures well defined.

It is noteworthy therefore that the drug discovery programme related to the novel P2T receptor antagonist, ARL67085 (Humphries et al 1995) was based on the identification of an opportunity to develop an improved antithrombotic agent using available data on structure–activity relationships for stable purine nucleotide ligands (Cusack 1993) and a body of biological knowledge related to ADP, ATP and platelet function that had been available for over two decades. This by no means belittles the significant achievement of the Fisons group, but rather points out that the recognition and *application* of existing knowledge to a specific therapeutic target was crucial for a successful outcome.

As knowledge in the area of P2 purinoceptor function continues to accumulate and as medicinal chemistry and screening efforts are focused on the various receptor targets, it is anticipated that novel P2 receptor ligands, both agonists and antagonists, will represent innovative therapeutic modalities for the treatment of disease states in a variety of tissue systems.

References

Abbracchio MP, Burnstock G 1994 Purinoceptors: are there families of P_{2X} and P_{2Y} purinoceptors? Pharmacol Ther 64:445–475

Abbracchio MP, Ceruti S, Burnstock G, Cattabeni F 1995 Purinoceptors on glial cells of the central nervous system: functional and pathological implications. In: Bellardinelli L, Pelleg A (eds) Adenosine and adenine nucleotides: from molecular biology to integrative physiology. Kluwer Acad, Norwell, MA, p 271–280

Arneric SP, Sullivan J, Williams W 1995 Neuronal nicotinic acetylcholine receptors. Novel targets for central nervous system therapeutics. In: Bloom FE, Kupfer DJ (eds) Psychopharmacology: the fourth generation of progress. Raven, New York, p 95–110

Barnard EA, Webb TE, Simon J, Kunapuli SP 1996 The diverse series of recombinant P2Y purinoceptors. In: P2 purinoceptors: localization, function and transduction mechanisms. Wiley, Chichester (Ciba Found Symp 198) p 166–188

Boarder MR, Weisman GA, Turner JT, Wilkinson GF 1995 G protein-coupled P_2 purinoceptors: from molecular biology to functional response. Trends Pharmacol Sci 16:133–139

Boeynaems J-M, Communi D, Pirotton S, Motte S, Parmentier M 1996 Involvement of distinct receptors in the actions of extracellular uridine nucleotides. In: P2 purinoceptors: localization, function and transduction mechanisms. Wiley, Chichester (Ciba Found Symp 198) p 266–277

Boucher RC, Knowles MR, Olivier KN, Bennett W, Mason SJ, Stutts MJ 1995 Mechanisms and therapeutic actions of uridine triphosphate in the lung. In: Bellardinelli L, Pelleg A (eds) Adenosine and adenine nucleotides: from molecular biology to integrative physiology. Kluwer Acad, Norwell, MA, p 525–532

Bowman C, Harvey AL 1995 The discovery of drugs. Proc R Coll Physicians Edinb 25:5–24

Brake AJ, Wagenbach MJ, Julius D 1994 New structural motif for ligand-gated ion channels defined by an ionotropic ATP receptor. Nature 371:519–523

Burnstock G 1996 P2 purinoceptors: historical perspective and classification. In: P2 purinoceptors: localization, function and transduction mechanisms. Wiley, Chichester (Ciba Found Symp 198) p 1–34

Changeux J-P, Galzi JL, Devillers-Thiery A, Betrand D 1992 The functional architecture of the acetylcholine nicotinic receptor explored by affinity labeling and site-directed mutagenesis. Quart Rev Biophys 25:395–432

Chen CC, Akopian AN, Sivilotti L, Colquhoun D, Burnstock G, Wood JN 1995 A P_{2X} receptor expressed by a subset of sensory neurons. Nature 377:428–431

Cusack NJ 1993 P_2 receptor: subclassification and structure–activity relationships. Drug Dev Res 28:244–252

Desai MC, Zukermann RN, Moos WH 1994 Recent advances in the generation of chemical diversity libraries. Drug Dev Res 33:174–188

Di Masi J, Hansen RW, Graboswki HG, Lasagna L 1991 Cost of innovation in the pharmaceutical industry. J Health Econ 10:107–142

Di Virgilio F 1995 The purinergic P2Z receptor: an intriguing role in immunity, inflammation and cell death. Immunol Today 16:524–528

Ehrlich YH, Snyder RM, Kornecki E, Garfield MG, Lenox RH 1988 Modulation of neuronal signal transduction systems by extracellular ATP. J Neurochem 50:295–301

Evans RJ, Lewis C, Buell G, Valera S, North RA, Surprenant A 1995 Pharmacological characterization of heterologously expressed ATP-gated cation channels (P_{2X} purinoceptors). Mol Pharmacol 48:178–183

Fesik SW 1991 NMR studies of molecular complexes as a tool in drug design. J Med Chem 34:2937–2945

Fukunaga AF, Miyamoto TA, Kikuta Y, Kaneko Y, Ichinohe T 1995 Role of adenosine and adenosine triphosphate as anesthetic adjuvants. In: Bellardinelli L, Pelleg A (eds) Adenosine and adenine nucleotides: from molecular biology to integrative physiology. Kluwer Acad, Norwell, MA, p 511–523

Greer J, Erickson JW, Baldwin JJ, Varney MD 1994 Application of three-dimensional structures of protein target molecules in structure-based drug design. J Med Chem 37:1035–1054

Gordon EM, Barratt RW, Dower WJ, Fodor SPA, Gallop MA 1994 Applications of combinatorial technologies. II. Combinatorial organic synthesis, library screening strategies and future directions. J Med Chem 37:1385–1401

Harden TK, Boyer JL, Nicholas TA 1995 P_2-purinergic receptors: subtype-associated signaling responses and structure. Annu Rev Pharmacol Toxicol 35:541–579

Headrick JP, Bendall MR, Faden AI, Vink R 1994 Dissociation of adenosine levels from bioenergetic state in experimental brain trauma: potential role in secondary injury. J Cereb Blood Flow Metab 14:853–861

Hoyer, D, Clarke DE, Fozard JR et al 1994 International Union of Pharmacology classification of receptors for 5-hydroxytryptamine (serotonin). Pharmacol Rev 46:157–203

Humphrey PPA, Feniuk W, Marriott AS, Tanner RJN, Jackson MR, Tucker ML 1991 Preclinical studies on the anti-migrane drug sumatriptan. Eur Neurol 31:282–290

Humphries RG, Robertson MJ, Leff P 1991 A novel series of P2T purinoceptor antagonists: definition of the role of ADP in arterial thrombosis. Trends Pharmacol Sci 16:179–181

Jacobson KA, Fischer B, Malliard M et al 1995 Novel ATP agonists reveal receptor heterogeneity within P_{2X} and P_{2Y} subtypes. In: Bellardinelli L, Pelleg A (eds) Adenosine and adenine nucleotides: from molecular biology to integrative physiology. Kluwer Acad, Norwell, MA, 149–166

Kennedy C, Leff P 1995 How should P_{2X} purinoceptors be classified pharmacologically? Trends Pharmacol Sci 16:168–174

King BF, Wang S, Ziganshin AU, Ziganshina LE, Spedding M, Burnstock G 1994 2′-2′-Pyridylisatogen potentiates ATP agonism at a recombinant P_{2Y1} purinoceptor. Br J Pharmacol 113:60P

Lazarowski ER, Harden TK 1994 Identification of a uridine nucleotide-sensitive G-protein-linked receptor that activates phospholipase C. J Biol Chem 269:11830–11836

Maruyama M, Sato K, Shimoji K 1979 Hypotensive anesthesia using ATP (translated from Japanese). J Clin Anesth 3:279–284

Murphy DE, Schneider J, Boehm D, Lehmann J, Williams M 1987 Binding of [^{3}H]CPP (3(carboxypiperazin-4-yl)propyl-1-phosphonic acid): a selective high affinity ligand for N-methyl-D-aspartate receptors. J Pharm Exp Ther 240:778–784

Neary JT, Norenberg MD 1992 Signalling by extracellular ATP: physiological and pathological considerations in neuronal–astrocytic interactions. Prog Brain Res 94:145–151

Neary JT, Rathbone MP, Cattabeni F, Abbracchio MP, Burnstock G 1996 Trophic actions of extracellular nucleotides and nucleosides on glial and neuronal cells. Trends Neurosci 19:13–18

Palea S, Artibani W, Ostardo E, Trist DG, Pietra C 1993 Evidence for purinergic neurotransmission in human urinary bladder affected by interstitial cystitis. J Urol 150:2007–2012

Parr CE, Sullivan DM, Paradiso AM et al 1994 Cloning and expression of a human P_{2U} nucleotide receptor, a target for cystic fibrosis therapy. Proc Natl Acad Sci USA 91:3275–3279

Rapaport E 1993 Anticancer activities of adenine nucleotides in tumor bearing hosts. Drug Dev Res 28:428–431

Rees DC 1993 Non-peptide ligands for neuropeptide receptors. Annu Rep Med Chem 28:59–68

Sawynok J, Sweeney MI 1989 The role of purines in nociception. Neuroscience 32: 557–569

Segerdahl M, Ekblom A, Sandelin K, Wickman M, Sollevi A 1995 Peroperative adenosine infusion reduces the requirement for isoflurane and postoperative analgesics. Anesth Analg 80:1145–1149

Simon J, Webb TE, Barnard EA 1995 Characterization of a P_{2Y} purinoceptor in the brain. Pharmacol Toxicol 76:302–307

Surprenant A, Buell G, North RA 1995 P_{2X} receptors bring new structure to ligand-gated ion channels. Trends Neurosci 18:224–229

Valera S, Hussy N, Evans RJ et al 1994 A new class of ligand-gated ion channel defined by P_{2X} receptor for ATP. Nature 371:516–519 van Rhee AM, Fischer B, van Galen PJM, Jacobson KA 1995 Modelling the P2Y purinoceptor using rhodopsin as a template. Drug Design Deliv 13:133–154

Weisbach J, Moos WH 1995 Diagnosing the decline of major pharmaceutical research laboratories: a prescription for drug companies. Drug Dev Res 34:243–259

Williams M 1995 Purinoceptors in central nervous system function. Targets for therapeutic intervention. In: Bloom FE, Kupfer DJ (eds) Psychopharmacology, the fourth generation of progress. Raven, New York, p 643–655

Willams M, Jacobson KA 1990 Radioligand binding assays. In: Williams M (ed) Adenosine and adenosine receptors. Humana, Clifton, NJ, p 17–55

Williams M, Jacobson KA 1995 P_2 purinoceptors: advances and therapeutic opportunities. Expert Opin Invest Drugs 4:925–934

Williams M, Gordon EM 1996 Drug discovery: an overview. In: Krogsgaard-Larsen P, Liljefors T, Madsen U (eds) A textbook of drug design and development, 2nd edn. Harwood Academic, Chur, p 1–36

Williams M, Giordano T, Elder RA, Reiser HJ, Neil GL 1993 Biotechnology in the drug discovery process: strategic and management issues. Med Res Rev 13:399–448

Williams M, Deecher D, Sullivan J 1995 Drug receptors. In: Wolff ME (ed) Burger's medicinal chemistry and drug discovery, 5th edn, vol 1: Principles and practice. Wiley Interscience, New York, p 349–397

Zimmermann H 1994 Signalling via ATP in the nervous system. Trends Neurosci 17:420–426

Summing-up

Geoffrey Burnstock

Department of Anatomy and Developmental Biology, University College London, Gower Street, London WC1E 6BT, UK

This has been a balanced and harmonious meeting: we have had honest debate without personal animosity and there has been little secrecy. The symposium has identified several unresolved issues and I expect that all of us have ideas for experiments where we think we can resolve those issues. I hope that new friendships might have been made through the meeting and through the new collaborations developed. The best way to move things forward is to find people with skills in adjacent areas and work with them; there's nothing more pleasant than doing this.

Molecular biological approaches are making enormous inroads into this field. Purinoceptors are a relatively recent development compared with what is known about receptors to other neurotransmitters, so we shall have much to do, but we are now moving and learning, and a lot more progress can be expected during the next few years. But I hope we never forget that we want to use this information about molecular structure to find out more about function, about pathophysiology, and to help us develop new compounds. We are desperately short of good antagonists that can be used *in vivo*. It is nice that the drug industry is represented here and their contribution has been particularly strong. I hope and I think that the members of the drug industry present see the advantage of linking with top academic labs around the world and taking advantage of their expertise to work together to produce new drugs of therapeutic value. I also hope that we will go on to look at more pathological situations. My guess is that there are going to be some surprises there.

I have personal views about some of the areas that I think will grow in the next few years. I'm very keen on the trophic roles of purines: these are going to be very important, particularly in early development. In a way this is related to the roles of ATP in apoptosis and in the development of antitumour strategies. I also hope that people begin to do some work on plasticity of expression of purinoceptors in different situations, and begin to look for the genes that control the expression of different kinds of purinoceptors in different conditions. I find the evolutionary side absolutely fascinating. I am amazed at the things that are hidden in the literature, material that I'm trying to bring together in a Biological Review that I am writing with Maria Abbracchio.

And then there is the therapeutic side. As more companies get interested in purinoceptors, this will grow. There were many companies interested in the adenosine receptor 10 years ago. I think that the work with ATP receptors may prove more fruitful.

I have really enjoyed this meeting and I have really appreciated the outstanding contributions from so many of you.

Index of contributors

Non-participating co-authors are indicated by asterisks. Entries in bold type indicate papers; other entries refer to discussion contributions.

Indexes compiled by Liza Weinkove.

Subject index

Other Ciba Foundation Symposia:

No. 151 **Neurobiology of incontinence**
Chairman: C. D. Marsden
1990 ISBN 0 471 92687 6

No. 179 **The molecular basis of smell and taste transduction**
Chairman: F. Margolis
1993 ISBN 0 471 93946 3

No. 190 **Somatostatin and its receptors**
Chairman: S. Reichlin
1995 ISBN 0 471 95382 2

No. 196 **Growth factors as drugs for neurological and sensory disorders**
Chairman: J. Kessler
1996 ISBN 0 471 95727 6